T0205700

NANOPOLYMERS AND MODERN MATERIALS

Preparation, Properties, and Applications

NANOPOLYMERS AND MODERN MATERIALS

Preparation, Properties, and Applications

Edited by

Oleg V. Stoyanov, DSc, A. K. Haghi, PhD, and Gennady E. Zaikov, DSc

Apple Academic Press Inc.
3333 Mistwell Crescent
Oakville, ON L6L 0A2
Canada

Apple Academic Press Inc.
9 Spinnaker Way
Waretown, NJ 08758
USA

First issued in paperback 2021

Exclusive worldwide distribution by CRC Press, a member of Taylor & Francis Group

ISBN 13: 978-1-77463-277-2 (pbk)
ISBN 13: 978-1-926895-47-5 (hbk)

Library of Congress Control Number: 2013945480

Library and Archives Canada Cataloguing in Publication

Nanopolymers and modern materials: preparation, properties, and applications/edited by Oleg V. Stoyanov, DSc, A.K. Haghi, PhD, and Gennady E. Zaikov, DSc.

Includes bibliographical references and index.
ISBN 978-1-926895-47-5
1. Nanofibers. 2. Nanocomposites (Materials). 3. Polymerization. 4. Nanotechnology.
I. Stoyanov, Oleg V., author, editor of compilation II. Haghi, A. K., author, editor of compilation III. Zaikov, G. E. (Gennadi™i Efremovich), 1935-, author, editor of compilation

TA418.9.N35N38 2013 620.1'92 C2013-904997-5

Apple Academic Press also publishes its books in a variety of electronic formats. Some content that appears in print may not be available in electronic format. For information about Apple Academic Press products, visit our website at **www.appleacademicpress.com** and the CRC Press website at **www.crcpress.com**

ABOUT THE EDITORS

Oleg V. Stoyanov, DSc

Oleg V. Stoyanov, DSc is Head and Chair of Industrial Safety and Professor and Chair of Technology of Processing Plastic and Composite Materials (TPPCM) at Kazan State Technological University in Russia. He is a world-renowned scientist in the field of chemistry and the physics of oligomers, polymers, composites, and nanocomposites.

A. K. Haghi, PhD

A. K. Haghi, PhD, holds a BSc in urban and environmental engineering from University of North Carolina (USA); a MSc in mechanical engineering from North Carolina A&T State University (USA); a DEA in applied mechanics, acoustics and materials from Université de Technologie de Compiègne (France); and a PhD in engineering sciences from Université de Franche-Comté (France). He is the author and editor of 65 books as well as 1000 published papers in various journals and conference proceedings. Dr. Haghi has received several grants, consulted for a number of major corporations, and is a frequent speaker to national and international audiences. Since 1983, he served as a professor at several universities. He is currently Editor in-Chief of the *International Journal of Chemoinformatics and Chemical Engineering* and *Polymers Research Journal* and on the editorial boards of many international journals. He is also a faculty member of University of Guilan (Iran) and a member of the Canadian Research and Development Center of Sciences and Cultures (CRDCSC), Montreal, Quebec, Canada.

Gennady E. Zaikov, DSc

Gennady E. Zaikov, DSc, is Head of the Polymer Division at the N. M. Emanuel Institute of Biochemical Physics, Russian Academy of Sciences, Moscow, Russia, and Professor at Moscow State Academy of Fine Chemical Technology, Russia, as well as Professor at Kazan National Research Technological University, Kazan, Russia. He is also a prolific author, researcher, and lecturer. He has received several awards for his work, including the the Russian Federation Scholarship for Outstanding Scientists. He has been a member of many professional organizations and on the editorial boards of many international science journals.

ABOUT THE EDITORS

CONTENTS

LIST OF CONTRIBUTORS

M. H. Moghadam Abatari
Department of Mathematics, Faculty of Mathematical Sciences, University of Guilan, Rasht, Iran

O. M. Alekseeva
N. M. Emanuel Institute of Biochemical Physics, Russian Academy of Sciences, Kosygin Str. 4, 119334 Moscow, Russia, E-mail: olgavek@yandex.ru

V. A. Babkin
403343 SF VolgSABU, c. Mikhailovka, Region Volgograd s. Michurina 21, E-mail: sfi@reg.avtlg.ru

S. A. Bekusarova
Gorsky State Agrarian University, Kirov Str. 37, Vladikavkaz, Republic of North Ossetia Alania, Russia, E-mail: ggau@globalalania.ru

I. S. Belostotskaja
N. M. Emanuel Institute of Biochemical Physics, Russian Academy of Sciences, Kosygin Str. 4, Moscow 119334, Russia

V. I. Binyukov
Federal State Budget Institution of Science, N. M. Emanuel Institute of Biochemical Physics, RAS, ul. Kosygina 4, Moscow 119334, Russian Federation

N. A. Bome
Tyumen State University, Semakova Str. 1, 625003 Tyumen, Russia, E-mail: president@utmn.ru

E. B. Burlakova
N. M. Emanuel Institute of Biochemical Physics, Russian Academy of Sciences, Kosygin Str. 4, Moscow 119334, Russia

A. K. Haghi
University of Guilan, Rasht, Iran, E-mail: Haghi@Guilan.ac.ir

M. Hasanzadeh
Department of Textile Engineering, Amirkabir University of Technology, Tehran, Iran
Department of Chemical Engineering, Imam Hossein Comprehensive University, Tehran, Iran

A. L. Iordanskii
N. N. Semenov Institute of Chemical Physics, street Kosygina 4, RAS 119991, Moscow, Russia

N. V. Khokhriakov
Izhevsk State Agricultural Academy, Basic Research and Educational Center of Chemical Physics and Mesoscopy, Udmurt Scientific Center, Ural Division, Russian Academy of Science, Izhevsk 426000, Russia

N. G. Khrapova
N. M. Emanuel Institute of Biochemical Physics, Russian Academy of Sciences, Kosygin Str. 4, Moscow 119334, Russia

G. A. Korablev
Izhevsk State Agricultural Academy, Basic Research and Educational Center of Chemical Physics and Mesoscopy, Udmurt Scientific Center, Ural Division, Russian Academy of Science, 426000 Izhevsk, Russia, E-mail: korablev@udm.net

R. Yu. Kosenko
N. N. Semenov Institute of Chemical Physics, street Kosygina 4, RAS 119991, Moscow, Russia

G. V. Kozlov
Dagestan State Pedagogical University, Yaragskii Str. 57, Makhachkala 367003, Russian Federation

G. V. Luschenko
PhD Student, North Caucasus Research Institute of Mountain and Foothill Agriculture, Williams's Str. 1, Mikhailovskoye 391502, Republic of North Ossetia Alania, Russia

G. M. Magomedov
Dagestan State Pedagogical University, Yaragskii Str. 57, Makhachkala 367003, Russian Federation

B. Hadavi Moghadam
Department of Textile Engineering, Amirkabir University of Technology, Tehran, Iran

L. I. Matienko
Federal State Budget Institution of Science, N. M. Emanuel Institute of Biochemical Physics, RAS, ul. Kosygina 4, Moscow 119334, Russian Federation, E-mail: matienko@sky.chph.ras.ru

E. M. Mil
Federal State Budget Institution of Science, N. M. Emanuel Institute of Biochemical Physics, RAS, ul. Kosygina 4, Moscow 119334, Russian Federation

V. M. Misin
N. M. Emanuel Institute of Biochemical Physics, Russian Academy of Sciences, Kosygin Str. 4, Moscow 119334, Russia

L. A. Mosolova
Federal State Budget Institution of Science, N. M. Emanuel Institute of Biochemical Physics, RAS, ul. Kosygina 4, Moscow 119334, Russian Federation

G. A. Nikiforov
N. M. Emanuel Institute of Biochemical Physics, Russian Academy of Sciences, Kosygin Str. 4, Moscow 119334, Russia

A. A. Olkhov
Moscow State Academy of Chemical Technology of M. V. Lomonosov, prosp.Vernadskogo 86, Moscow 119571, Russia, E-mail: aolkhov72@yandex.ru

P. G. Pronkin
N. M. Emanuel Institute of Biochemical Physics, Russian Academy of Sciences, ul. Kosygina 4, Moscow 119334, Russia

V. N. Shtol'ko
N. M. Emanuel Institute of Biochemical Physics, Russian Academy of Sciences, Kosygin Str. 4, Moscow 119334, Russia

Yu. S. Simonova
Moscow State Academy of Chemical Technology of M. V. Lomonosov, prosp.Vernadskogo 86, Moscow 119571, Russia

A. S. Tatikolov
N. M. Emanuel Institute of Biochemical Physics, Russian Academy of Sciences, ul. Kosygina 4, Moscow 119334, Russia, E-mail: tatikolov@sky.chph.ras.ru

F. T. Tzomatova
PhD Student, North Caucasus Research Institute of Mountain and Foothill Agriculture, Williams's Str. 1, Mikhailovskoye 391502, Republic of North Ossetia Alania, Russia

Yu. G. Vasiliev
Izhevsk State Agricultural Academy, Basic Research and Educational Center of Chemical Physics and Mesoscopy, Udmurt Scientific Center, Ural Division, Russian Academy of Science, Izhevsk 426000, Russia

L. I. Weisfeld
Senior Researcher, N.M. Emanuel Institute of Biochemical Physics, Russian Academy of Sciences, Kosygina Str. 4, Moscow 119334, Russia, E-mail: chembio@sky.chph.ras.ru

Kh. Sh. Yakh'yaeva
Dagestan State Pedagogical University, Yaragskii Str. 57, Makhachkala 367003, Russian Federation

G. E. Zaikov
Federal State Budget Institution of Science, N. M. Emanuel Institute of Biochemical Physics, RAS, ul. Kosygina 4, Moscow 119334, Russian Federation, E-mail: chembio@sky.chph.ras.ru
N. M. Emanuel Institute of Biochemical Physics, RAS, Kosygin Str. 4, Moscow 119991, Russia
Russian Academy of Sciences, Moscow, Russia
N.N. Semenov Institute of Chemical Physics, street Kosygina 4, RAS119991, Moscow, Russia

D. S. Zakharov
403343 SF VolgSABU, c. Mikhailovka, Region Volgograd s. Michurina 21, E-mail: sfi@reg.avtlg.ru

LIST OF ABBREVIATIONS

AFD	Average fiber diameter
ANN	Artificial neural network
ANOVA	Analysis of variance
AO	Antioxidants
BSA	Bovine serum albumin
CA	Contact angle
CCD	Central composite design
CEP	Copolymer of ethylene with polypropylene
CEVA	Copolymer of ethylene with vinyl acetate
CRAC	Ca^{2+}-release-activated Ca^{2+} channels
CVD	Chemical vapor deposition
DMF	Dimethylformamide
DMPC	Dimyristoilphosphatidylcholine
DSC	Differential scanning calorimetry
DSK	Differential scanning microcalorymetry
EAC	Ascetic Ehrlich carcinoma
EAC	Ehrlich ascetic carcinoma
EC	Ethyl cellulose
EEET	Electronic excitation energy transfer
EHD	Electrohydrodynamic
EPR	Electron paramagnetic resonancespectroscopy
ER	Endoplasmic reticulum
FSE	Free surface energy
FTIR	Fourier transform infrared spectroscopy
FWHM	Full width at half maximum intensity
GPC	Gel-penetrating chromatography
HBP	Hyperbranched polymer
HPET	Alkali-treated PET fabric
HPPE	High-pressure polyethylene
LBM	Lattice Boltzmann method
LPO	Lipid peroxide oxidation
MFI	Melt flow index
MP	N-Methylpirrolidon-2
mRNA	Messenger ribonucleic acid
MTD	Much as possible transferable dose

NMR	Nuclear magnetic resonance
ODE	Ordinary differential equation
PABA	Para-aminobenzoic acid
PEH	α-Phenyl ethyl hydro peroxide
PEK	Polyethyleneketone
PET	Poly ethylene terephthalate
PHB	Polyhydroxybutyrate
PP	Polypropylene
PPO	2,6-Dimethyl 1,4-phenylene oxide
PS	Polystyrene
PTFE	Polytetrafluorethylene
RNA	Ribonucleic acid
RSM	Response surface methodology
SEI	Spatial-energy exchange interactions
SEM	Scanning electron microscope
SEP	Copolymer of ethylene with propylene
SEP	Spatial-energy parameter
SSD	Supersmall doses effect
UCM	Upper-convected Maxwell

PREFACE

Nanopolymer technology continues to grow at an ever-increasing rate. The use of nanopolymers and modern materials compounds in commercial applications is gaining momentum. Behind much of the hype surrounding nanopolymers and modern materials is a real opportunity for companies to produce superior, sustainable materials faster, cheaper, and more efficiently. This new book provides data on the rapid growth of nanopolymers and demonstrates, via a range of case studies, their use in different industry sectors. Nanotechnology and modern materials is transforming the transport, energy, packaging, healthcare, and construction sectors by offering improved properties from lighter weight materials, from flame retardant polymers, to functionalized surfaces that can enable anti-corrosive, anti-reflective, anti-microbial, anti-scratch composite surfaces or provide barrier properties.

This state-of-the art book provides empirical and theoretical research concerning nanopolymers and modern materials. Themes within the book include:

- a new class of polymer material with particular application in nanotechnology
- production of electrospun nanofibers and the importance of governing parameters
- new research on the influence of nanofiller in composites
- optimization of the electrospinning process in the production of nanofibers along with mathematical models on the transport properties of electrospun-nanofibers
- quantum-chemical modeling
- application of bioactive substances and biochemichal treatments
- polymer processing and formation

This book covers all aspects of this fascinating and rapidly developing field that is already beginning to impact polymer users and researchers in most industry sectors. The book will update producers and users involved in polymers, both for components and films, with the latest thoughts and developments that will affect them. In this book, trends in the field of nanocomposites are also covered, along with information on the latest developments into composites.

This book features the characterization of nanocomposites, regulatory proposals and potential restraints, migration of the nano-component from polymers, recyclability, and the current patent landscape. Specific applications for nanopolymers, including nano-biopolymers, are presented and focus on their use in some key industry sectors such as transport, energy, and environmental applications.

The wide coverage makes this book an excellent reference book for academics, researchers, engineering professionals, and graduate students on this important and ever-emerging field.

— Oleg V. Stoyanov, DSc, Gennady E. Zaikov, DSc, and A. K. Haghi, PhD

CHAPTER 1

THE SELECTIVE CATALYTIC ALKYLARENS OXIDATIONS TO HYDRO PEROXIDES WITH DIOXYGEN, CATALYZED WITH NICKEL COMPLEXES: THE FORMATION OF SUPRAMOLECULAR NANOSTRUCTURES ON THE BASIS OF CATALYTIC ACTIVE HETEROLIGAND NICKEL COMPLEXES

L. I. MATIENKO, L. A. MOSOLOVA, V. I. BINYUKOV, E. M. MIL, and G. E. ZAIKOV

CONTENTS

1.1 INTRODUCTION

Technical and technological development demands the creation of new materials, which are stronger, more reliable and more durable, that is, materials with new properties. Up-to-date projects in creation of new materials go along the way of nanotechnology.

Nanotechnology can be referred to as a qualitatively new round in human progress. This is a wide enough concept, which can concern any area: information technologies, medicine, military equipment, robotics and so on. The concept of nanotechnology is narrowed and considered it with the reference to polymeric materials as well as composites on their basis.

The prefix 'nano' means that in the context of these concepts, the technologies based on the materials, elements of constructions and objects which are considered, and whose size makes 10^{-9}m. As fantastically as it may sound but the science reached the nanolevel long ago. Unfortunately, everything connected with such developments and technologies for now is impossible to apply to mass production because of their low productivity and high cost.

That means nanotechnology and nanomaterials are only accessible in research laboratories for now, but it is only a matter of time. What sort of benefits and advantages will the manufacturers have after the implementation of nanotechnologies in their manufactures and starting to use nanomaterials? Nanoparticles of any material have absolutely different properties than micro- or macroparticles. This results from the fact that alongside with the reduction of particles' sizes of the materials to nanometric sizes, physical properties of a substance change too. For example, the transition of palladium to nanocrystals leads to the increase in its thermal capacity to more than 1.5 times; it causes the increase of solubility of bismuth in copper to 4,000 times and the increase of self-diffusion coefficient of copper at room temperature to the order of 21.

Such changes in properties of substances are explained by the quantitative change of atoms' surface and volume ratio of individual particles, that is, by the high-surface area. Insertion of such nanoparticles in a polymeric matrix while using the apparently old and known materials gives a chance of receiving the qualitatively and quantitatively new possibilities in their use.

Nanocomposites based on thermoplastic matrix and containing natural, laminated inorganic structures are referred to as laminated nanocomposites. Such materials are produced on the basis of ceramics and polymers, however, with the use of natural laminated inorganic structures such as montmorillonite or vermiculite which are present, for example, in clays. A layer of filler ~1 nm thick is saturated with monomer solution and later polymerized. The laminated nanocomposites in comparison with initial polymeric matrix possess much smaller permeability for liquids and gases. These properties allow applying them to medical and food-processing industry. Such materials can be used in manufacturing of pipes and containers for the carbonated beverages.

These composite materials are ecofriendly, absolutely harmless to the person and possess fire-resistant properties. The derived thermoplastic laboratory samples have been tested and really confirmed those statements.

Nowadays, manufacturing technique of thermoplastic materials causes difficulties, notably dispersion of silicate nanoparticles in monomer solution. To solve this problem, it is necessary to develop the dispersion technique, which could be transferred from laboratory conditions to the industrial ones.

The advantages which the manufacturers can have if they decide to reorganize their manufacture for the use of such materials can be predicted even today. As these materials possess more mechanical and gas-barrier potential in comparison with the initial thermoplastic materials, their application in manufacture of plastic containers or pipes will lead to raw materials' saving by means of reduction of product thickness.

The improvement of physical and mechanical properties allows applying nanocomposite products under higher pressures and temperatures. For example, the problem of thermal treatment of plastic containers can be solved. Another example of the application of the valuable properties of laminated nanocomposites concerns the motor industry. American company General Motors manufactures the laminated nanocomposite material for the production of parts of the car body of Hummer H_2.

Ultrathin nanoclay surpasses talc and other fillers in its quality: it forms particles which are more rigid and less fragile at low temperatures and approximately for 20% easier. The application of such material allows lowering the weight of the car and raising the durability of its parts.

Another group of materials is metal containing nanocomposites. Thanks to the ability of metal particles to create the ordered structures (clusters), metal containing nanocomposites can possess a complex of valuable properties. The typical sizes of metal clusters from 1 to 10 nm correspond to their huge specific surface area. Such nanocomposites demonstrate the super-paramagnetizm and catalytic properties; therefore, they can be used while manufacturing semiconductors, catalysts, optical, and luminescent devices, and so on.

Such valuable materials can be produced in several ways, for example, by means of chemical or electrochemical reactions of isolation of metal particles from solutions. In this case, the major problem is not so much the problem of metal restoration but the preservation of its particles, that is, the prevention of agglutination and formation of large metal pieces.

For example, under laboratory conditions metal is deposited on the thin polymeric films capable to catch nano-sized particles. Metal can be evaporated by means of high energy and nano-sized particles can be produced, which then should be preserved. Metal can be evaporated by using explosive energy, high-voltage electric discharge or simply high temperatures in special furnaces.

The practical application of metal-containing nanocomposites (not going into details about high technologies) can involve the creation of polymers possessing some valuable properties of metals. For example, the polyethylene plate with the tenth fractions of palladium possesses the catalytic properties similar to the plate made of pure palladium.

An example of applying metallic composite is the production of packing materials containing silver and possessing bactericidal properties. By the way, some countries have already been applying the paints and the polymeric coverings with silver nanoparticles. Owing to their bactericidal properties, they are applied in public facilities (painting of walls, coating of handrails etc.).

At present, independent works on creation of concentrates containing stable silver nanoparticles are carried out parallel in Russia and the Ukraine. On addition of such concentrates to plastics, the latter obtain bactericidal properties therewith opening new prospects in the field of packing materials and water purification. In the near future, such concentrates can become accessible to mass manufacturers of products from plastics.

The technology of polymeric nanocomposites manufacture forges ahead; its development is directed to simplify and making the production processes cheap of composite materials with nanoparticles in their structure. For example, Hybrid Plastics/USA has developed **POSS® additives** (polyhedral oligomeric silsesquioxanes—mineral fillers based on silicon) in the form of crystalline solids, liquids and oils. More recently, these products were estimated in a range of 1,000 dollars per pound (0.454 kg) however, now they cost 50 dollars per pound. It is to note that nanotechnology is still a very young scientific field and is at the stage of primary development, as the basic part of information and knowledge is applied only in laboratory conditions. However, the nanotechnologies develop high rates; what seemed impossible yesterday, will be accessible to introduce on a commercial scale tomorrow.

The prospects in the field of polymeric composite materials upgrading are retained by nanotechnologies. Ever-increasing manufacturer's demand for new and superior materials stimulates the scientists to find new ways of solving tasks on the qualitatively new nanolevel.

The desired event of fast implementation of nanomaterials in mass production depends on the efficiency of cooperation between the scientists and the manufacturers in many respects. Today's high technology problems of applied character are successfully solved in close consolidation of scientific and business worlds.

Research of structure and catalytic activity of complexes of nickel in different reactions of oxidation with molecular O_2 causes increase in interest of researchers in connection with nickel as an essential catalytic cofactor of enzymes which are found in eubacteria, archaebacteria, fungi, and plants. These enzymes catalyze a diverse array of reactions that include both redox and nonredox chemistries [1].

Last year's development of researches in sphere of a homogeneous catalysis occurs in two directions—free radicals chain catalytic oxidation and a catalysis with complexes of the metals modeling action of enzymes. The majority of the reactions which are carried out commercially are based on auto-oxidation reactions. Low exits of products of oxidation counting on the spent hydrocarbon, caused by fast deactivation of the catalyst are the main obstacles for commercial application of majority of biomimetic systems. At the same time, conclusions about the mechanism of dioxygenases action and their models appear rather useful at treatment of the mechanism of catalysis with metal complexes in processes of oxidation of hydrocarbons by molecular O_2 [2, 3]. On the other hand, research results of catalysis of oxidation processes can be used for studying the mechanism of enzyme's operation.

The method was developed for enhancing the catalytic activity of transition metal complexes in the processes of alkylarens (ethyl benzene, cumene) oxidation with dioxygen to afford the corresponding hydro peroxides [2, 3] intermediates in the large-scale production of important monomers [4]. This method consists of introducing additional mono- or multidentate modifying ligands into catalytic metal complexes. The activity of systems $\{ML^1_n + L^2\}$ (M = Ni, Fe, L^1 = acac$^-$, L^2 = N-methylpirrolidon-2 (MP), HMPA, MSt (M = Na, Li, K), crown ethers or quaternary ammonium salts) is associated with the fact that during the ethyl benzene oxidation, the primary $(M^{II}L^1_2)_x(L^2)_y$ complexes and the real active catalytic hetero-ligand $M^{II}_xL^1_y(L^1_{ox})_z(L^2)_n(H_2O)_m$ ("A") complexes are formed to be involved in the oxidation processes [2, 3]. As it is established by us that the mechanism of formation of active complexes ("A") is similar to the action of ARD enzymes, complexes "A" can be considered as models of Ni^{II}-ARD [5].

On the other hand, unusual effect was discovered of significant increase in efficiency of binary systems $\{Ni^{II}(acac)_2+L^2\}$ (electron-donating ligand-modifier L^2 = MSt (M = Na, Li), MP (MP = N-methylpirrolidon-2), HMPA) as the selective catalyst of ethyl benzene oxidation into α-phenyl ethyl hydro peroxide (PEH), with the introduction in catalytic system of the third component—phenol. The advantage of these ternary systems is the long-term activity of the *in situ* formed complexes $Ni^{II}(acac)_2{\cdot}L^2{\cdot}PhOH$ [2, 3].

The most effective binary and triple catalytic systems as catalysts of ethyl benzene oxidation into α-phenyl ethyl hydro peroxide (PEH) with dioxygen on parameters ***C*** (***C*** = $(\Delta[RH]/[RH]_0)$% (at $S_{PEH} \geq 90\%$, $S_{PEH,max}$ = 90–99%; (L^2 = 18C6); $S_{PEH,max}$ = 90–97% (L^2 = Me_4NBr); $S_{PEH,max}$ = 90% (L^2=MP, NaSt)) and $[PEH]_{max}$ (mass%) are presented in Figure 1.1. As one can see, the best results were obtained in the case of catalysis with system $\{Ni^{II}(acac)_2$+ NaSt + PhOH$\}$ (Figure 1.1).

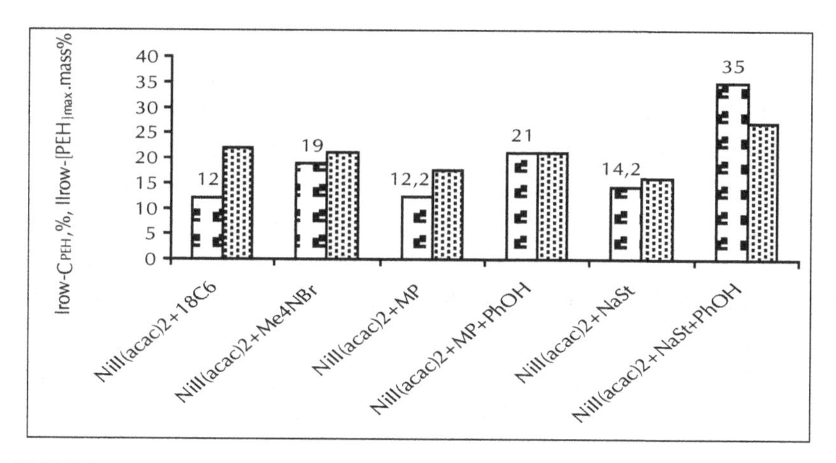

FIGURE 1.1 Values of conversion C_{PEH} (%), maximum values of hydro peroxide concentrations $[PEH]_{max}$ (mass %) in reactions of ethyl benzene oxidation in the presence of the best binary and triple catalytic systems {$Ni(II)(acac)_2+L^2$} and {$Ni(II)(acac)_2+L^2+PhOH$} (L^2 = 18C6, Me_4NBr, MP, NaSt), 120°C.

The preservation of high activity of catalysts–heteroligand complexes $Ni^{II}_xL^1_y(L^1_{ox})_z(L^2)_n(H_2O)_m$ ("A") and heterobinuclear heteroligand triple complexes $Ni^{II}(acac)_2$·NaSt·PhOH during oxidation seems to be due to the formation of the stable supramolecular structures on the basis of ("A"), or triple complexes concluding PhOH, with assistance of intermolecular H-bonds. This hypothesis is evidenced by AFM (Atomic Force Microscopy) data. Thus the new approach to the research of mechanism of catalysis has been offered with the use of AFM technique [6, 7].

Nanostructure science and supramolecular chemistry are fast evolving fields that are concerned with manipulation of materials that have important structural features of nanometer size (1nm–1μm) [8, 9]. Nature has been exploiting noncovalent interactions for the construction of various cell components. For instance, microtubules, ribosomes, mitochondria, and chromosomes use mostly hydrogen bonding in conjunction with covalently formed peptide bonds to form specific structures. The self-assembled systems and self-organized structures mediated by transition metals are considered in connection with increasing research interest in chemical transformations with use of these systems [10].

Hydrogen bonds vary enormously in bond energy from ~15–40 kcal/mol (for the strongest interactions) to less than 4 kcal/mol (for the weakest). It is proposed largely based on calculations that the strong hydrogen bonds have more covalent character, whereas, electrostatics are more important for weak hydrogen bonds, but the precise contribution of electrostatics to hydrogen bonding is widely debated [11]. Hydrogen bonds are important in non-covalent aromatic interactions, where π-electrons play the role of the proton acceptor, which is a very common phenomenon in chemistry and biology. They play an important role in the structures of proteins and DNAs, as well as in drug receptor binding and catalysis [3, 12].

The H-bonding can be a remarkably diverse driving force for the self-assembly and self-organization of materials. H-bonds are commonly used for the fabrication of supramolecular assemblies because they are directional and have a wide range of interaction energies that are tunable by adjusting the number of H-bonds, their relative orientation, and their position in the overall structure. The H-bonds in the center of protein helices can carry 20 kcal/mol due to cooperative dipolar interactions [13, 14].

One of the reasons of stability of heteroligand complexes $M^{II}{}_xL^1{}_y(L^1{}_{ox})_z(L^2)_n(H_2O)_m$ ($L^1{}_{ox} = CH_3COO^-$) ("A") and heteroligand triple complexes $Ni^{II}(acac)_2 \cdot L^2 \cdot PhOH$ in the conditions of oxidation may be a formation of intermolecular H-bonds [2, 3]. One used the AFM method in analytic aims, for research of possibility of the supramolecular nanostructures formation on the basis of nickel heteroligand complexes $Ni_2(AcO)_3(acac)L^2 \cdot 2H_2O$ (L^2 = MP) and binuclear heteroligand $Ni^{II}(acac)_2 \cdot NaSt \cdot PhOH$ with the assistance of H-bonding [6, 7].

1.2 MATERIALS AND METHODS

The AFM SOLVER P47/SMENA/with Silicon Cantilevers NSG11S (NT MDT) with curvature radius 10nm, tip height: 10–15μm and cone angle ≤22° in taping mode with resonant frequency = 150KHz was used.

As substrate the polished silicone surface (especially chemically modified) was used.

Waterproof modified silicone surface was exploited for the self-assembly-driven growth due to H-bonding of heteroligand complexes

$Ni_2(AcO)_3(acac)MP \cdot 2H_2O$ ("A") and $\{Ni^{II}(acac)_2 \cdot NaSt \cdot PhOH\}_n$ with silicone surface.

The saturated solution of complex ("A") in water was put on a surface, maintained some time, and then the water was removed from surface by means of special method—spin-coating process.

In the course of scanning of investigated samples, it has been found that the structures based on ("A") are fixed on a surface not strongly enough to spend measurements in contact mode. So, samples are well measured in taping mode. At that in additional experiments, it has been shown that at scanning there is a turning movement and orientation of investigated structures with cantilever.

The saturated chloroform ($CHCl_3$) solution of complex $Ni^{II}(acac)_2 \cdot NaSt \cdot PhOH$ (1:1:1) was put on a surface, and maintained for some time, and then solvent was removed from the surface by means of special method—spin-coating process.

In the course of scanning of investigated samples, it has been found that the structures are fixed on a surface strongly due to H-bonding. The self-assembly-driven growth of the supramolecular structures formation on the basis of complexes $Ni(II)(acac)_2 \cdot NaSt \cdot PhOH$ due to H-bonds and perhaps the other non covalent interactions were observed on silicone surface. One can watch these structures with big height and volume (Figure 1.2 and 1.3). In check experiments, it has been shown that for binary systems $\{Ni(II)(acac)_2+NaSt\}$, and $\{Ni(II)(acac)_2+PhOH\}$, the formation of the similar structures (height exceeding 2–10 nm) is not observed.

1.3 RESULTS AND DISCUSSION

1.3.1 ROLE OF H-BONDING IN STABILIZATION OF CATALYTIC COMPLEXES, $NI_X(ACAC)_Y(ACO)_Z(L^2)_N(H_2O)_M$ ("A")

It is well known that the transition metals β-diketonates are involved in various substitution reactions. Methine protons of chelate rings in β-diketonate complexes can be substituted by different electrophiles. Formally, these reactions are analogous to the Michael addition reactions [3].

This is a metal-controlled process of the C–C bond formation. The complex $Ni^{II}(acac)_2$ is the most efficient catalyst of such reactions [3].

It has been established that the electron-donating ligand L^2, axially coordinated to $M^{II}L^1_2$ (M = Ni, Fe, L^1 = $acac^-$) controls the formation of primary active complexes $M^{II}L^1_2 \cdot L^2$ and the subsequent reactions in the outer coordination sphere of these complexes. The coordination of an electron-donating extra-ligand L^2 with an $M^{II}L^1_2$ complex favorable for stabilization of the transient zwitter-ion $L^2[L^1M(L^1)^+O_2^-]$ enhances the probability of regioselective O_2 addition to the methine C–H bond of an acetylacetonate ligand, activated by its coordination with metal ions. The outer-sphere reaction of O_2 incorporation into the chelate ring depends on the nature of the metal. Transformation routes of a ligand $(acac)^-$ for Ni and Fe are various, but lead to formation of similar heteroligand complexes $M^{II}_xL^1_y(L^1_{ox})_z(L^2)_n(H_2O)_m$ (L^1_{ox} = CH_3COO^-) ("A") [2, 3]. For nickel complexes, the reaction of acac-ligand oxygenation follows a mechanism analogous to those of Ni^{II}-containing dioxygenase ARD [5] or Cu- and Fe-containing quercetin 2, 3-dioxygenases [15, 16]. Namely, incorporation of O_2 into the chelate acac-ring was accompanied by the proton transfer and the redistribution of bonds in the transition complex leading to the scission of the cyclic system (A) to form a chelate ligand OAc^-, acetaldehyde and CO (in the Criegee rearrangement) (Scheme 1).

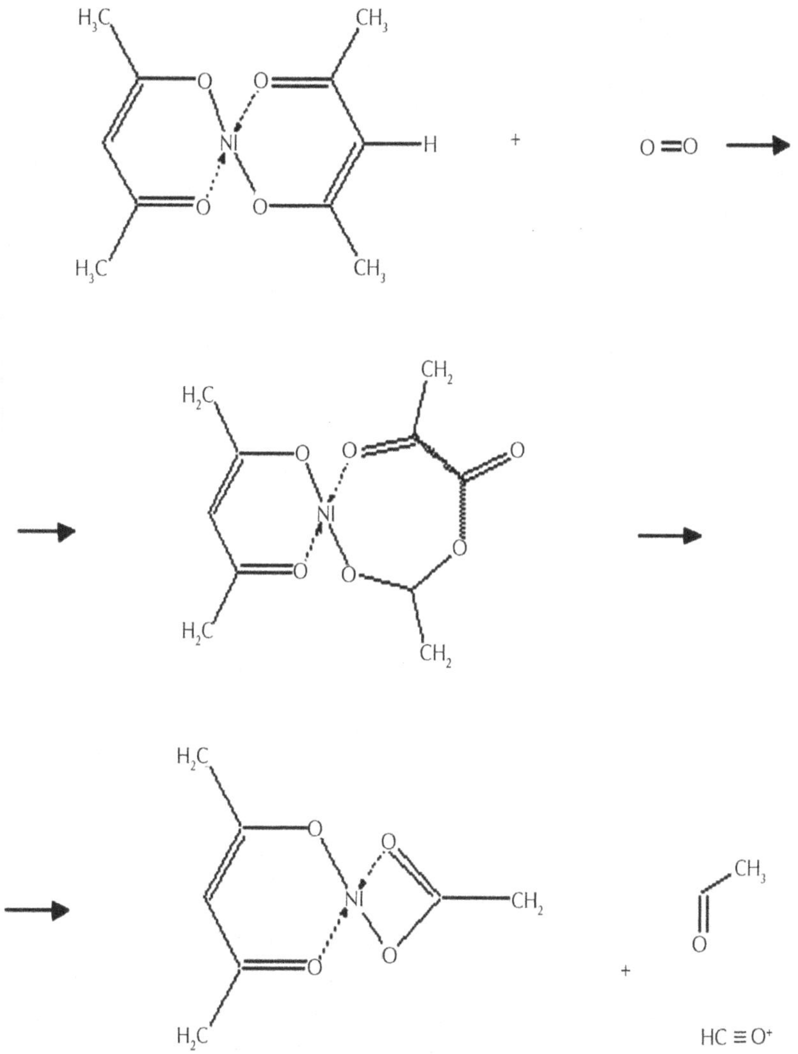

H_3C CH_3
O O
Ni H + O=O →
O O
H_3C CH_3

CH_2
H_2C
O O O
→ Ni →
O O O
H_2C CH_2

H_2C
O O CH_3
→ Ni CH_2
O O O
+
H_2C HC ≡ O^+

Scheme 1

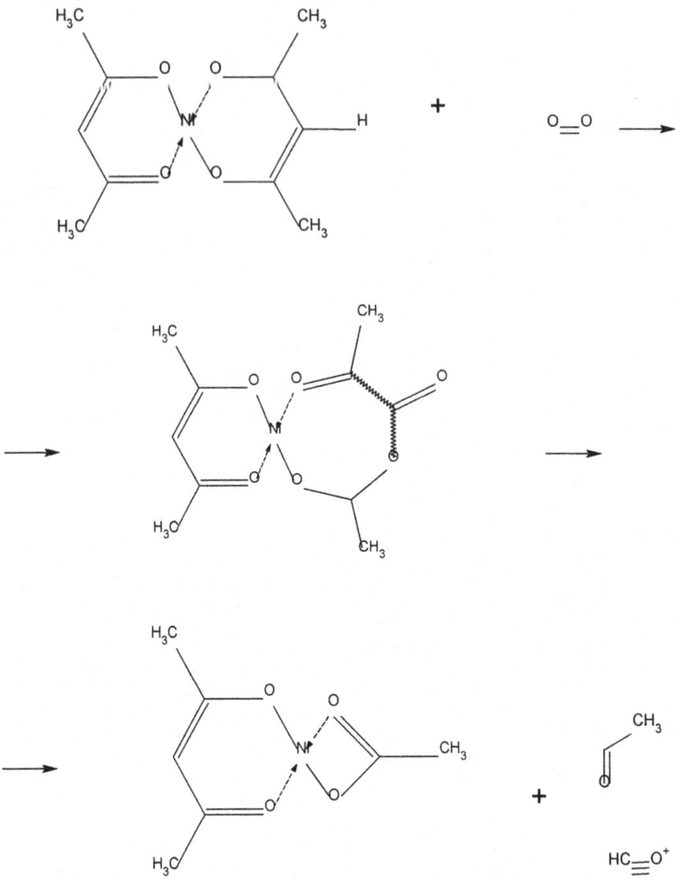

It is known that heteroligand complexes are more active in relation to reactions with electrophiles in comparison with homoligand complexes [3]. Thus the stability of heteroligand complexes $Ni_x(acac)_y(OAc)_z(L^2)_n(H_2O)_m$ with respect to conversion into inactive form $Ni(OAc)_2$ seems to be due to the formation of intermolecular H-bonds. The possibility of supramolecular nanostructures formation was investigated on the example of active complex of type ("A"), formed in the course of ethyl benzene oxidation, catalyzed with system $\{Ni^{II}(acac)_2 + MP\}$. Complex has been synthesized by us and its structure has been defined with mass spectrometry, electron and IR spectroscopy and element analysis [2, 3]. The certain structure of a complex $Ni_2(OAc)_3(acac)MP \cdot 2H_2O$ ("A") ($L^2 = MP$, x = 2, y = 1, z = 3, n = 1, m = 2) predicted on the basis of the kinetic data [2, 3].

Prospective structure of the complex $Ni_2(OAc)_3(acac)MP{\cdot}2H_2O$ ("A") is presented with Scheme 2.

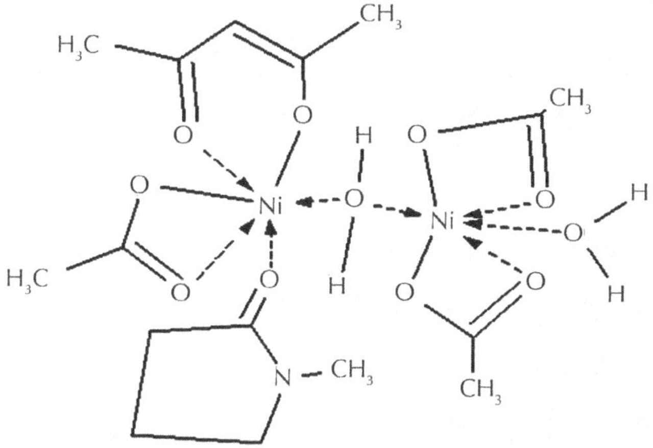

Scheme 2

On the basis of the knowledge from the literature facts, it was possible to assume that binuclear heteroligand complexes $Ni_2(OAc)_3(acac){\cdot}MP{\cdot}2H_2O$ are capable to form macro structures with the assistance of intermolecular H-bonds (H_2O–MP, H_2O–acetate (or $acac^-$) group) [17, 18]. The association of $Ni_2(OAc)_3(acac)MP{\cdot}2H_2O$ to supramolecular structures as a result of H-bonding are demonstrated in Figure 1.4–1.6.

In Figures 1.4–1.6 three-dimensional and two-dimensional AFM image (10×10 μm, 2×2 μm) of the structures formed at drawing of a uterine solution on a surface of modified silicone are presented (μm = micron (micrometer) = 10^3 nm). It is visible that the majority of the generated structures have rather similar form of three almost merged spheres.

As it is possible to see in Figure 1.4a and 1.6, except particles with the form reminding three almost merged spheres (Figure 1.4b), there are also structures of more simple form (with the height approximately equal to 3–4 nm). In Figure 1.6c, profile of one of the particles of the size 3–4 nm on height is presented.

The distribution histogram (Figure 1.4b) shows that the greatest numbers of particles are particles of size 3–4 nm on height.

It is important to notice that for all structures the sizes in plane XY do not depend on height of Z. They make about 200 nm along a shaft passing through two big spheres, and about 150 nm along a shaft crossing the big and small spheres. But all structures are different with respect to height from the minimal 3–4 nm to ~20–25 nm for maximal values. In distribution on height there is a small quantity of particles with maximum height 20–25 nm and considerably smaller quantity with height to 35 nm (Figure 1.4b).

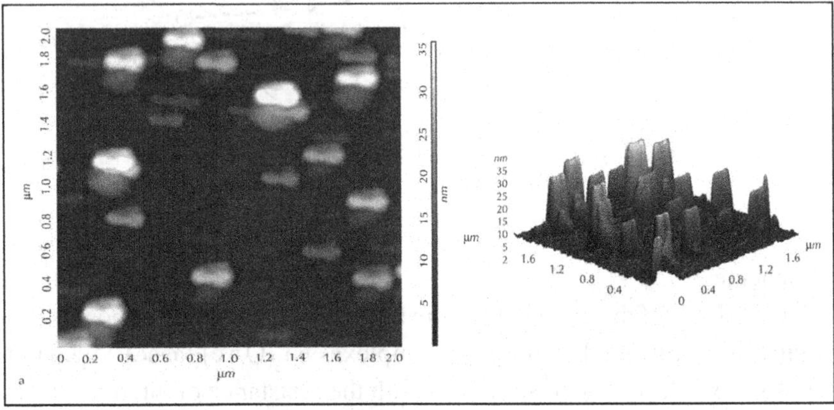

FIGURE 1.4 The AFM two **a** and three-dimensional **b** images of nanoparticles on the basis of $Ni_2(OAc)_3(acac)MP{\cdot}2H_2O$, formed on the surface of modified silicone.

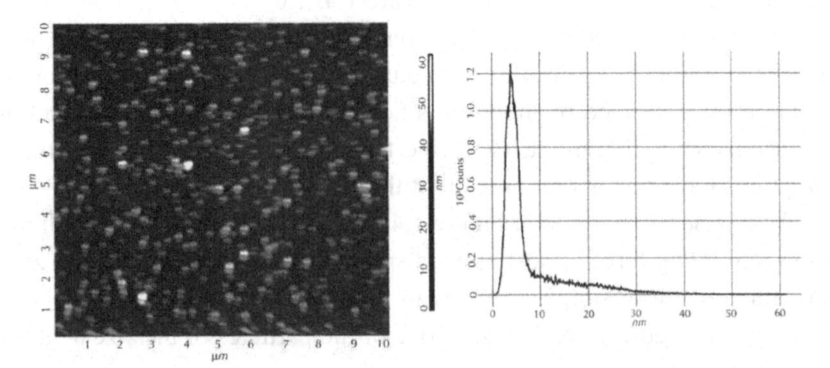

FIGURE 1.5 AFM image of structures (on the basis of $Ni_2(OAc)_3(acac)MP{\cdot}2H_2O$) on the modified silicone surface 10 × 10 μm (at the left). The distribution histogram of height of nanoparticles (to the right).

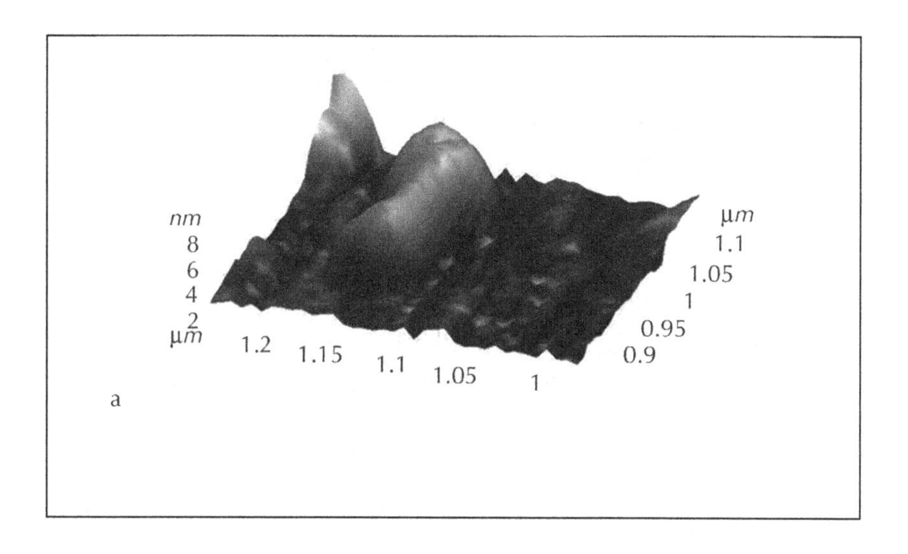

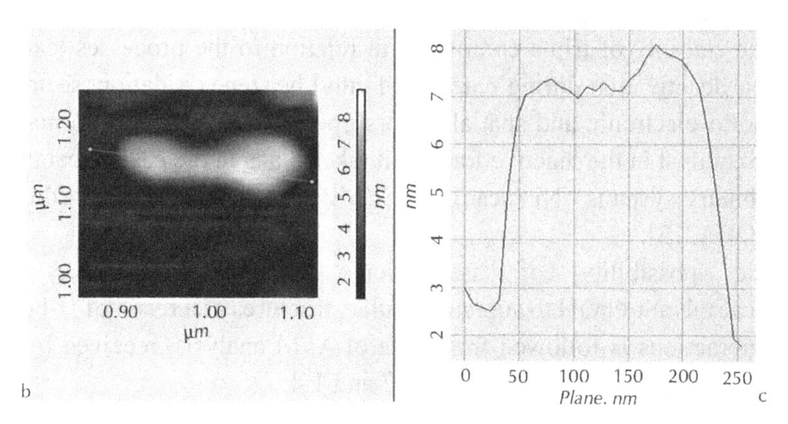

FIGURE 1.6 The AFM three **a** and two-dimensional **b** image and profile of the structure **c** (on the basis of $Ni_2(OAc)_3(acac)MP{\cdot}2H_2O$) with minimum height along the greatest size in plane XY.

Thus, it is shown what the self-assembly-driven growth seems to be due to H-bonding of binuclear heteroligand complex $Ni_2(OAc)_3(acac){\cdot}MP{\cdot}2H_2O$ ("A") with a surface of modified silicone, and further due to directional intermolecular H-bonds, apparently at participation of H_2O molecules, $acac^-$, OAc^- groups, MP [17, 18] (see Scheme 2).

Role of H-bonding in stabilization of triple catalytic complexes $Ni^{II}(acac)_2{\cdot}L^2{\cdot}PhOH$. Nanoparticles $\{Ni^{II}(acac)_2{\cdot}L^2{\cdot}PhOH\}_n$ formation as evidence of possibility of association of $Ni^{II}(acac)_2{\cdot}L^2{\cdot}PhOH$ complexes in supra molecular nanostructures in the course of the ethyl benzene oxidation.

Earlier H-bonding interactions were established in mechanism of formation of catalytic complexes on $Ni^{II}(acac)_2{\cdot}L^2{\cdot}PhOH$ (L^2 = NaSt(MP)) in the course of ethyl benzene oxidation in the presence of triple systems $\{Ni^{II}(acac)_2+L^2+PhOH\}$[3]. Besides the high activity (and stability) of triple complexes on $Ni^{II}(acac)_2{\cdot}L^2{\cdot}PhOH$ (L^2 = NaSt) as catalysts of the ethyl benzene oxidation into PEH (it is expressed in substantial growth of degree of conversion of RH and the yields of PEH) seems to be due to the formation of supramolecular structures $\{Ni^{II}(acac)_2{\cdot}NaSt{\cdot}PhOH\}_n$ in the course of ethyl benzene oxidation as a result of intra- and intermolecular (phenol-carboxylate) H-bonds [18].

The stability of triple complexes in relation to the processes leading to their deactivation during reaction of ethyl benzene oxidation, seems to be due to electronic and spatial factors operating against transformation of the catalyst in the inactive form as it takes place in the case of catalysis with binary systems $\{Ni^{II}(acac)_2+NaSt(MP)\}$ at the absence of additives of PhOH [2, 3].

The possibility of association of triple complexes on $Ni^{II}(acac)_2{\cdot}NaSt{\cdot}PhOH$ to supramolecular structures as a result of H-bonding interactions is followed from data of AFM analysis, received by us. Results are presented in the Figure 1.7 and 1.8.

Figure 1.7 and 1.8 demonstrated three-dimensional and two-dimensional AFM image (30 × 30 and 10 × 10 μm) of the structures on the basis of triple complexes $Ni^{II}(acac)_2{\cdot}NaSt{\cdot}PhOH$, formed at drawing of a uterine solution on a surface of modified silicone.

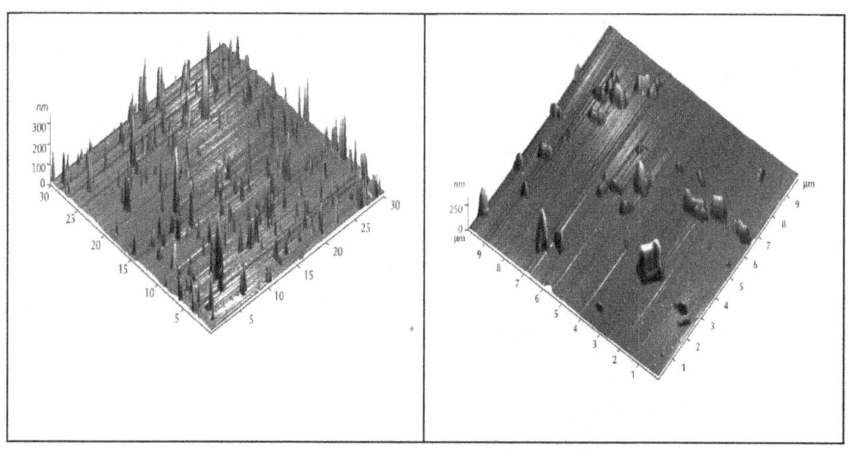

FIGURE 1.7 The AFM three-dimensional image of the structures formed on a surface of modified silicone on the basis of triple complexes $Ni^{II}(acac)_2{\cdot}NaSt{\cdot}PhOH$; **a** 30 30μm and **b** 10 × 10μm.

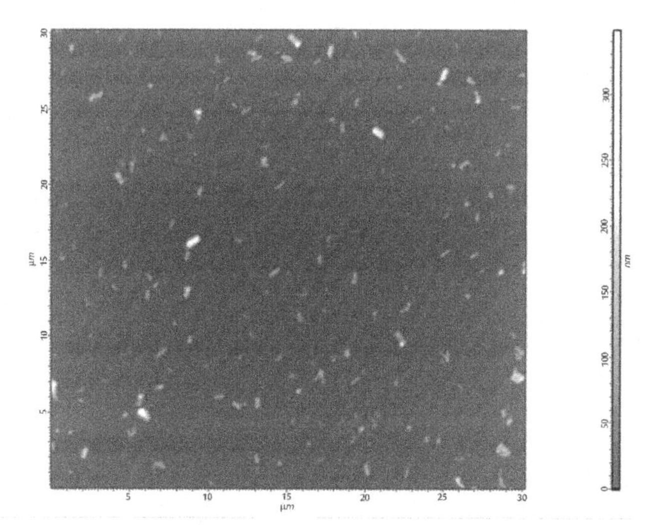

FIGURE 1.8 The AFM two-dimensional image of nanoparticles on the basis Ni(II)$(acac)_2{\cdot}NaSt{\cdot}PhOH$, formed on the surface of modified silicone.

In the Figure 1.9, the histogram of height mean values of the AFM images of nanoparticles on the basis of complexes $Ni^{II}(acac)_2{\cdot}NaSt{\cdot}PhOH$ is presented. As one can see, structures are various at heights ranging from 25 to ~250–300 nm for maximal values. The distribution histogram shows that the greatest numbers are particles of the mean size of 50–100 nm at height.

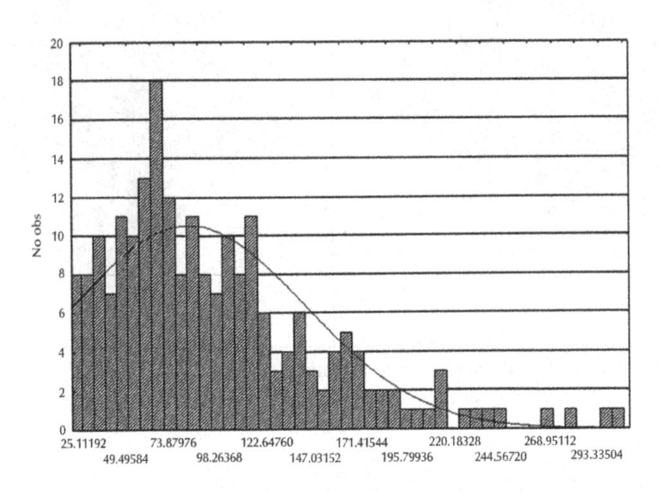

FIGURE 1.9 Histogram of height mean values (nm) of the AFM images of nanostructures based on $Ni^{II}(acac)_2$·NaSt·PhOH formed on the surface of modified silicone.

Table 1.1 shows the mean values of area, volume, height, width, length of nanoscale structures on the basis of triple complexes on $Ni^{II}(acac)_2$·NaSt·PhOH formed on the surface of modified silicone.

TABLE 1.1 The mean values of area, volume, height, length, width of the AFM nanoparticle images on the basis of $Ni^{II}(acac)_2$·NaSt·PhOH formed on the surface of modified silicone.

Variable	Mean values	Confidence	
		-95.000%	+95.000%
Area (μm^2)	0.13211	0.11489	0.14933
Volume (μm^3)	14.11354	11.60499	16.62210
Height (*Z*) (nm)	80.56714	73.23940	87.89489
Length (μm)	0.58154	0.53758	0.62549
Width (μm)	0.19047	0.17987	0.20107

An interesting fact was revealed that the length of the formed nanoparticles in the XY plane exceeds the width of nanoparticles about three times (Table 1.1).

1.4 CONCLUSION

1. Thus, it is shown with use of AFM method the self-assembly-driven growth due to H-bonding of heteroligand complex $Ni_2(AcO)_3(acac)$ MP·$2H_2O$ ("A") with a surface of modified silicone, and further due to directional inter-molecular H-bonds, apparently at participation of H_2O molecules, acac, acetate groups, MP [17-19]. The role of van Der Waals-attractions and π–π bonding seems to take into account.
   H-bonding seems to be one of the factors responsible for the stability of real catalysts—eteroligand complexes $Ni_x(acac)_y(AcO)_z(L^2)_n(H_2O)_m$, in the course of alkylarens (ethyl benzene, cumene) oxidation by dioxygen into hydro peroxide (intermediates in the large-scale production of important monomers) at the presence of catalytic systems $\{Ni^{II}(acac)_2 + L^2\}$.
   As it is established that mechanism of complex "A" formation is analogous to Acireductone dioxygenase Ni^{II}-ARD action, the received results can be used in interpretation of biology effects in Ni^{II}-ARD operation. The transformation of catalytic complexes into supramolecular nanostructures in assistance of intermolecular H-bonding may be one of the factors of regulation of Ni^{II}-ARD activity in plants and living organisms.
2. With the use of AFM method it is shown the self-assembly-driven growth on the basis of binuclear heteroligand triple complexes $Ni^{II}(acac)_2$·NaSt·PhOH due to H-bonding with the modified silicone surface, and further the supramolecular nanostructures formation $\{Ni^{II}(acac)_2\cdot NaSt\cdot PhOH\}_n$ due to directional intermolecular (phenol-carboxylate) H-bonds [18], and, possibly, other non-covalent interactions (van Der Waals-attractions, π–π bonding).

These data support the very probable supramolecular structure appearance on the basis of heterobinuclear heteroligand triple complexes

$Ni^{II}(acac)_2$·NaSt·PhOH in the course of the ethyl benzene oxidation with dioxygen, catalyzed by three-component catalytic system {$Ni^{II}(acac)_2$+NaSt+PhOH} and therefore the high values of the conversion of the ethyl benzene into PEH at the selectivity S_{PEH} preservation at the level not below S_{PEH} = 90% in this process.

KEYWORDS

- **H-Bonds**
- **Nanoparticles**
- **Palladium**
- **Polymeric Matrix**
- **π–π Bonding**

REFERENCES

1. Li, Y. & Zamble, D. B. (2009). *Chemical Reviews, 109,* 4617.
2. Matienko, L. I. (2007). Chapter 2. In D'Amore, A. & Zaikov, G. (Eds.), *Reactions and properties of monomers and polymers* (p. 21), New York: Nova Science Publishers, Inc.
3. Matienko, L. I., Mosolova, L. A., & Zaikov, G. E. (2010). *Selective catalytic hydrocarbons oxidation. new perspectives* (p. 158). New York: Nova Science Publishers, Inc.
4. Weissermel, K. & Arpe, H. J. (1997). *Industrial Organic Chemistry*, 3rd ed., translated by Lindley C. R. (p. 427). New York: VCH.
5. Dai, Y., Pochapsky, Th. C., & Abeles, R. H. (2001). *Biochemistry, 40,* 6379.
6. Matienko, L. I., Binyukov, V. I., Mosolova, L. A., & Mil, E. M. (2011). *Polymer Yearbook 2011* (p. 221). New York: Nova Science Publisher, Inc.
7. Matienko, L. I., Binyukov, V. I., Mosolova, L. A., & Zaikov, G. E. (2011). *Polymer Research Journal, 5,* 151.
8. Leninger, St., Olenyuk, B., & Stang, P. J. (2000). *Chemical Reviews, 100,* 853.
9. Stang, P. J., & Olenyuk, B. (1997). *Accounts of Chemical Research, 30,* 502.
10. Beletskaya, I., Tyurin, V. S., Tsivadze, A. Yu., Guilard, R., & Stem, Ch. (2009). *Chemical Reviews, 109,* 1659.
11. Saggu, M., Levinson, N. M., & Boxer, S. G. (2011). *Journal of the American Chemical Society, 133,* 17414.
12. Ma, J. C. & Dougherty, D. A. (1997). *Chemical Reviews, 97,* 1303.

13. Drain, C. M., Varotto, A., & Radivojevic, I. (2009). *Chemical Reviews, 109,* 1630.
14. Chu, Cheng-Che, Raffy, G., Ray, D., Del Guerzo, A., Kauffmann, B., Wantz, G., Hirsch, L., & Bassani, D. M. (2010). *Journal of the American Chemical Society, 132,* 12717.
15. Gopal, B., Madan, L. L., Betz, S. F., & Kossiakoff, A. A. (2005). Biochemistry, *44,* 193.
16. Balogh-Hergovich, É., Kaizer, J., & Speier, G. (2000). *Journal of Molecular Catalysis. A, Chemical, 159,* 215.
17. Basiuk, E. V., Basiuk, V. V., Gomez-Lara, J., & Toscano, R. A. (2000). *Journal of Inclusion Phenomena and Macrocyclic Chemistry, 38,* 45.
18. Mukherjee, P., Drew, M. G. B., Gómez-Garcia, C. J., & Ghosh, A. (2009). *Inorganic Chemistry, 48,* 4817.
19. Dubey, M., Koner, R. R., & Ray, M. (2009). *Inorganic Chemistry, 48,* 9294.

CHAPTER 2

POLYMER PROCESSING AND FORMATION: PART I

KH. SH. YAKH'YAEVA, G. V. KOZLOV, G. M. MAGOMEDOV, and G. E. ZAIKOV

CONTENTS

2.1 THE TEMPERATURE DEPENDENCE OF SELF-ADHESION BONDING STRENGTH FOR MISCIBLE POLYMERS

At present, it has been generally acknowledged that in self-adhesion bonding the mechanical properties depend on two factors, namely: the macromolecular entanglement formation in the boundary layer and the macromolecular coils interdiffusion kinetics [1]. However, as a rule, an experimental result of treatment of self bonded layers may give rise to only qualitative analysis. Instead, the present communication purpose is the development of quantitative structural model for self-adhesion bonding strength using the fractal analysis notions.

2.1.1 EXPERIMENTAL

Amorphous polystyrene (PS, $M_w = 23 \times 10^4$, $M_w/M_n = 2.84$) and poly (2, 6-dimethyl 1, 4-phenylene oxide) (PPO, $M_w = 44 \times 10^3$, $M_w/M_n = 1.91$) obtained from Dow Chemical and General Electric (USA), respectively, were used. The films of polymers with thickness about 100 microns were prepared by an extrusion method. The PS-PPO interfaces were healed during 10min within the temperature range of 343–396 K at pressure of 0.8 MPa. The lap shear strength of bonded polymers was conducted at temperature 293 K by means of an Instron tensile tester, model 1130 at a cross-head speed of 3×10^{-2} m/s.

2.1.2 RESULTS AND DISCUSSION

In Figure 2.1, the experimental dependence of shear strength, τ_b, of PS-PPO bonding on the bonding temperature is shown. The authors [1] explained the indicated dependence by faster interdiffusion of macromolecular coils at increasing temperature. In a previous work, the shear strength, τ_b, was found to obey the following equation:

$$\ln \tau_b = N_i - 16.6D_f + 20.5 \quad (1)$$

where, N_i is the number of coils intersecting along the boundary layer, physically describing the macromolecular entanglements, density D_f is the macromolecular coil dimension.

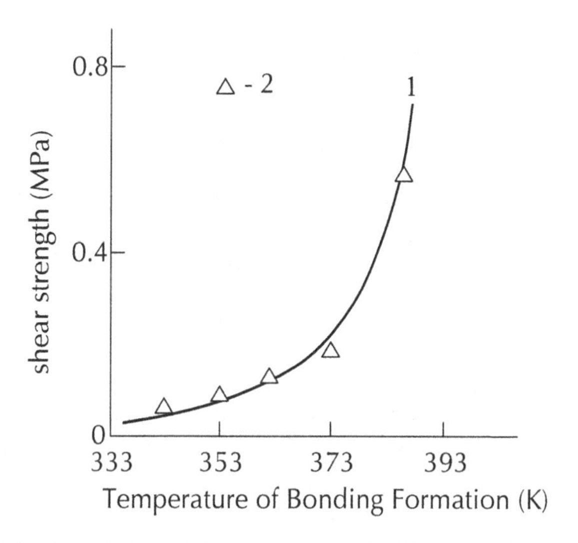

FIGURE 2.1 The dependences of shear strength τ_b of self-adhesion bonding PS-PPO as function of the bonding formation temperature T. *Line*: the calculation according to the equation (1). *Symbols*: experimental data [1].

The equation (1) takes into consideration both the factors influencing the shear strength, τ_b, namely: the number of coils intersections, N_i, that takes into account the macromolecular entanglement formation, and the coil dimension ($-16.6\ D_f$) that accounts for the interdiffusion weakening at increasing D_f.

N_i can be determined according to the following fractal relationship [2]:

$$N_i \sim R_g^{D_{f1}+D_{f2}-d} \quad (2),$$

where, R_g is macromolecular coil gyration radius, D_{f_1} and D_{f_2} are fractal dimension of coils structure, forming self-adhesion bonding, d is the dimension of Euclidean space, in which a fractal is considered (it is obvious that in our case $d = 3$).

For the estimation of the parameter D_f, the following approximated technique will be used. As it is known that the correlation between D_f and structure dimension d_f of linear polymers in the condensed state obey the following equation:

$$D_f = \frac{d_f}{1.5}. \tag{3}$$

with d_f estimation conducted according to the following formula [3]:

$$d_f = 3 - 6\left(\frac{\phi_{cl}}{SC_\infty}\right)^{1/2} \tag{4},$$

where. φ_{cl} is relative fraction of local order domains (clusters), S is the macromolecule cross-section area, C_∞ is the characteristic ratio.

The estimation of φ_{cl} was performed according to the following percolation relationship [3]:

$$\varphi_{cl} = 0.03\left(T_g - T\right)^{0.55} \tag{5},$$

where, T_g and T are glass transition and the bonding temperatures, respectively.

It was already found that for PS $C_\infty = 9.8$ and $S = 54.8$ Å^2 and for PPO $C_\infty = 3.8$ and $S = 27.9$ Å^2 [3]. The macromolecular coil gyration radius was calculated further as follows [3, 4, 5, and 6]:

$$R_g = l_0\left(\frac{C_\infty M_w}{6m_0}\right)^{1/2} \tag{6},$$

where, l_0 is the length of the main chain skeletal bond, which is equal to 0.154nm for PS and 0.541nm for PPO, m_0 is the molar mass per backbone bond ($m_0 = 52$ for PS and $m_0 = 25$ for PPO [4]).

Let us note an important methodological aspect. R_g was calculated according to the equation (6) where C_∞ was considered a variable that can be calculated according to the following relationship [3]:

$$C_\infty = \frac{2d_f}{d(d-1)(d-d_f)} + \frac{4}{3}. \tag{7}$$

In Figure 2.1 the comparison of experimental data and the theoretical curve obtained on the basis of the indicated method (the solid curve) is reported. Based on equation (1), the $\tau_b(T)$ calculation gives rise to the following considerations. Firstly, the mean R_g value for PS and PPO was used as macromolecular coil gyration radius. Secondly, at $T = 386$ K, that is, where $T > T_g$ for PS the value $d_f = 2.95$, that is, the greatest one for real solids [3], was used to calculate C according to the equation (7). As it follows from the data of Figure 2.1, a good correspondence of theory and experiment was obtained that confirms the technique reliability.

2.1.3 CONCLUSION

Hence, from the analysis of the results, the shear strength, τ_b, is influenced by a number of factors, namely: the glass transition T_g of polymers and the bonding temperature, T, the molecular characteristics, C_∞, S, l_0, m_0, and the molecular weight of polymers M_w.

KEYWORDS

- **Amorphous polystyrene**
- **Gyration radius**
- **Interdiffusion kinetics**
- **Lap shear strength**
- **Self-adhesion bonding**

REFERENCES

1. Boiko, Yu. M., & Prud'homme, R. E. (1998). *Journal Polymer Science. Part B: Polymer Physics, 36,* 567–572.
2. Vilgis, T. A. (1988). *Physica A, 153,* 341–354.
3. Kozlov, G. V., & Zaikov, G. E. (2004). *Structure of the Polymer Amorphous State* (p. 465). Leiden: Brill Academic Publishers.
4. Grassia, L., & D'Amore, A. (2009). *Journal of Rheology, 53(2),* 339–356.
5. Schnell, R., Stamm, M., & Creton, C. (1998). *Macromolecules,* 31, 2284–2292.
6. Basile, A., Greco, F., Mader, A., & Carrà, S. (2003). *Plastics, Rubber and Composites,* 32(8–9), 340–344.

CHAPTER 3

POLYMER PROCESSING AND FORMATION: PART II

S. A. BOGDANOVA, O. R. SHASHKINA, V. P. BARABANOV,
G. P. BELOV, G. E.ZAIKOV, and O. V. STOYANOV

CONTENTS

3.1 INFLUENCE OF FREE SURFACE ENERGY OF POLYMERS ON FORMATION OF THEIR INTERPHASE CONTACT WITH OXYETHYLATED ISONONYLPHENOL

The addition products of ethylene oxide to organic compounds have found applications in the processes of producing and processing polymers as surfactants (e.g., as emulsifiers of synthetic latexes), technological additives (to improve the operational characteristics of the epoxide and composite materials, decrease in the external stresses), vulcanizate activators of rubber blends, and antistatics. The possibility of applying these compounds to intensify the wetting and increase in the adhesive and sealing compositions coated on the polymer surface in the production processes of facing polyurethane foams and paint-and-lacquer compositions is of special interest. Oxyethylated compounds are chemically inert, nontoxic, and stable at elevated temperatures and in the aggressive media, do not cause the corrosion of the equipment. The liquid aggregate state at a particular number of oxyethylated groups (*n* oxyethylation degree) promotes the introduction of these additives into the adhesives, melts, and oligomers. Previously, in our investigations, it was shown that the adducts of ethylene oxide and alkylphenols (oxyethylated isononylphenols) take part in the formation of the surface layer of liquid reactive oligomer systems, and composites on their basis [1-3] influence the work of adhesion of oligomers to various substrates [4] and modify the surface properties of the polymers during coating from the solutions [5]. At the same time, the mechanism that determines the character of the contacting-phase interaction in the presence of these additives is not adequately investigated. The study of the combined influence of the oxyethylene group content in the molecules of adducts and surface energy characteristics of the polymers that differ in the nature and content of the functional groups in the surface layer on the adhesive interaction of the oxyethylated compounds with the polymers, will promote the directed selection of additives that regulate the surface phenomena in polymer materials.

The purpose of this work was to study the wetting and adhesion in systems of oxyethylated isononylphenols, that is, surface of solid polymers in the wide range of polarity and parameters.

3.1.1 EXPERIMENTAL

Polyolefins and their copolymers with vinyl acetate and monoxide of carbon, polytetrafluorethylene, polyethylene terephthalate, and cellulose triacetate were used as objects of investigation. Samples of high-pressure polyethylene (HPPE) of 15303-003 and 16803-070 trademarks of OJSC Kazan'orgsintez production (MFI 0.3 and 7.0g/10 min respectively) were obtained by pressing at 150–170°C and pressure of 3.4MPa according to GOST State Standard 16337–77. HPPE-153 additionally contained Irganox 10-10 phenol stabilizer in the amount of 0.1%. To produce thermally oxidized HPPE, samples were held in the oven (forced circulation of air at 473 K) for various lengths of time (15, 30, 60, 90, 180 min).

In the work, polypropylene (PP) of 01030 trademark (70% degree of crystallinity, MFI 3g/10 min) was used. The polypropylene film samples were obtained by pressing the granulated material at 210°C (storage time was 5 min, cooling time was 2 min). The sample of the copolymer of ethylene with polypropylene (CEP) of 02035 trademark (MFI 3.5 g/10 min) were obtained by pressing at 180°C (exposure time was 5 min, cooling time was 2min). The copolymer of ethylene with vinyl acetate (CEVA) of the 11104-030 trademark (Specifications 6-05-1636–97, 5–7% content of vinyl acetate groups) produced by OJSC Sevilen production (Kazan) was used. The samples were obtained by pressing at 140 ± 5°C according to GOST State Standard 12019–66.

In work, we applied a biaxial-oriented film of polyethylene terephthalate (GOST State Standard 24234–80, thickness = 110 μm, and p = 1.4 g/cm^3), as well as a film of cellulose triacetate (deacetylation degree 60.2–62.2%, mean viscosity molecular weight = 1 x 10^{-5}) obtained at OJSC Tasma were applied. Alternating copolymers of ethylene and carbon monoxide (polyethyleneketone, PEK) were synthesized at the Institute of Problems of Chemical Physics of Russian Academy of Sciences using the method of catalytic copolymerization by applying the catalytic system $Pd(CH_3COO)_2$–$Ph_2(CH_2)3PPh_2$–CF_3COOH at a ratio of monomers 50:50 [6]. Depending on the synthesis conditions, copolymers with the various molecular weights (M) determined by the viscosimetric method in meta-cresol at 298K were obtained. The molecular weight characteristics of copolymers of ethylene with carbon monoxide calculated by Mark–Kuhn–

Hou-wink equation with the coefficients $K = 4.31$ and a = 0.75 determined [7] based on the data of vapor phase osmometry and gel-penetrating chromatography (GPC) are given above. The films were prepared from solution in metacresol by watering on the glass substrate with the following evaporation of metacresol at a temperature of about 473 K. The final removal of the solvent from the films was performed at 293 K under vacuum to a constant weight.

η (dl/g)	$M\eta \times 10^{-4}$
0.34	0.65
1.05	2.8
1.18	3.3
3.35	13.1
5.4	44.1
8.4	24.7

In this work, polytetrafluoroethylene (PTFE) was used in the form of sheets made of fluoroplastic-4 of the P trademark (specifications 95-2467–93) made by OJSC Kirovo-Chepetskii Khimkombinat.

Oxyethylated isononylphenols (neonols A F, specifications 2483-077-05766801–98) were obtained at OJSC Nizhnekamskneftekhim. The degree of oxyethylation varied from 3 to 12. The interfacial angle of wetting (inflow wetting angle) was determined by the sitting-drop method at 293 (±1) K. The measurements were performed in the cell with a hydraulic gate using a KM-8 cathetometer equipped with a micrometer attachment. No less than 10 drops of the same volume were coated by a microsyringe on the surface of the polymers. The root-mean-square deviation was 1.

To estimate the roughness of the investigated surfaces, the profilograms on a P-203 profilograph-pro-filometer were analyzed. The root-mean-square deviation of the profile was applied as the main parameter of roughness and determined according to GOST State Standard 25142–82. We calculated the roughness coefficient K_r, which was considered during the determination of the interfacial angle by the Ventsel–Deryagin equation [8]. For all investigated surfaces, K_r is in the range of 1.085–1.001.

The free surface energy (FSE) of the polymers and its constituents were determined by the graphical method based on the values of the interfacial angle of the surfaces using test liquids based on the Fowkes concept and Owens–Wendt equation [9, 10]. Freshly distilled water, glycerol, formamide, dimethylformamide, dimethylsulfoxide, methylene iodide, phenol, and aniline were used as test liquids. The FSE of the samples was determined immediately before performing the investigations. Considering that the films of high-molecular compounds are characterized by the capacity for long-term storage in the adsorbed molecules of low-molecular compounds, the repeated applications of the same polymer samples were completely excluded.

The acidity parameter of the polymer surface *D* was determined by the method in [11]. The surface tension of the oxyethylated isononylphenols on the boundary with the air was determined in the thermostated cell via the method of retraction using a Wilhelmi glass plate at 293 (±1)K. The infrared spectra of HPPE samples used as films were investigated on a Specord IP-75IK spectrometer and identified by the absorption bands at 1306, 1896, 1376, 909, and 888 cm^{-1} [12].

3.1.2 RESULTS AND DISCUSSION

The free surface energy of the polymers (γ_s) is a highly sensitive parameter that significantly influences the intensity of the interfacial interactions in polymer systems. The acid–base constituent of the free surface energy (γ_s^{ab}) which is closely connected with the presence of the functional groups in the surface layer, that is, active centers that promote adhesive interaction is especially informative. To investigate the interactions of the oxyethylated isononylphenols with the low-energy surfaces, the polymers with sequential increase in γ_s^{ab} were chosen. Their surface energy characteristics and parameters of acidity of the surface are presented in Table 1. Free surface energy of polymers (γ_s), their acid–base (γ_s^{a}) and dispersion (γs) constituents, and acidity parameter *D*As can be seen in Table 3.1 that the polytetrafluorethylene, which is often used as the standard non polar surface has the smallest value of γs.

Polymer	γ_s, mN/M	γ^{ab}_s, mN/M	γ^d_s, mN/M	D,(mN/m)$^{0.5}$
Polytetrafluor-ethylene (PTFE)	18.3	0.09	18.2	–3.8
Copolymer of ethylene with propylene (CEP)	30.7	0.7	30.0	1.3
Polypropylene (PP)	28.6	1.5	27.1	0.1
High-pressure polyethylene HPPE-153	29.2	2.2	27.0	–2.2
High-pressure polyethylene HPPE-168	29.3	4.2	25.1	–1.6
Polyethylene terephthalate (PETP)	43.5	7.9	35.6	7.8
Copolymer of ethylene with vinyl acctate (CEV)	38.4	9.0	29.4	2.3
Cellulose triac-etate (CTA)	43.9	14.2	29.7	5.5
Polyethylenek-etone (PEK)	42.3	22.7	19.6	3.4

fab *fab*

High values of γs were noted for the case where the copolymer of ethylene alternates with carbon monoxide (polyethyleneketone), the special characteristics of which are conditioned by the inclusion of a polar comonomer in a polyolefin chain. The values of the acidity parameter indicate that in the surface layer of the investigated polymers either the donor $(D > 0)$ or acceptor $(D < 0)$ character of the functional groups dominates. As a whole, the surface energy characteristics of other investigated polymer materials correspond to the literature data [13].

The intensity of the adhesive interaction of liquid with the surface can be judged by the work of adhesion (W_a), which is calculated by the Dupre–Young equation

$$W_a = \gamma_l(1 + \cos\theta),$$

where, γ_l is the surface tension of liquid and θ is the interfacial angle of the solid phase of this liquid [8].

To calculate the work of adhesion of oxyethylated isononylphenols to the surface of the investigated polymers, the results of the determination of their surface tension on the boundary with air obtained by the modified method of Wilhelmi and data of determination of the surface equilibrium interfacial angle via method of "sitting drop" were used by us.

The equilibrium values of the interfacial angles were determined based on the preliminary kinetic investigations. An analysis showed that the degree of oxyethylation responsible for the length of the polar fragment of the adduct molecule influences on the wetting parameter. The dependence of the wetting of polytetrafluorethylene (PTFE), copolymer of ethylene with propylene (SEP), and polypropylene (PP) surfaces on the degree of oxyethylation n is shown in Figure 3.1. Here, it is obvious that as n increases, the wetting decreases nonlinearly, which inidcates the prevailing realization of the lipophilic interfacial interactions of polymers and adducts. For PTFE (Figure 3.1, curve 1) in the region of $n = 6–9$, a region is observed with almost constant values of cosθ; in the region of $n = 8$, the value of cosθ even increases slightly. The wetting of polypropylene (PP) and copolymer of ethylene with propylene (SEP) surfaces depending on n is presented in Figure 1 by curves 2 and 3, respectively. It is obvious that the emergence of polar groups on the surface layer of the polymer leads to an increase in wetting, which indicates that the polar fragments of the molecules of the oxyethylated compounds are included in the interfacial interactions. Furthermore, as the length of the polar chain increases, the observed decrease in the wetting ability of the adducts is less significant than on the PTFE surface.

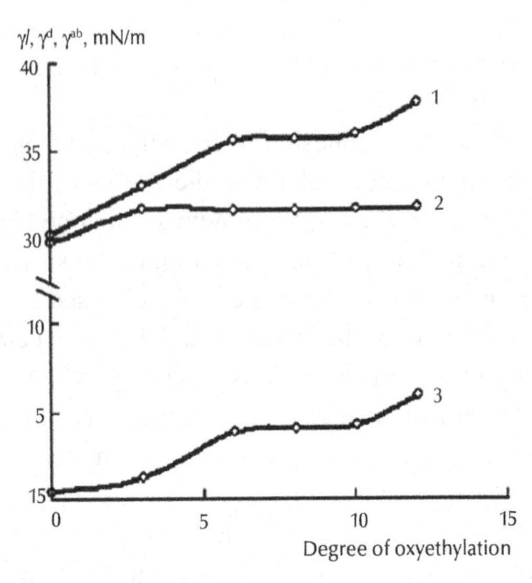

FIGURE 3.1 Dependence of surface tension of (*1*) oxyethylated isononylphenols, (*2*) its dispersion, and (*3*) acid–base constituents on degree of oxyethylation

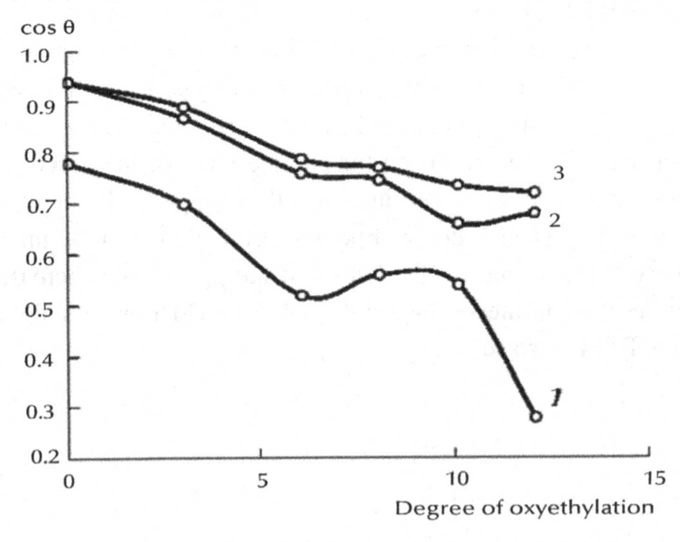

FIGURE 3.2 Dependence of wetting on degree of oxyethylation for oxyethylated isononylphenols on surfaces of (*1*) polytetrafluorethylene, (*2*) copolymer of ethylene with propylene, and (*3*) polypropylene.

The oxidation processes have a γ significant influence on the surface properties of the polymers. It should be taken into account during the selection of surfactants and other additives based on the ethylene oxide adducts that interact with the polymer material surface, especially during their long exploitation. On the surface of thermally oxidized polyethylene, a significant amount of ketones, carboxyl, ester, and other groups are found that influence the polar constituent values of the free surface energy. Figure 3.2 presents the dependence of energy characteristics of the HPPE surface on the time of thermal oxidation. As can be seen from Figure 3.2, the increase in γ_s is conditioned by the increase in the polar constituent, which is evidence of the emergence of the polar groups in the surface layer, and γ_s^d hardly changes. As was noted above, the basic HPPE trademarks (153 and 168) that we used differ in their molecular weight and the presence of 0.1% stabilizer in HPPE-153, which is reflected in the quantitative differences of the FSE value. The unstabilized HPPE-168 sample has a value of $\gamma_s^{a\,b}$ = 4.2 mN/m, while stabilized HPPE-153 has a value of $\gamma_s^{a\,b}$ = 2.2 mN/m.

Thermally oxidized HPPE samples were investigated using infrared spectroscopy. Using the relative intensities, the degree of crystallinity, branching, and relative optical density were calculated for all samples. The results of Table 1 are evidence of an increase in the degree of branching [CH_3] for HPPE-168; consequently, HPPE-168 quickly underwent destruction compared to the data using γ_s^p for both trademarks. The crystallinity coefficient (*k*) is almost unchanged, since polyethylene is crystallized quickly even at small temperatures and it is known that its boundary layer is amorphized [14]. During heating the content of CO groups in a polymer changes which is characterized by the relative optical density D_{co}/D_b, where D_b is a basic optical density, that is, unoxidized PE.

The results of estimating the wetting parameters of a high-pressure polyethylene surface by oxyethylated isononylphenols are presented in Figure 3.3 and 3.4.

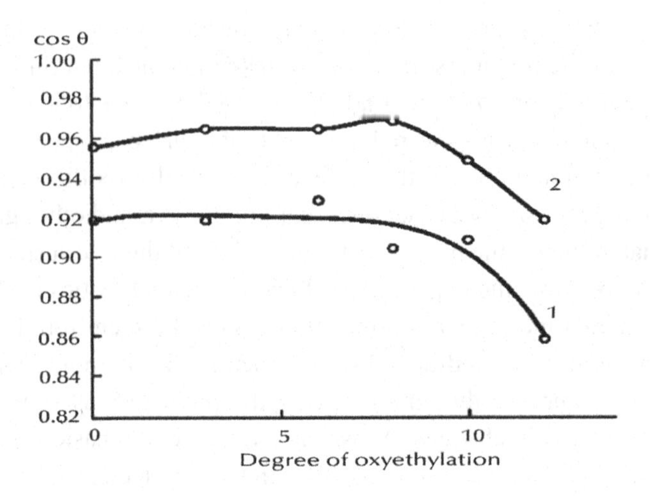

FIGURE 3.3 Dependence of wetting on degree of oxyethylation for oxyethylated isononylphenols on surfaces of (*1*) polyethylenes HPPE-153 and (*2*) thermally oxidized HPPE-153.

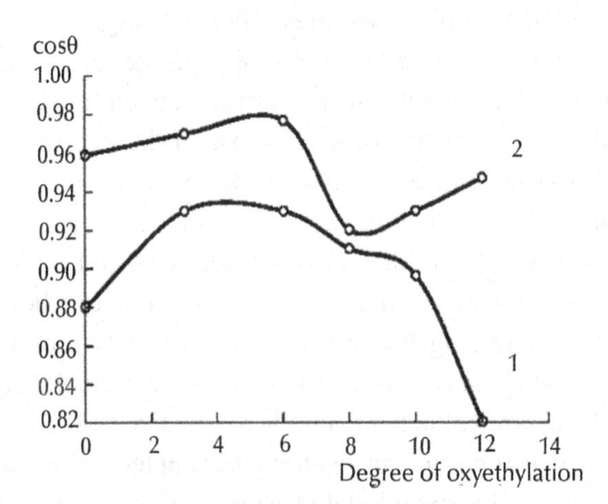

FIGURE 3.4 Dependence of wetting on the degree of oxyethylation for oxyethylated isononylphenols on surfaces of (*1*) poly ethylenes HPPE-168 and *(2)* thermally oxidized HPPE-168.

The influence of the degree of oxyethylation on the interaction between the adduct and the surface is revealed in different ways for unoxidized surfaces, as well as surfaces subjected to thermal oxidation. For HPPE-

153 (Figure 3.3), as the degree of oxyethylation rises the wetting ability remains high and almost constant, slightly decreasing only for adducts with $n > 8$–9. If the polar groups emerge on the surface as a result of oxidation, the character of the dependence hardly changes, although the wetting decreases. For HPPE-168 (Figure 3.4), the value of $\gamma_s^{a\,b}$ which is not subjected to the thermal oxidation of the samples, increases upto 4.2 mN/m. This is evidence of the emergence of a significant amount of polar groups during the formation of the HPPE sample from melt in the surface layer before thermal oxidation is carried out by air oxygen. As a whole, an insignificant decrease in wetting has been noted as the amount of the polar groups of adduct increases; furthermore, the dependence is stepwise with a break at $n = 8$. For thermally oxidized HPPE-168 ($y_s = 10$ mN/m), a sharper change in cost occurs, that is, a decrease in the region of $n = 8$ and a rise in wetting at high degrees of n.

As the degree of surface oxidation rises,, an increase occurs in the acidity parameter D (Figure 3.5).

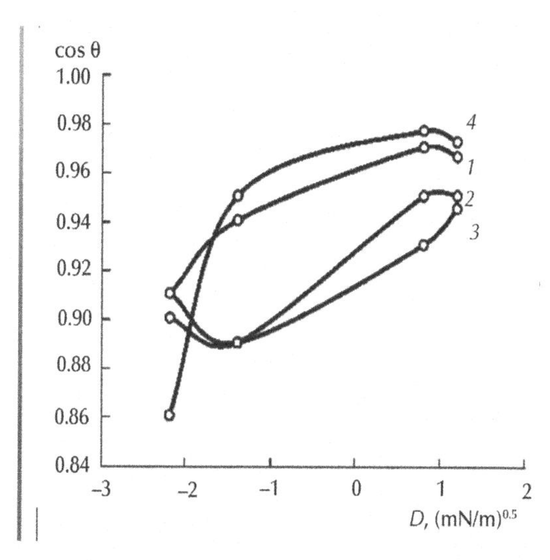

FIGURE 3.5 Dependence of wetting of oxyethylated isononylphenols on D parameter of thermally oxidized surfaces of HPPE-153: (*1*) $n = 6$; (*2*) $n = 8$; (*3*) $n = 10$; (*4*) $n = 12$.

The initial sample (HPPE-153) has the value $D = -2.2$ (mN/m)$^{0.5}$, which indicates the presence of the main groups, that is, proton acceptors in the surface layer in particular carbonyl. With growth in *ys* during oxidization, the surface strongly favors acid properties. This is probably connected with the emergence of OH and COOH groups. With an increase in the value of D, oxyethylated alkylphenol with $n = 12$ demonstrates the highest wetting ability in relation to the polymer. It is obvious that the wetting increases as a result of the donor–acceptor interactions of ester oxygen with the adduct molecule and acid groups present in the surface layer.

The obtained data shows that the wetting on the non polar surface is conditioned by the dispersion forces. To analyze the influence of y_s of HPPE on cosθ for oxidized samples, it is convenient to use the values of change in the cosine of the interfacial angle as follows:

$$\Delta\cos\theta = (\cos\theta - \cos\theta_0),$$

where, $\cos\theta_0$ is the cosine of the equilibrium interfacial angle of the given nonionic surfactant on the polyethylene surface, which was not subjected to thermooxidative destruction, and cosθ is the equilibrium value of the interfacial angle cosine on the oxidized surface under consideration.

The dependence of the change in Δcosθ on the acid–base constituent of FSE for OE nonylphenols is given in Figure 3.6.

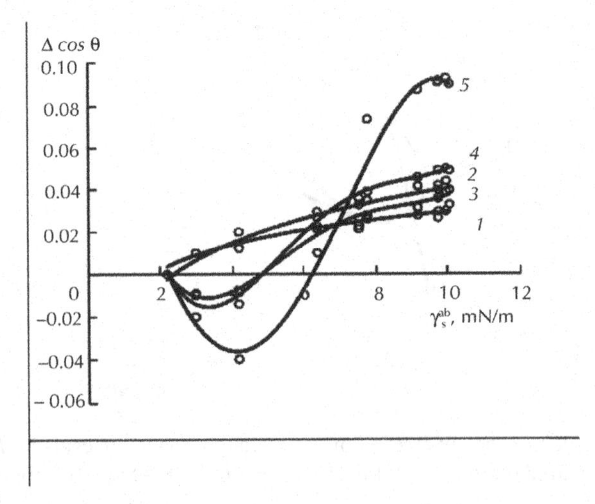

FIGURE 3.6 Dependence of cosθ change of oxyethylated isononylphenols on acid–base constituent of free surface energy of HPPE: (*1*) $n = 3$; (*2*) $n = 6$; (*3*) $n = 8$; (*4*) $n = 10$; (*5*) $n = 12$.

The results show that for OE nonylphenols with low n, an insignificant increase in the wetting ability was noted. As for adducts with high n ($n > 6$), the region of negative values, which indicates the decrease in the wetting ability for non oxidized HPPE is noted with low γ_s^{ab}. With increasing FSE and its acid–base constituent, the wetting improves, and for $n = 12$, it reaches almost complete spreading.

In the case when FSE γ_s^{ab} increases, the polar regions of the surfactant also enter into the interaction with the surface, which obtains this ability after $n = 6$ and under the condition of the presence of polar groups in the surface boundary layer.

Figure 3.7 presents the dependence of wetting on n for oxyethylated isononylphenols on surfaces with the greatest values of polar constituent in the explored series of polymers; they spread almost completely over the entire range of variations in the number of the oxyethylene groups. Apparently, the wetting is conditioned not only by dispersion forces, but also and, to a great extent, by the polar interactions. The correlation between the acid–base constituent of the free surface energy and the value of the interfacial angle remains. Thus, the highest value of cosθ was obtained for a surface with $\gamma_s^{ab} = 24$ mN/m.

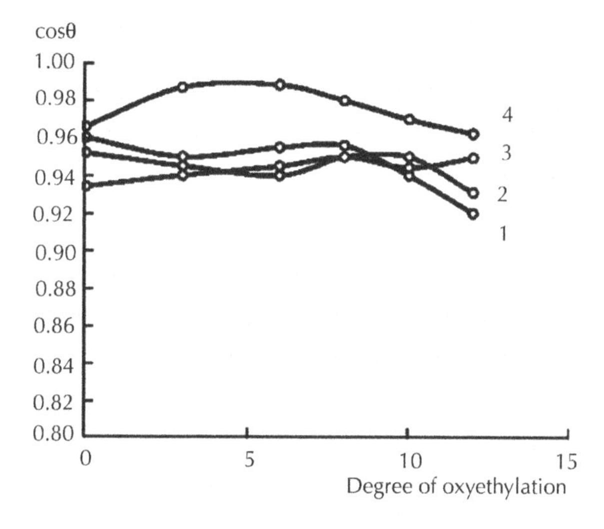

FIGURE 3.7 Dependence of wetting on degree of oxyethylation for oxyethylated isononylphenols on surfaces of polymers (*1*) CEVA, (*2*) CTA, (*3*) PETP, and (*4*) PEK.

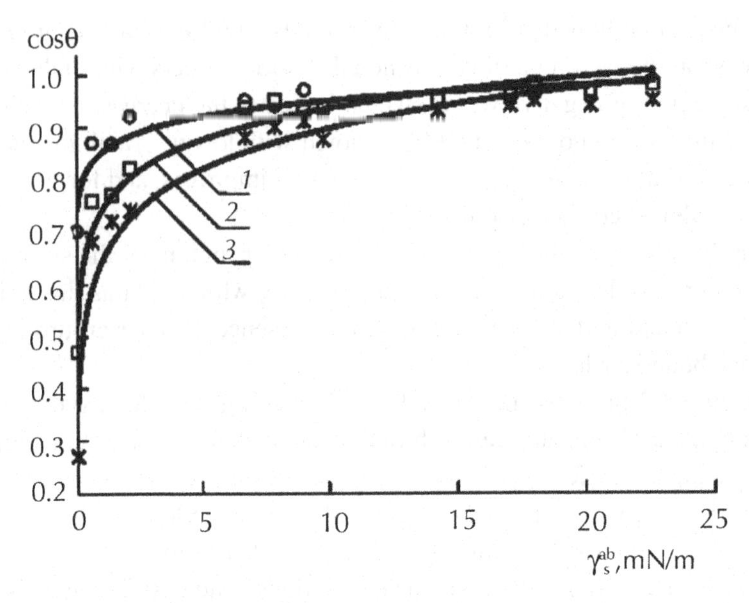

FIGURE 3.8 Dependence of wetting of oxyethylated isononylphenols on acid–base constituent of free surface energy of polymers: (*1*) $n = 3$; (*2*) $n = 6$; (*3*) $n = 12$

Figure 3.8 shows the generalized dependence of the wetting ability of oxyethlated iso-nonylphenols in relation to the surfaces of the studied polymers depending on the acid–base constituent of the free surface energy of the solid surface. Analysis shows that for lower-energy surfaces (γ_s^a ^{b} lies in the range of 0.09–4 mN/m), it is necessary to differentiate the approach to select an adduct with a particular degree of oxyethylation to use as modifying and surfactant additives. For surfaces with high contents of polar groups, oxyethylated isononylphenols show variability in this effect.

Using the considered results of the wetting estimation and the equilibrium values of the surface tension of oxyethylated isononylphenols at the boundary with air determined by the Wilhelmi method, we calculated the work of adhesion (W_a). The obtained data are given in Table 3.2.

TABLE 3.2 Work of adhesion (W_a) of oxyethylated isononylphenols to polymer surfaces

Polymers	W_a (mN/m), at oxyethylation degree				
	3	6	8	10	12
PTFE	54.0	48.3	49.1	54.2	46.8
CEP	59.0	56.2	55.5	58.4	61.5
PP	59.0	56.2	55.9	61.1	62.9
HPPE-153	60.8	61.0	59.0	67.1	68.1
HPPE-153 (oxidized)	62.0	62.5	62.1	68.5	70.3
HPPE-168					
HPPE-168	61.0	61.4	59.1	66.7	66.6
(oxidized)	62.1	62.5	58.8	66.7	70.7
CEV					
PETP	62.4	62.3	61.7	67.8	69.9
CTA	61.0	61.8	61.6	67.9	71.7
PEK	61.6	61.7	61.6	68.6	70.7
	62.9	63.1	62.6	69.1	71.4

Based on the values of γ_s^d of the free surface energy of polymers and γ_l^d of the surface tension of oxyethylated isononylphenols, it can be judged that the significant contribution to the work of adhesion are made by the dispersion forces; however, the data of Table 3.2 show that the acid–base non dispersion constituent of the work of adhesion significantly influences the inter action with surface. As n increases, the differences in the adhesion to PTFE and polethyleneketones, which in the investigated range of γ_s^{ab} values, lies in the region of the lowest and highest values, are already significant. On non polar surfaces, as n increases, the work of adhesion as a whole decreases, and on the surface of the polymers for which γ_s^{ab} values are high, it increases. The significant increase in W_a occurs at high (8–12) n.

Based on the investigations of the adhesive interactions of the contacting phases, it can be concluded that the adhesion of oxyethylated compounds with a low content of oxyethylene groups to polymers for which γ_s^{a} ^{b} is small, is conditioned by the dispersion forces caused by the interaction of the non polar hydrocarbon radicals. As the polar groups emerge on the surface layer of the polymer and the length of the polar oxyethylene chain increases, the oxyethylated groups are included in the adhesive interactions with the surface. High adhesion was noted in polymers that differ in their positive values of *D* acidity parameter.

3.1.3 CONCLUSION

The established connection between the value of the free surface energy of the polymers and intensities of the adhesive interaction, which is determined by the combined influence of the polar oxyethylene chain length and the presence of polar groups of various character in the polymer surface layer promote the scientifically based selection of modifying additives for the effective influence on the surface properties of polymer systems.

KEYWORDS

- **Free surface energy**
- **Mark–Kuhn–Hou-wink equation**
- **Oxyethylated compounds**
- **Polyolefins**
- **Thermo-oxidative destruction**

REFERENCES

1. Slobozhaninova, M. V., Bogdanova, S. A., Bara-banov, V. P., & Deberdeev, R. Ya. (2004). *Lakokras Mater Ikh Pri-men, 7,* 29–31.
2. Ebel', A. O., Bogdanova, S. A., Slobozhaninova, M. V. et al. (2001). In *Structure and Dynamics of Molecular Systems: Collected Papers,* Yoshkar-Ola, Khimiya: Moscow. Vol 8, pp. 155– 158.
3. Slobozhaninova , M. V., Bogdanova , S. A., & Barabanov, V. P. (2005). *Effect of oxyethylated alkylphenol additives on surface energy characteristics of cured epoxy compositions.* Vest: KGGU.

4. Slobozhaninova, M. V., Bogdanova, S. A., Bara-banov V. P., & Lakokras, O. V. (2007). *Mater. Ikh Primen. 1–2,* 68–73.
5. Sautina, N. V., Bogdanova, S. A., Barabanov, V. P. (2009). Khimiya: Moscow.
, No. 2, 77–83.
6. Belov, G. P. (1998). *Vysokomol. Soedin., Ser. B, 40(3),* 503–517.
7. Belov, G. P., Prudskova, T. N., & Urman, Ya. G. et al. (1998). *Vysokomol. Soedin., Ser. A, 40(4),* 571–575.
8. Summ, B. D. (2007). *Fundamentals of colloid chemistry: a textbook,* (ed. 2nd). Akademiya: Moscow [in Russian].
9. Fowkes, F. M. (1983). In K. L. Mittal (Ed.), *Physicochemical aspects of polymer surfaces* (Vol. 2, pp. 583–595). New York: Plenum.
10. Owens, D. K., Wsndt, R. C. (1969). *Journal of Applied Polymer Science, 13*(8), 1740–1748.
11. Berger, E. J. (1990). *Journal of Adhesive Science and Technology, 4*(5), 373–391.
12. Kazitsyna , L. A., & Kupletskaya, N. B. (1971). *Application of UV, IR, and NMR Spectroscopy in Organic Chemistry.* Moscow: Vysshaya Shkola [in Russian].13. Kinlok, E. (1991). *Adhesion and adhesives: Science and technology.*Mir: Moscow, 1991) (in Russian).
14. Pugachevich, P. P., Beglyarov, E. M., & Lavygin, I. A. (1982). *Surface Phenomena in Polymers.* Khimiya: Moscow.

CHAPTER 4

POLYMER PROCESSING AND FORMATION: PART III

V. A. PANKRATOV, O. A. SDOBNIKOVA, and N. S. SHMAKOVA

CONTENTS

4.1 SURFACE-ACTIVE QUATERNARY AMMONIUM SALTS IN POLYMER PROCESSING

Quaternary ammonium salts (QAS) as the main representatives of cationic surfactants are widely used in different branches of industry. However, they are practically not used in polymer processing, except vulcanization of the rubber [1]. Meanwhile, high surface activity, a variety of structure of hydrophobic radicals and cationic centers make it possible to use QAS for modification of different plastics, especially on the basis of filled polymers. The antimicrobial activity typical of all quaternary salts gives the advantages in developing polymer films for food products.

A range of cationic surfactants produced in Russia is quite limited. Only water and alcoholic solutions or pastes of alkyl trimethyl- and alkyl dimethyl benzyl ammonium halogenides are sold at reasonable prices. In polymer blends, however, all components should generally be crystalline or powdered. Thus, we synthesized a group of new, previously undescribed salts to conduct different tests.

Most of QAS contain one hydrophobic radical (generally alkyl) and one cationic center. We synthesized the salts containing two hydrophobic radicals on the same or different nitrogen atoms and also including oxymethylene group in radicals or in a bridge between the cationic centers.

Mono-quaternary salts were obtained by alkylation of secondary and tertiary amines or nitrogenous heterocycles with higher alkyl bromides and chloromethyl ethers of higher alcohols.

$$2\,RBr + R'_2NH \longrightarrow [R_2NR'_2]^+ Br^- \quad (I)$$

$$ROCH_2Cl + RNR'_2 \longrightarrow [ROCH_2\text{—}N(R')_2R]^+ Cl^- \quad (II)$$

$$ROCH_2Cl + NC_5H_5 \longrightarrow [ROCH_2\text{—}NC_5H_5]^+ Cl^- \quad (III)$$

Bis-quaternary salts were obtained by alkylation of tertiary diamines or by reaction of dibromoalkanes, bis(chloromethyl) ethers of glycols, and chlorex with higher tertiary amines:

$$2RBr + R'_2N(CH_2)_nNR'_2 \;\text{or}\; 2\,R{-}NR'_2 + Br(CH_2)_nBr \longrightarrow \left[R{-}\overset{R'}{\underset{R'}{N}}{-}(CH_2)_n{-}\overset{R'}{\underset{R'}{N}}{-}R \right]^{2+} 2Br^- \quad (IV)$$

$$2\,ROCH_2Cl + R'_2N(CH_2)_nNR'_2 \longrightarrow \left[ROCH_2{-}\overset{R'}{\underset{R'}{N}}{-}(CH_2)_n{-}\overset{R'}{\underset{R'}{N}}{-}CH_2OR \right]^{2+} 2Cl^- \quad (V)$$

$$2\,R{-}NR'_2 + ClCH_2O(CH_2CH_2O)_nCH_2Cl \longrightarrow \left[R{-}\overset{R'}{\underset{R'}{N}}{-}CH_2O(CH_2CH_2O)_n{-}CH_2\overset{R'}{\underset{R'}{N}}{-}R \right]^{2+} 2Cl^- \quad (VI)$$

$$2\,R{-}NR'_2 + ClCH_2CH_2OCH_2CH_2Cl \longrightarrow \left[R{-}\overset{R'}{\underset{R'}{N}}{-}CH_2CH_2OCH_2CH_2\overset{R'}{\underset{R'}{N}}{-}R \right]^{2+} 2Cl^- \quad (VII)$$

Synthesized salts represent white solid crystalline substances. Most of the salts (except I) are soluble in water; the solutions form an abundant foam. In order to extend the field of application of the new QAS, the surface tension of their water solutions and bactericidal activity were measured. The main characteristics are shown in Table 4.1.

TABLE 4.1 The structure and properties of quaternary ammonium salts I-IV

No	Reference designation	R	R' (R'_2NH)	n	Tm (°C)	Surface tension of 0.5% water solution	Bactericidal dilutions upon exposure for 20 min at 20°C	
							S. aureus	*E. coli*
1.	I a	$C_{10}H_{21}$	C_2H_5	-	88	-	-	-
2.	I b	$C_{12}H_{25}$	C_2H_5	-	59	-	-	-
3.	I c	$C_{15}H_{31}$	C_2H_5	-	144	-	-	-
4.	I d	$C_{16}H_{33}$	C_2H_5	-	72	-	-	-
5.	I e	$C_{12}H_{25}$	(piperidine)	-	150	-	-	-
6.	I f	$C_{12}H_{25}$	(morpholine)	-	150	-	-	-
7.	I g	$C_{15}H_{31}$	(morpholine)	-	165	-	-	-
8.	II a	$C_{12}H_{25}$	(morpholine)	-	176	-	-	-
9.	II b	$C_{12}H_{25}$	(piperidine)	-	126	-	-	-
10.	III	$C_{14}H_{29}$	-	-	56	33	1:10000	1:1000
11.	IV a	$C_{11}H_{23}$	CH_3	2	165	29	-	-
12.	IV b	$C_{12}H_{25}$	CH_3	2	167	27.9	-	-
13.	IV c	$C_{14}H_{29}$	CH_3	2	143	33.9	-	-
14.	IV d	$C_{16}H_{33}$	CH_3	2	140	-	-	-
15.	IV e	$C_{12}H_{25}$	C_2H_5	2	182	30.5	-	-
16.	IV f	$C_{16}H_{33}$	C_2H_5	2	125	-	-	-
17.	IV g	$C_{12}H_{25}$	CH_3	1	162	29.5	-	-
18.	IV h	$C_{12}H_{25}$	(piperidine)	2	126	33.2	-	-

TABLE 4.1 *(Continued)*

19.	IV i	$C_{12}H_{25}$	(piperidine)	3	150	36.2	-	-
20.	IV j	$C_{12}H_{25}$	C_2H_5	1	-	29	-	1:3000
21.	IV k	$C_{12}H_{25}$	C_2H_5	4	196	40	-	-
22.	IV l	$C_{12}H_{25}$	(piperidine)	4	168	36	-	-
23.	IV m	$C_{12}H_{25}$	(morpholine)	4	189	40	-	1:10000
24.	IV n	$C_{12}H_{25}$	CH_3	6	185	26	-	-
25.	IV o	$C_{12}H_{25}$	(piperidine)	6	204	38.5	-	1:3000
26.	IV p	$C_{12}H_{25}$	(morpholine)	6	230	32.9	-	1:3000
27.	IV q	$C_{12}H_{25}$	C_2H_5	8	119	41	-	1:3000
28.	V a	$C_{12}H_{25}$	CH_3	2	-	32	-	-
29.	V b	$C_{12}H_{25}$	C_2H_5	2	-	34.6	-	-
30.	VI a	$C_{10}H_{21}$	CH_3	0	-	38.4	1:2000	1:2000
31.	VI b	$C_{10}H_{21}OCH_2$	CH_3	0	-	35	1:10000	1:2000
32.	VI c	$C_{10}H_{21}OCH_2$	C_2H_5	0	-	34.2	1:1000	1:1000
33.	VI d	$C_{10}H_{21}$	CH_3	1	-	38.4	1:2000	1:1000
34.	VI e	$C_{16}H_{33}$	CH_3	1	56	49.1	1:2000	1:1000
35.	VI f	$C_{10}H_{21}OCH_2$	CH_3	2	-	35.1	1:2000	1:10000
36.	VI g	$C_{10}H_{21}$	CH_3	2	-	44.5	1:10000	1:20000
37.	VI h	$C_{10}H_{21}$	C_2H_5	2	-	36.7	1:1000	1:2000
38.	VII i	$C_{10}H_{21}$	CH_3	-	-	29.1	1:1000	1:2000
39.	VII j	$C_{16}H_{33}$	CH_3	-	80	38.7	1:2000	1:2000

The cellulose ethers and esters and polyethylene were studied as polymer subjects. The cellulose ethers and esters are used in developing ecologically clean packaging materials. In order to improve biodegradability and lower the price of a film, it was proposed to use starch as filler; however, its addition notably increases viscosity of melt that deteriorates the processibility of a material.

We studied the effect of QAS additives on rheological properties of the filled and unfilled samples of cellulose esters and ethers—diacetyl cellulose (DAC) and ethyl cellusose (EC).

The experiments were carried out by the capillary viscometry method using the IIRT Instrument. The specific test parameters were selected for each polymer material.

The preliminary results showed that the character of the melt flow did not change on incorporation of cationic surfactants (both mono- and bis-quaternary) into polymer matrix; however, effective viscosity notably decreased.

It is known that QAS demonstrate the effect in very small concentrations, and they are used in too high quantities in many technological processes. Using diacetyl cellulose plasticized with 35% of triacetin as an example, we found that the optimum addition of bis-quaternary salt IVn was as low as 0.1% (Figure 4.1).

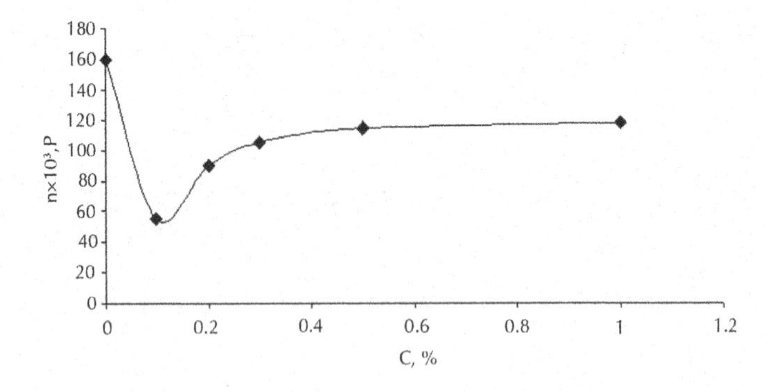

FIGURE 4.1 The viscosity of the melt of DAC modified by the additive IVn.

Such a low quantity of an additive lowers the requirements to the cost of a modifier. Thus, we considered that it was expedient to test all types of salts that we synthesized (Figure 4.2).

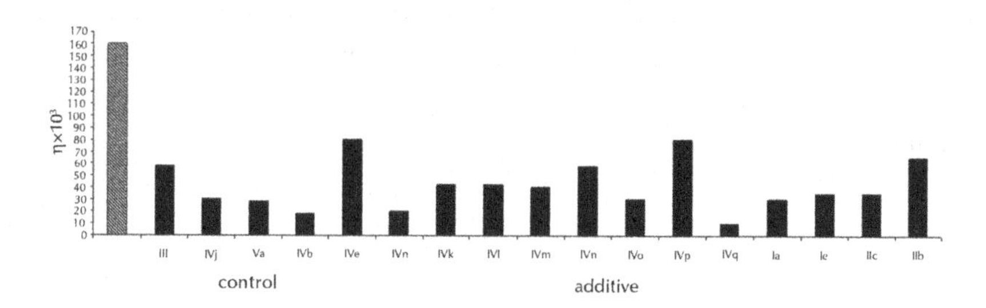

FIGURE 4.2 The viscosity of the melt of DAC with 0.1% of different additives.

Regrettably, we did not manage to establish the regular dependence of viscosity decrease upon the structure of the tested samples of QAS, but the high effectiveness of relatively cheap chlorides (IIc, Va and, of course, III) indicates the reality of their practical application.

Figure 4.3 shows the high effectiveness of the AMPIK (III) preparation.

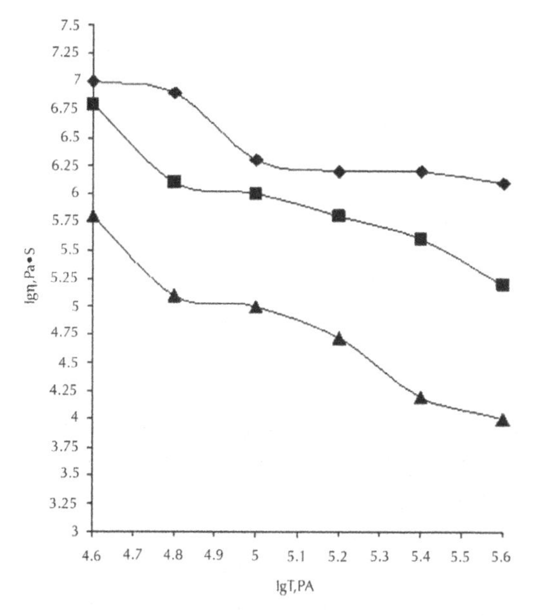

FIGURE 4.3 Flow curves for DAC with surfactant additives.♦ DAC; ■ DAC + IVn; ▲ DAC + III.

The modification of filled compositions is of the utmost interest. Figure 4.4 demonstrates the decrease in the melt viscosity of the ethyl cellulose-triacetin-starch (100:35:50) system by more than an order on the incorporation of 0.1% III.

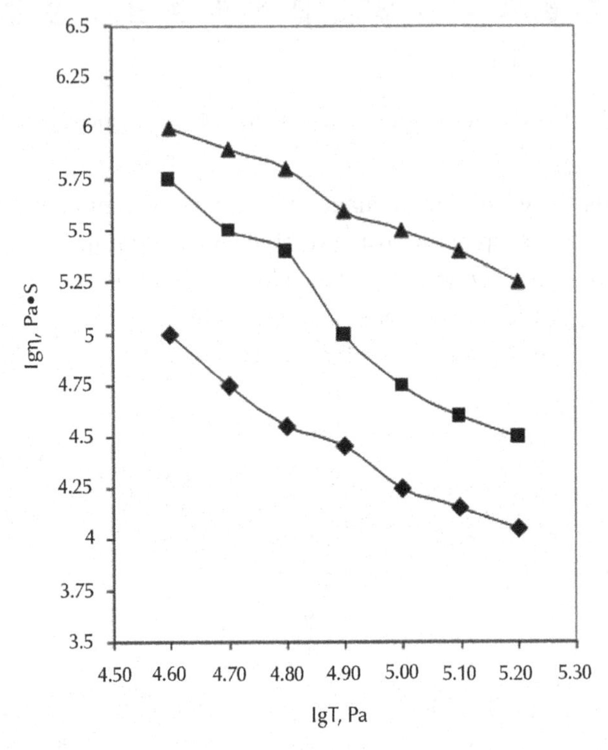

FIGURE 4.4 Flow curves for filled ethyl cellulose (EC). ■ EC + plasticizer; ▲ EC + plasticizer + filler; ♦ EC + plasticizer + filler + III.

The rheological properties of polyethylene are much less affected by QAS. The technology of polyethylene films production using an extruder with circular head does not allow to obtain a homogeneous film with low content of a surfactant. An increase in melt flow index (MFI) in the presence of high concentrations of an additive was observed in some films. MFI for polymer compositions is shown in Table 4.2.

TABLE 4.2 MFI of polymer compositions.

Composition	MFI (g/10min)
HDPE	0.8
HDPE + 0.5% CTAB	1.0
HDPE + 2% IVb	1.2

The small amounts of QAS additives also improve the strength properties of polymer compositions. The comparative studies of the samples of plasticized diacetyl cellulose with the mono-quaternary (III) and bis-quaternary (IVn) salt additives using the tearing machine demonstrated the significant increase in both elongation and breaking stress at rupture (Figure 4.5).

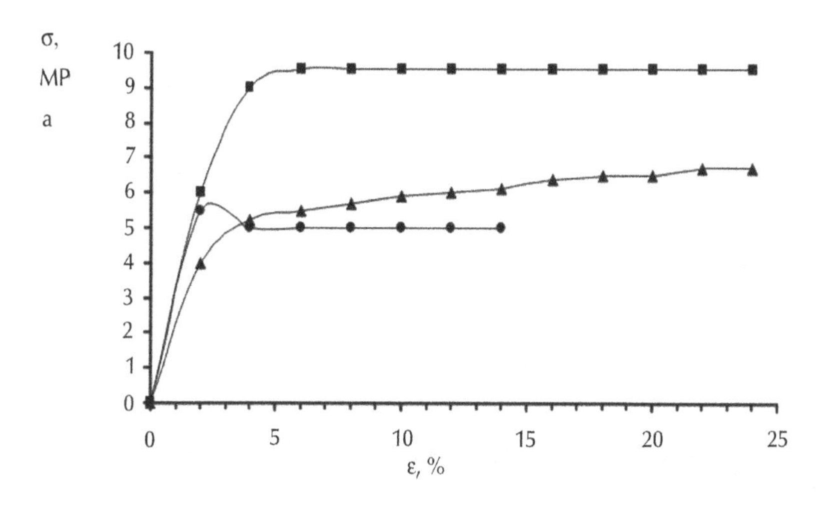

FIGURE 4.5 Tension curves for the samples of DAC. ● control DAC; ▲ DAC modified with III; ■ DAC modified with IVn

The physical and mechanical properties of polyethylene are less influenced by QAS. Mono-quaternary and bis-quaternary salts therewith show the effect in higher concentrations (Figure 4.6 and 4.7).

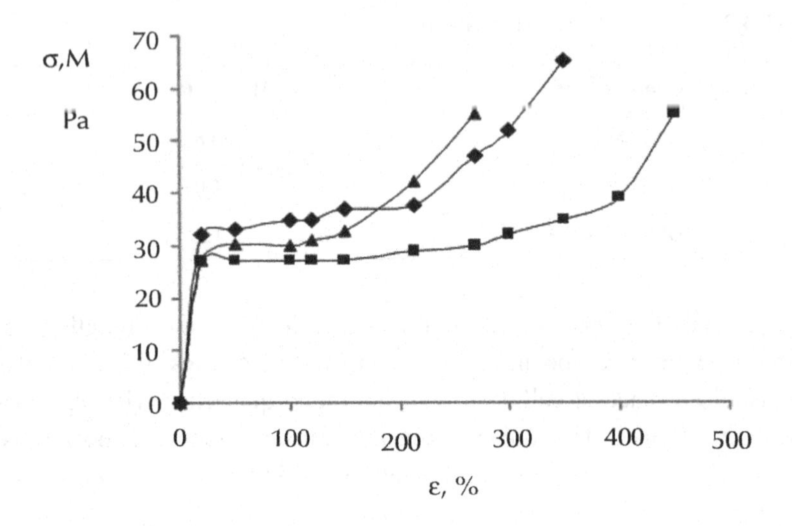

FIGURE 4.6 Tension curves for films from HDPE. ▲ Pure HDPE; ♦ HDPE + 0.5% CTAB; ■ HDPE + 3% CTAB.

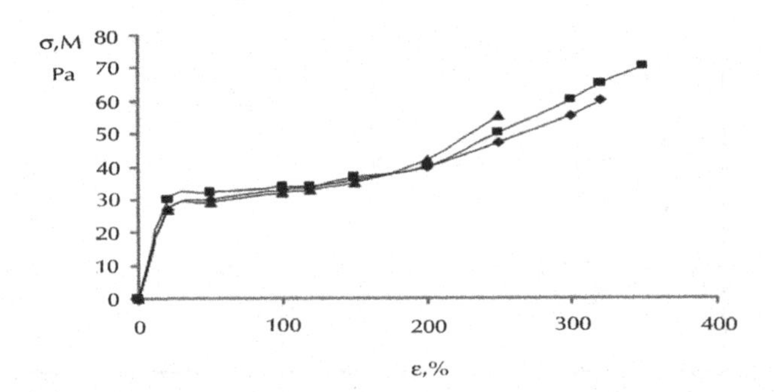

FIGURE 4.7 Tension curves for films from HDPE. ▲ Pure HDPE; ♦ HDPE + 0.5% IVd; ■ HDPE + 3% IVd.

Both mono- and bis-quaternary ammonium salts significantly improve elongation at rupture. The strength of polymer materials therewith does not decrease.

In order to evaluate the possibility of use of the modified films for food products packaging, the study on the detection of the modifying additives migration from a film into environment was carried on.

The water extracts from polymer films were the subjects of research; the exposure time was 31 days. Two methods were used for the qualitative detection of the additives migration.

First method: The qualitative reaction on the halogen ion using 2% solution of $AgNO_3$. The qualitative reaction did not reveal the presence of the halogen ion in the water extracts.

Second method: Biotesting on protozoan microorganisms [2]. *Infusoria Tetrahymena pyriformis* were used as test-microorganisms. The shapes and the character of movements of infusoria placed into the water extracts were assessed using a microscope. The observations were made after 1 and 2hr. On the basis of the observations, it can be concluded that there is no migration of the additives from the obtained polymer materials into the model medium.

4.2 CONCLUSION

The examination of fungicidal activity of the polymer films was conducted according to the method of the laboratory mold resistance testing. The method establishes the presence of fungicidal properties in a material [3]. Samples (ø25 mm) are cut from a film, placed into a culture medium and inoculated with spores. The test duration is 28 days. A material has fungicidal properties if a zone of the absence of fungal growth is seen on a culture medium around a sample or fungal growth with scores 0 and 1 on a six-point scale is observed on the surface and the edges of a sample. Table 4.3 presents the results of the test.

TABLE 4.3 The results of the test

Composition	Growth on a surface	Score
HDPE	35% affect	4
HDPE + 0.5% CTAB	Growth on the edges of a sample	1
HDPE + 3% CTAB	Growth on the edges of a sample	1
HDPE + 0.5% ED-120	Growth on the edges of a sample	1

TABLE 4.3 *(Continued)*

HDPE + 2% ED-120	Growth on the edges of a sample	1
HDPE + 0.5% ED-160	Growth on the edges of a sample	1
HDPE + 2% ED-160	Growth on the edges of a sample	1

The test demonstrated the pronounced fungicidal activity of the polymer compositions.

KEYWORDS

- **Ammonium halogenides**
- **Fungicidal activity**
- **Quaternary ammonium salts**
- **Rheological properties**
- **Viscometry method**

REFERENCES

1. Pankratov, V. A., Romanova, T. V., Fomin, A. G., & Fonsky, D. Yu. (1995). *The influence of triethyl alkyl ammonium bromides on sulphur vulcanization of diene rubbers* (N 3). Moscow: Kauchuk I Resina.
2. Tcheremnyh, E. G., Simbireva, E. I. (2009). *Infusoria Taste Food* (N 1). Moscow: Khimiya I Zhizn.
3. Snezhko, A. G., Kulaeva, G. V., Dontsova, E. P. (2006). *Materials with antimicrobial activity. PNILPMiPP MGUPB* (N5). Moscow: Meat Technologies.

CHAPTER 5

TRANSFORMATION OF HIGH-ENERGY BONDS IN ATP

G. A. KORABLEV, N. V. KHOKHRIAKOV, G. E. ZAIKOV, and YU.G. VASILIEV

CONTENTS

5.1 INTRODUCTION

With the help of spatial energy concept it is demonstrated that the formation and change of high-energy bonds in ATP take place at the functional transitions of valence-active orbitals of the system “phosphorus-oxygen”.

These values of energy bonds are in accord with experimental and quantum-mechanical data.

5.2 METHODS AND MATERIALS

5.2.1 SPATIAL-ENERGY PARAMETER

During the interaction of oppositely charged heterogeneous systems the certain compensation of volume energy of interacting structures takes place which leads to the decrease in the resulting energy (e.g., during the hybridization of atomic orbitals). But this is not the direct algebraic deduction of the corresponding energies. The comparison of multiple regularities of physical, chemical and biological processes allows assuming that in such and similar cases the principle of adding the reciprocals of volume energies or kinetic parameters of interacting structures is executed.

Lagrangian equation for the relative movement of the system of two interacting material points with the masses m_1 и m_2 in coordinate x is as follows:

$$M_{red}\, x'' = -\frac{\partial U}{\partial x}, \text{where} \frac{1}{m_{red}} = \frac{1}{m_1} + \frac{1}{m_2} \qquad (1), (1a)$$

here U—mutual potential energy of material points; m_{red}—reduced mass. Herein $x'' = a$ (system acceleration).

For the elementary areas of interactions Δx can be accepted: $\frac{\partial U}{\partial x} \approx \frac{\Delta U}{\Delta x}$.

Then: $m_{red} a\Delta x = -\Delta U$; $\qquad \frac{1}{1/(a\Delta x)} \cdot \frac{1}{(1/m_1 + 1/m_2)} \approx -\Delta U$

or: $\frac{1}{1/(m_1 a\Delta x)+1/(m_2 a\Delta x)} \approx -\Delta U$

Since the product $m_i a\Delta x$ by its physical sense equals the potential energy of each material point ($-\Delta U_i$), then: $\frac{1}{\Delta U} \approx \frac{1}{\Delta U_1} + \frac{1}{\Delta U_2}$ (2)

Thus the resulting energy characteristic of the system of two interacting material points is found by the principle of adding the reciprocals of initial energies of interacting subsystems.

"The electron with the mass m moving about the proton with the mass M is equivalent to the particle with the mass: $m_{red} = \frac{mM}{m+M}$" [1].

Therefore, modifying the equation (2) It can be assumed that the energy of atom valence orbitals (responsible for interatomic interactions) can be calculated [2] by the principle of adding the reciprocals of some initial energy components based on the equations:

$$\frac{1}{q^2/r_i} + \frac{1}{W_i n_i} = \frac{1}{P_E} \text{ or } \frac{1}{P_0} = \frac{1}{q^2} + \frac{1}{(Wrn)_i}; \; P_E = P_0/r_i \qquad (3), (4), (5)$$

where: W_i—orbital energy of electrons [3]; r_i—orbital radius of i orbital [4]; $q=Z^*/n^*$—by [5, 6], n_i—number of electrons of the given orbital, Z^* and n^*—nucleus effective charge and effective main quantum number, r—bond dimensional characteristics.

The value P_0 is called a spatial-energy parameter (SEP), and the value P_E—effective P-parameter (effective SEP). Effective SEP has a physical sense of some averaged energy of valence orbitals in the atom and is measured in energy units, for example, in electron-volts (eV).

The values of P_0-parameter are tabulated constants for electrons of the given atom orbital.

For SEP dimensionality:

$$[P_0] = [q^2] = [E] \cdot [r] = [h] \cdot [v] = \frac{kgm^3}{s^2} = J \cdot m,$$

where [E], [h] and [υ]—dimensionalities of energy, Planck's constant and velocity.

The introduction of P-parameter should be considered as further development of quasi-classical concepts with quantum-mechanical data on atom structure to obtain the criteria of phase-formation energy conditions. For the systems of similarly charged (e.g. orbitals in the given atom) homogeneous systems the principle of algebraic addition of such parameters is preserved:

$$\Sigma P_E = \Sigma(P_0 / r_i); \cdot \Sigma P_E = \frac{\Sigma P_0}{r} \quad (6), (7)$$

$$\text{or: } \Sigma P_0 = P_0^{'} + P_0^{''} + \ldots; \cdot r\Sigma P_E = \Sigma P_0 \quad (8), (9)$$

Here P-parameters are summed up by all atom valence orbitals.

To calculate the values of P_E-parameter at the given distance from the nucleus either the atomic radius (R) or ionic radius (r_I) can be used instead of r depending on the bond type.

Let us briefly explain the reliability of such an approach. As the calculations demonstrated the values of P_E-parameters equal numerically (in the range of 2%) the total energy of valence electrons (U) by the atom statistic model. Using the known correlation between the electron density (β) and intra-atomic potential by the atom statistic model [7], the direct dependence of P_E-parameter can be obtained on the electron density at the distance r_i from the nucleus. The rationality of such technique was proved by the calculation of electron density using wave functions by Clementi [8] and comparing it with the value of electron density calculated through the value of P_E-parameter.

The modules of maximum values of the radial part of Ψ-function were correlated with the values of P_0-parameter and the linear dependence between these values was found. Using some properties of wave function as applicable to P-parameter, the wave equation of P-parameter with the formal analogy with the equation of Ψ-function was obtained [9].

5.2.2 WAVE PROPERTIES OF P-PARAMETERS AND PRINCIPLES OF THEIR ADDITION

Since P-parameter has wave properties (similar to Ψ'-function), the regularities of the interference of the corresponding waves should be mainly fulfilled at structural interactions.

The interference minimum, weakening of oscillations (in antiphase) occurs if the difference of wave move (Δ) equals the odd number of semi-waves:

$$\Delta = (2n+1)\frac{\lambda}{2} = \lambda\left(n+\frac{1}{2}\right), \text{ where n} = 0, 1, 2, 3, \ldots \tag{10}$$

As applicable to P-parameters this rule means that the interaction minimum occurs if P-parameters of interacting structures are also "in antiphase"—either oppositely charged or heterogeneous atoms (e.g., during the formation of valence-active radicals CH, CH_2, CH_3, NO_2 ..., etc) are interacting.

In this case P-parameters are summed by the principle of adding the reciprocals of P-parameters – equations (3, 4).

The difference of wave move (Δ) for P-parameters can be evaluated via their relative value $\left(\gamma = \frac{P_2}{P_1}\right)$ of relative difference of P-parameters (coefficient α) which at the interaction minimum produce an odd number:

$$\gamma = \frac{P_2}{P_1} = \left(n+\frac{1}{2}\right) = \frac{3}{2}; \frac{5}{2} \ldots$$

When $n = 0$ (main state) $\frac{P_2}{P_1} = \frac{1}{2}$ (11)

It should be pointed out that for stationary levels of one-dimensional harmonic oscillator the energy of these levels $\varepsilon = h\nu\left(n+\frac{1}{2}\right)$, therefore in quantum oscillator, in contrast to the classical one, the least possible energy value does not equal zero.

In this model the interaction minimum does not provide zero energy corresponding to the principle of adding reciprocals of P-parameters – equations (3, 4).

The interference maximum, strengthening of oscillations (in phase) occurs if the difference of wave move equals the even number of semi-waves:

$$\Delta=2n\frac{\lambda}{2}=\lambda n \text{ or } \Delta=\lambda(n+1).$$

As applicable to P-parameters the maximum interaction intensification in the phase corresponds to the interactions of similarly charged systems or systems homogeneous by their properties and functions (e.g., between the fragments or blocks of complex inorganic structures, such as CH_2 and NNO_2 in octogene).

And then:

$$\gamma=\frac{P_2}{P_1}=(n+1) \quad (12)$$

By the analogy, for "degenerated" systems (with similar values of functions) of two-dimensional harmonic oscillator the energy of stationary states:

$$\varepsilon = h\nu(n + 1)$$

By this model the interaction maximum corresponds to the principle of algebraic addition of P-parameters—equations (6–8). When $n=0$ (main state) we have $P_2=P_1$, or: the interaction maximum of structures occurs if their P-parameters are equal. This concept was used [2] as the main condition for isomorphic replacements and formation of stable systems.

5.2.3 EQUILIBRIUM-EXCHANGE SPATIAL-ENERGY INTERACTIONS

During the formation of solid solutions and in other structural equilibrium-exchange interactions the unified electron density should be established in the contact spots between atoms-components. This process is accompanied by the re-distribution of electron density between valence areas of both particles and transition of a part of electrons from some external spheres into the neighboring ones.

It is obvious that with the proximity of electron densities in free atoms-components the transition processes between the boundaries atoms of particles will be minimal thus contributing to the formation of a new structure. Thus the task of evaluating the degree of such structural interactions in many cases comes down to comparative assessment of electron density of valence electrons in free atoms (on the averaged orbitals) participating in the process.

Therefore the maximum total solubility evaluated via the structural interaction coefficient α is defined by the condition of minimal value of coefficient α which represents the relative difference of effective energies of external orbitals of interacting subsystems:

$$\alpha = \frac{P'_o/r_i' - P''_o/r_i''}{(P'_o/r_i' + P''_o/r_i'')/2} 100\% \quad \text{or} \quad \alpha = \frac{P'_S - P''_S}{P'_S + P''_S} 200\% \qquad (13), (14),$$

where P_S—structural parameter is found by the equation:

$$\frac{1}{P_S} = \frac{1}{N_1 P_E'} + \frac{1}{N_2 P_E''} + \ldots \qquad (15),$$

here N_1 and N_2—number of homogeneous atoms in subsystems.

The nomogram of the dependence of structural interaction degree (ρ) upon the coefficient α, the same for the wide range of structures, was prepared by the data obtained.

Isomorphism as a phenomenon is usually considered as applicable to crystalline structures. But obviously the similar processes can also take place between molecular compounds where the bond energies can be assessed via the relative difference of electron densities of valence orbitals of interacting atoms. Therefore the molecular electronegativity is rather easily calculated via the values of corresponding P-parameters.

In complex organic structures the main role in intermolecular and intramolecular interactions can be played by separate "blocks" or fragments considered as "active" areas of the structures. Therefore it is necessary to identify these fragments and evaluate their spatial-energy parameters. Based on wave properties of P-parameter, the total P-parameter of each element should be found following the principle of adding the reciprocals of initial P-parameters of all the atoms. The resulting P-parameter of the fragment block or all the structure is calculated following the rule of algebraic addition of P-parameters of their constituent fragments.

Apparently, spatial-energy exchange interactions (SEI) based on leveling the electron densities of valence orbitals of atoms-components have in nature the same universal value as purely electrostatic coulomb interactions and complement each other. Isomorphism known from the time of E. Mitscherlich (1820) and D.I. Mendeleev (1856) is only a special demonstration of this general natural phenomenon.

The quantitative side of evaluating the isomorphic replacements both in complex and simple systems rationally fits into P-parameter methodology. More complicated is the problem of evaluating the degree of structural SEI for molecular structures, including organic ones. Such structures and their fragments are often not completely isomorphic to each other. Nevertheless, SEI is going on between them and its degree can be evaluated either semi-quantitatively numerically or qualitatively. By the degree of isomorphic similarity all the systems can be divided into three types:

I Systems mainly isomorphic to each other – systems with approximately the same number of heterogeneous atoms and cumulatively similar geometric shapes of interacting orbitals.

II Systems with organic isomorphic similarity – systems which:

1) either differ by the number of heterogeneous atoms but have cumulatively similar geometric shapes of interacting orbitals;

2) or have certain differences in the geometric shape of orbitals but have the same number of interacting heterogeneous atoms.

III Systems without isomorphic similarity – systems considerably different both by the number of heterogeneous atoms and geometric shape of their orbitals.

Taking into account the experimental data, all SEI types can be approximately classified as follows:

SYSTEMS I

1. $\alpha < (0–6)\%$; $\rho = 100\%$. Complete isomorphism, there is complete isomorphic replacement of atoms-components;
2. $6\% < \alpha < (25–30)\%$; $\rho = 98 – (0–3)\ \%$. There is wide or unlimited isomorphism.
3. $\alpha > (25–30)\%$; no SEI.

SYSTEMS II

1. $\alpha < (0–6)\%$;
   (a) There is reconstruction of chemical bonds that can be accompanied by the formation of a new compound;
   (b) Cleavage of chemical bonds can be accompanied by a fragment separation from the initial structure but without adjoinings and replacements.
2. $6\% < \alpha < (25–30)\%$; the limited internal reconstruction of chemical bonds is possible but without the formation of a new compound and replacements.
3. $\alpha > (20–30)\%$; no SEI.

SYSTEMS III

I. $\alpha < (0–6)\%$;
   (a) The limited change in the type of chemical bonds in the given fragment is possible, there is an internal re-grouping of atoms

without the cleavage from the main molecule part and replacements;

(b) The change in some dimensional characteristics of the bond is possible;

II. $6\% < \alpha < (25–30)\%$; A very limited internal re-grouping of atoms is possible;

III. $\alpha > (25–30)\%$; no SEI.

When considering the above systems, it should be pointed out that they can be found in all cellular and tissue structures in some form but are not isolated and are found in spatial-time combinations.

The values of α and ρ calculated in such a way refer to a definite interaction type whose nomogram can be specified by fixed points of reference systems. If the universality of spatial-energy interactions in nature is taken into account, such evaluation can have the significant meaning for the analysis of structural shifts in complex bio-physical and chemical processes of biological systems.

Fermentative systems contribute a lot to the correlation of structural interaction degree. In this model the ferment role comes to the fact that active parts of its structure (fragments, atoms, ions) have such a value of P_E-parameter which equals the P_E-parameter of the reaction final product. That is the ferment is structurally "tuned" via SEI to obtain the reaction final product, but it will not enter it due to the imperfect isomorphism of its structure.

The important characteristics of atom-structural interactions (mutual solubility of components, chemical bond energy, energy of free radicals, etc.) for many systems were evaluated following this technique [10, 11].

5.2.4 CALCULATION OF INITIAL DATA AND BOND ENERGIES

Based on the equations (3–5) with the initial data calculated by quantum-mechanical methods [3–6] the values of P_0-parameters are calculated for the majority of elements being tabulated, constant values for each atom valence orbital. Mainly covalent radii—by the main type of the chemical bond of interaction considered were used as a dimensional characteristic for calculating P_E-parameter (Table 5.1). The value of Bohr radius and the value of atomic ("metal") radius were also used for hydrogen atom.

In some cases the bond repetition factor for carbon and oxygen atoms was taken into consideration [10]. For a number of elements the values of P_E-parameters were calculated using the ionic radii whose values are indicated in column 7. All the values of atomic, covalent and ionic radii were mainly taken by Belov–Bokiy, and crystalline ionic radii – by Batsanov [12].

The results of calculating structural P_S-parameters of free radicals by the equation (15) are given in Table 5.2. The calculations are done for the radicals contained in protein and amino acid molecules (CH, CH_2, CH_3, NH_2 etc), as well as for some free radicals formed in the process of radiolysis and dissociation of water molecules.

The technique previously tested [10] on 68 binary and more complex compounds was applied to calculate the energy of coupled bond of molecules by the equations:

$$\frac{1}{E}=\frac{1}{P_s}=\frac{1}{\left(P_E\frac{n}{K}\right)_1}+\frac{1}{\left(P_E\frac{n}{K}\right)_2};P_E\frac{n}{K}=P \qquad (16), (17)$$

where n—bond average repetition factor, K—hybridization coefficient which usually equals the number of registered atom valence electrons.

Here the P-parameter of energy characteristic of the given component structural interaction in the process of binary bond formation.

"Non-valence, non-chemical weak forces act ... inside biological molecules and between them apart from strong interactions" [13]. At the same time, the orientation, induction and dispersion interactions are used to be called Van der Waals. For three main biological atoms (nitrogen, phosphorus and oxygen) Van der Waals radii numerically equal approximately the corresponding ionic radii (Table 5.3).

It is known that one of the reasons of relative instability of phosphorus anhydrite bonds in ATP is the strong repulsion of negatively charged oxygen atoms. Therefore it is advisable to use the values of P-parameters calculated via Van der Waals radii as the energy characteristic of weak structural interactions of biomolecules (Table 5.3).

TABLE 5.1 P-parameters of atoms calculated via the bond energy of electrons

Atom	Valence electrons	W (eV)	r_i (Å)	q_0^2 (eVÅ)	P_0 (eVÅ)	R (Å)	P_0/R (eV)
						0.5292	9.0644
H	$1S^1$	13.595	0.5295	14.394	4.7985	0.28	17.137
						R_I^-=1.36	3.525
	$2P^1$	11.792	0.596	35.395	5.8680	0.77	7.6208
						0.67	8.7582
	$2P^2$	11.792	0.596	35.395	10.061	0.77	13.066
C						0.67	15.016
	$2S^2$				14.524	0.77	18.862
	$2S^2+2P^2$				24.585	0.77	31.929
					24.585	0.67	36.694
	$2P^1$	15.445	0.4875	52.912	6.5916	0.70	9.4166
	$2P^2$				11.723	0.70	16.747
	$2P^3$				15.830	0.70	22.614
N						0.55	28.782
	$2S^2$	25.724	0.521	53.283	17.833	0.70	25.476
	$2S^2+2P^3$				33.663	0.70	48.09

TABLE 5.1 *(Continued)*

	$2P^1$	17.195	0.4135	71.383	6.4663	0.66	9.7979
	$2P^1$					R_I=1.36	4.755
	$2P^1$					R_I=1.40	4.6188
	$2P^2$	17.195	0.4135	71.383	11.858	0.66	17.967
						0.59	20.048
O						R_I=1.36	8.7191
						R_I=1.40	8.470
	$2P^4$	17.195	0.4135	71.383	20.338	0.66	30.815
						0.59	34.471
	$2S^2$	33.859	0.450	72.620	21.466	0.66	32.524
	$2S^2+2P^4$				41.804	0.66	63.339
						0.59	70.854
	$4S^1$	5.3212	1.690	17.406	5.929	1.97	3.0096
Ca	$4S^2$				8.8456	1.97	4.4902
	$4S^2$					R^{2+}=1.00	8.8456
	$4S^2$					R^{2+}=1.26	7.0203

TABLE 5.1 *(Continued)*

	$3P^1$	10.659	0.9175	38.199	7.7864	1.10	7.0785
	$3P^1$					R^{3-}=1.86	$P_э$=4.1862
P	$3P^3$	10.659	0.9175	38.199	16.594	1.10	15.085
	$3P^3$					R^{3-}=1.86	8.0215
	$3S^2$+$3P^3$				35.644	1.10	32.403
	$3S^1$	6.8859	1.279	17.501	5.8568	1.60	3.5618
Mg	$3S^2$				8.7787	1.60	5.4867
						R^{2+}=1.02	8.6066
	$4S^1$	6.7451	1.278	25.118	6.4180	1.30	4.9369
Mn	$4S^1$+$3d^1$				12.924	1.30	9.9414
	$4S^2$+$3d^2$				22.774	1.30	17.518
		4.9552	1.713	10.058	4.6034	1.89	2.4357
Na	$3S^1$					R^{1+}_I=1.18	3.901
						R^{1+}_I=0.98	4.6973
	$4S^1$	4.0130	2.612	10.993	4.8490	2.36	2.0547
K						R^{1+}_I=1.45	3.344

TABLE 5.2 Structural P_S-parameters calculated via the bond energy of electrons

Radicals, molecule fragments	$P_i^{'}(eV)$	$P_i^{''}(eV)$	$P_s(eV)$	Orbitals
OH	9.7979	9.0644	4.7080	O $(2P^1)$
	17.967	17.138	8.7712	O $(2P^2)$
H_2O	2·9.0644	17.967	9.0227	O $(2P^2)$
CH_2	17.160	2·9.0644	8.8156	C $(2S^1 2P^3{}_r)$
	31.929	2·17.138	16.528	C $(2S^2 2P^2)$
CH_3	15.016	3·9.0644	9.6740	C $(2P^2)$
	40.975	3·9.0644	16.345	C $(2S^2 2P^2)$
CH	31.929	12.792	9.1330	C $(2S^2 2P^2)$
NH	16.747	17.138	8.4687	N$(2P^2)$
	19.538	17.132	9.1281	N$(2P^2)$
NH_2	19.538	2·9.0644	9.4036	N$(2P^2)$
	28.782	2·17.132	18.450	N$(2P^3)$

TABLE 5.2 *(Continued)*

CO–OH	8.4405	8.7710	4.3013	C(2P[illegible])
C=O	15.016	20.048	8.4405	C(2P[illegible])
C=O	31.929	34.471	16.576	O(2P[illegible])
CO=O	36.694	34.471	17.775	O(2P[illegible])
C–CH_3	17.435	19.694	9.2479	–
C–NH_2	17.435	18.450	8.8844	–
CO–OH	12.315	8.7712	5.1226	C($2S^2$[illegible]P^2)
$(HP)O_3$	23.122	23.716	11.708	O(2[illegible]^2) P($3S^2$[illegible]P^3)
$(H_3P)O_4$	17.185	17.244	8.6072	O(2[illegible]^1) P(3[illegible]^1)
$(H_3P)O_4$	31.847	31.612	15.865	O(2[illegible]^2) P($3S^2$[illegible]P^3)

TABLE 5.2 *(Continued)*

H_2O	2·4.3623	8.7191	4.3609	O(2P²) r =1.36 Å
H_2O	2·4.3623	4.2350	2.8511	O(2P²) r=1.40
$C\text{-}H_2O$	2.959	2.8511	1.4520	–
$(C\text{-}H_2O)_3$ Lactic acid	–	–	1.4520·3= 4.3563	–
$(C\text{-}H_2O)_6$ Glucose	–	–	1.4520·6= 8.7121	–

Bond energies for P and O atoms were calculated taking into account Van der Waals distances for atomic orbitals: $3P^1$ (phosphorus)-$2P^1$ (oxygen) and for $3P^3$ (phosphorus)-$2P^2$ (oxygen). The values of E obtained slightly exceeded the experimental; reference ones (Table 5.4). But for the actual energy physiological processes, for example: during photosynthesis, the efficiency is below the theoretical one, being about 83%, in some cases [14, 15].

Perhaps the electrostatic component of resulting interactions at anion-anionic distances is considered in such a way. Actually the calculated value of 0.83 E practically corresponds to the experimental values of bond energy during the phospholyration and free energy of ATP in chloroplasts.

Table 5.4 contains the calculations of bond energy following the same technique but for stronger interactions at covalent distances of atoms for the free molecule P--O (sesquialteral bond) and for the molecule P=O (double bond). The sesquialteral bond was evaluated by introducing the coefficient $n = 1.5$ with the average value of oxygen P_E-parameter for single and double bonds. The average breaking energy of the corresponding chemical bonds in ATP molecule obtained in the frameworks of semi-empirical method PM3 with the help of software GAMESS [16] are given in column 11 of Table 5.4 for comparison. The calculation technique is detailed in [17].

The calculated values of bond energies in the system K-C-N being close to the values of high-energy bond P□O in ATP demonstrate that such structure can prevent the ATP synthesis.

When evaluating the possibility of hydrogen bond formation, such

value of $n =$ is taken into account in which $K = 1$, and the value $n =$ 3.525/17.037 characterizes the change in the bond repetition factor when transiting from the covalent bond to the ionic one.

5.2.5 FORMATION OF STABLE BIOSTRUCTURES

At equilibrium-exchange spatial-energy interactions similar to isomorphism the electrically neutral components do not repulse but approach each other and form a new composition whose α in the equations (13, 14).

This is the first stage of stable system formation by the given interaction type which is carried out under the condition of approximate equality of component P-parameters: $P_1 \approx P_2$.

Hydrogen atom, element No 1 with the orbital $1S^1$ determines the main criteria of possible structural interactions. Four main values of its P-parameters can be taken from Tables 5.1 and 5.3:

1. For strong interactions: $P_E^{''}$ = 9.0644eV with the orbital radius 0.5292Å and $P_E^{'''}$ = 17.137eV with the covalent radius 0.28Å.
2. For weaker interactions: $P_E^{'}$ = 4.3623eV and P_E = 3.6352eV with Van der Waals radii 1.10Å and 1.32Å. The values of P-parameters $P' : P'' : P'''$ relates as 1:2:4. In accordance with the concepts, such values of interaction P-parameters define the normative functional states of biosystems, and the intermediary can produce pathologic formations by their values.

The series with approximately similar values of P-parameters of atoms or radicals can be extracted from the large pool of possible combinations of structural interactions (Table 5.5). The deviations from the initial, primary values of P-parameters of hydrogen atom are in the range ± 7%.

The values of P-parameters of atoms and radicals given in the Table define their approximate equality in the directions of interatomic bonds in polypeptide, polymeric and other multi-atom biological systems.

In ATP molecule these are phosphorus, oxygen and carbon atoms, polypeptide chains – CO, NH, and CH radicals. In Table 5.5 you can also see the additional calculation of their bond energy taking into account the sesquialteral bond repetition factor in radicals C⩦O и N⩦H.

On the example of phosphorus acids it can be demonstrated that this approach is not in contradiction with the method of valence bonds which explains the formation peculiarities of ordinary chemical compounds. It is demonstrated in Table 5.6 that this electrostatic equilibrium between the oppositely charged components of these acids can correspond to the structural interaction for H_3PO_4 $3P^1$ orbitals of phosphorus and $2P^1$ of oxygen, and for HPO_3 $3S^23P^3$ orbitals of phosphorus and $2P^2$ of oxygen. Here it is stated that P-parameters for phosphorus and hydrogen subsystems are added algebraically. It is also known that the ionized phosphate groups are transferred in the process of ATP formation that is apparently defined for phosphorus atoms by the transition from valence-active $3P^1$ orbitals to $3S^23P^3$ ones that is four additional electrons will become valence-active. According to the experimental data the synthesis of one ATP molecule is connected with the transition of four protons and when the fourth proton is being transited the energy accumulated by the ferment reaches its threshold [18, 19]. It can be assumed that such proton transitions in ferments initiate similar changes in valence-active states in the system P-O. In the process of oxidating phospholyration the transporting ATP-synthase uses the energy of gradient potential due to $2H^+$-protons which, in the given model for such a process, correspond to the initiation of valence-active transitions of phosphorus atoms from $3P^1$ to $3P^3$-state.

In accordance with the equation (17) it can be assumed that in stable molecular structures the condition of the equality of corresponding effective interaction energies of the components by the couple bond line is fulfilled by the following equations:

$$\left(P_E \frac{n}{K}\right)_1 \approx \left(P_E \frac{n}{K}\right)_2 \rightarrow P_1 \approx P_2 \tag{18}$$

And for heterogeneous atoms (when $n_1 = n_2$):

$$\left(P_E/K\right)_1 \approx \left(P_E/K\right)_2 \tag{18a}$$

In phosphate groups of ATP molecule the bond main line comprises phosphorus and oxygen molecules. The effective energies of these atoms by the bond line calculated by the equation (18) are given in Tables 5.4 and 5.5, from which it is seen that the best equality of P_1 and P_2 parameters is fulfilled for the interactions $P(3P^3)$ – 8.7337eV and $O(2P^2)$ – 8.470eV that is defined by the transition from the covalent bond to Van der Waals ones in these structures.

The resulting bond energy of the system P-O for such valence orbitals and the weakest interactions (maximum values of coefficient K) is 0.781eV (Table 5.4). Similar calculations for the interactions $P(2P^1)$ – 4.0981eV and $O(2P^1)$ – 4.6188eV produce the resulting bond energy 0.397eV.

The difference in these values of bond energies is defined by different functional states of phosphorous acids HPO_3 and H_3PO_4 in glycolysis processes and equals 0.384eV that is close to the phospholyration value (0.34–0.35eV) obtained experimentally.

Such ATP synthesis is carried out in anaerobic conditions and is based on the transfer of phosphate residues onto ATP via the metabolite. For example: ATP formation from creatine phosphate is accompanied by the transition of its NH group at ADP to NH_2 group of creatine at ATP.

TABLE 5.3 Ionic and Van der Waals radii (Å)

Atom	Ionic radii			Van der Waals radii		
	Orbital	R_I	P_E/K (eV)	R_B	Orbital	P_E/K (eV)
H	$1S^1$	$R^- = 1.36$	3.525	1.10	$1S^1$	4.3623
		$r = 0.5292$	9.0644	1.32		3.6352
N	$2P^3$	$R^{3-} = 1.48$	10.696/3 = 3.5653	1.50	$2P^1$	4.3944/1
				1.50	$2P^3$	10.553/3 = 3.5178
				1.50	$2S^22P^3$	22.442/5 = 4.4884
P	$3P^3$	$R^{3-} = 1.86$	8.9215/3 = 2.9738	1.9	$3P^1$	4.0981/1
				1.9	$3P^3$	8.7337/3 = 2.9112
				1.9	$3S^23P^3$	18.760/5 = 3.752
O	$2P^2$	$R^{2-} = 1.40$	8.470/2 = 4.2350	1.40	$2P^1$	4.6188/1
				1.50	$2P^1$	4.3109/1
		$R^{2-} = 1.36$	8.7191/2 = 4.3596	1.40	$2P^2$	8.470/2 = 4.2350
				1.50	$2P^2$	7.9053/2 = 3.9527
C	$2S^22P^2$	$d^*/2 = 3.2/2 = 1.6$	15.365/4 = 3.841	1.7	$2P^1$	3.4518/1
				1.7	$2P^2$	5.9182/2 = 2.9591
				1.7	$2S^22P^2$	14.462/4 = 3.6154

d*—Contact distance between C–C atoms in polypeptide chains [13].

TABLE 5.4 Bond energy (eV)

Atoms, structures, orbitals	Bond	Component 1		Component 2		Component 3		Calculation	E [13, 14, 15]	E [16, 17]	Remarks
		P_E (eV)	n/K	P_E (eV)	n/K	P_E (eV)	n/K	E			
1	2	3	4	5	6	7	8	9	10	11	12
								6.277			
P---O				70.854	1.5/6				6.1385		
	Cov.	32.403	1.5/5			6.14		6.024			PO free molecule
$3S^23P^3$-$2S^22P^4$				63.339	1.5/6				6.14		
								<6.15>			
H_2O	Cov.	2 × 9.0624	1/1	17.967	1/6			2.570	2.476		Decay
$1S^1$-$2P^2$	Cov.	2 × 9.0624	1/1	20.048	2/2			9.520		10.04	of one molecule
H_3PO_4	Cov.	3×9.0624	1/1	32.405	1/5	4 × 17.967	1/2	4.8779	4.708		
1	2	3	4	5	6	7	8	9	10	11	12
C---O											
	Cov.	7.6208	1.125/2	9.7979	1/1			4.2867			
($2P^1$-$1S^1$)											
C–N											
	Cov.	7.6208	1/4	9.4166	1/5			0.9471			
$2P^1$-$2P^1$											

TABLE 5.4 *(Continued)*

C⋯N $2P^1$-$2P^1$	Cov.	7.6208	1.125/4	9.4166	1.1667/5			1.0898	0.870
K–C–N $4S^2$-$2P^1$-$2P^1$	Cov.	2.0547	1/1	7.6208	1/4	9.4166	1/5	0.648	
($C–H_2O$)-($C–H_2O$)	VdW	1.4520	1/1	1.4520	1/1			0.726	
C⋯O	Cov.	31.929	1.125/4	20.048	1/2			4.7367	
$2S^22P^2$-$2P^2$		31.929	1/4	20.042	1/2			4.4437	
N⋯H	Cov.	9.4166	1.1667/1	9.0644	1/1			4.9654	
$2P^1$-$1S^1$		9.4166	1/1	9.0644	1/1			4.6186	
C–H $2P^1$-$1S^1$	Cov.	13.066	1/2	9.0644	1/1			3.797	3.772
C–H $2P^2$-$1S^1$		13.066	1/2	17.137	1/1			4.7295	

TABLE 5.4 *(Continued)*

N–H_2 $2P^3$-$1S^1$	Cov.	22.614	1/3	2× 9.0644	1/1		5.3238			
-H···O		3.525	3.525/17.037	4.6188	1/6		0.3730	0.3742		Hydrogen bond
P=O $3P^3$-$2P^2$	Cov.	15.085	2/3	20.042	2/2		6.6970	6.504	6.1385	Free molecule
P–O $3P^3$-$2P^2$	VdW	8.7337	1/5	8.470	1/6		0.781	0.670		ΔG ATP
P–O $3P^1$-$2P^1$	Cov.	7.0785	1/1	9.7979	1/1		4.1096	4.2059	4.2931	
P–O $3P^1$-$2P^1$	VdW	4.0981	1/5	4.6188	1/6		0.3970	0.34-0.35		Phosphol-yration

TABLE 5.5 Bio-structural spatial-energy parameters (eV)

Series No	H	C	N	O	P	CH	CO	NH	Glucose	Lactic acid	OH	Remarks
I	9.0644	8.7582	9.4166	9.7979	8.7337	9.1330	8.4405	8.4687	8.7121			Strong interaction
	$(1S^1)$	$(2P^1)$	$(2P^1)$	$(2P^1)$	$(3P^3)$	$(2S^22P^2–1S^1)$	$(2P^2–2P^2)$	$(2P^2–1S^1)$	$2P^2–$		8.7710	
		9.780						9.1281	$–(1S^1–2P^2)$			
		$(2P^1)$						$(2P^2–1S^1)$				
II	17.132	17.435	16.747	17.967	18.760	C and H blocks	16.576	N and H				Strong interaction
	$(1S^1)$	$(2S^12P^1)$	$(2P^2)$	$(2P^2)$	$(3S^23P^3)$		$(2S^22P^2-2P^4)$	blocks				

TABLE 5.5 *(Continued)*

III	(4.3623)	3.8696	4.3944	4.3109	4.0981	4.7295	4.4437	4.6186		4.3563	4.7084	Weak
	($1S^1$)	($2P^2$)	($2P^1$)	($2P^1$)	($3P^1$)		4.7367	4.9654		$2P^2$–		interaction
				4.6188						($1S^1$– $2P^2$)		
				($2P^1$)								
IV	3.6352	3.4518	3.5178	4.2350	4.0981	4.7295	4.4437	4.6186				Effective
	($1S^1$)	($2P^1$)	($2P^3$)	($2P^2$)	($3P^1$)		4.7367	4.9654				bond energy
		3.6154		3.6318	3.752							
		($2S^2 2P^2$)		($2P^4$)	($3S^2 3P^3$)							

TABLE 5.6 Structural interactions in phosphorus acids

Molecule	Component 1			Component 2			$\alpha=(\Delta P/<P>)*100\%$
	Atom	Orbitals	$P=P_1+P_2$ (eV)	Atom	Orbitals	P(eV)	
$(H_3P)O_4$	H_3P	$1S^1$-$3P^1$	4.3623*3+ 4.0981 =17.185	O_4	$2P^1$	4.3109*4=17.244	0.34
		$1S^1$-$(3S^23P^3)$	4.3623*3+18.760 =31.847	at r=1.50Å	$2P^2$	7.9053*4=31.612	0.74
$(HP)O_3$	HP	$1S^1$-$(3S^23P^3)$	4.3623+18.760 =23.122	O_3	$2P^2$	7.9053*3=23.716	2.54
				at r=1.50Å			

From Table 5.4 it is seen that the change in the bond energy of these two main radicals of metabolite is 5.3238 – 4.9654 = 0.3584eV—taking the sesquialteral bond N--H into account (as in polypeptides) and 5.3238 – 4.6186 = 0.7052eV—for the single bond N–H. This is one of the intermediary results of the high-energy bond transformation process in ATP through the metabolite. From Tables 5.4 and 5.6 it can be concluded that the phosphorous acid H_3PO_4 can have two stationary valence-active states during the interactions in the system P-O for the orbitals with the values of P-parameters of weak and strong interactions, respectively. This defines the possibility for the glycolysis process to flow in two stages. At the first stage, the glucose and H_3PO_4 molecules approach each other due to similar values of their P-parameters of strong interactions (Table 5.2). At the second stage, H_3PO_4 P-parameter in weak interactions 4.8779 eV (Table 5.4) in the presence of ferments provokes the bond (H_2O–C)–(C–H_2O) breakage in the glucose molecule with the formation of two molecules of lactic acid whose P-parameters are equal by 4.3563 eV. The energy of this bond breakage process equalled to 0.726 eV (Table 5.4) is realized as the energy of high-energy bond it ATP.

According to the reference data about 40% of the glycolysis total energy, that is about 0.83 eV, remains in ATP.

By the hydrolysis reaction in ATP in the presence of ferments $\left(HPO_3 + H_2O \rightarrow H_3PO_4 + E\right)$ for structural P_S-parameters (Table 5.2) E = 11.708 + 4.3609 – 15.865 = 0.276 eV.

It is known that the change in the free energy (ΔG) of hydrolysis of phosphorous anhydrite bond of ATP at pH = 7 under standard conditions is 0.311 – 0.363 eV. But in the cell the ΔG value can be much higher as the ATP and ADP concentration in it is lower than under standard conditions. Besides, the ΔG value is influenced by the concentration of magnesium ions which is the acting co-ferment in the complex with ATP. Actually Mg^{2+} ion has the P_E-parameter equalled to 8.6066eV (Table 5.1) which is very similar to the corresponding values of P-parameters of phosphorous and oxygen atoms.

The quantitative evaluation of this factor requires additional calculations.

5.3 CONCLUSIONS

1. Bond energies of some biostructures have been calculated following P-parameter and quantum-mechanical techniques.
2. High-energy bonds in ATP are formed in the system P-O under functional transitions of their valence-active states.
3. The data obtained agree with the experimental ones.

KEYWORDS

- **High-energy Bonds**
- **Orbital Energy**
- **Phosphorus-Oxygen**
- **P-Parameter**
- **Spatial-energy**

REFERENCES

1. Eyring, H., Walter, J., Kimball, & G. E. (1948). *Quantum chemistry* (p. 528). Moscow: L.
2. Korablev, G. A. (2005). *Spatial-energy principles of complex structures formation* (p. 426). Netherlands: Brill Academic Publishers and VSP.
3. Fischer, C. F. (1972). *Average-Energy of configuration: Hartree-Fock Results for the atoms helium to radon, 4,* 301–399.
4. Waber, J. T. & Cromer, D. T. (1965). Orbital Radii of Atoms and Ions. *Journal of Chemical Physics, 42*(12), 4116–4123.
5. Clementi, E. & Raimondi, D. L. (1963). Atomic Screening constants from S.C.F. Functions. *Journal of Chemical Physics, 38*(11), 2686–2689.
6. Clementi, E. & Raimondi, D. L. (1967). Atomic Screening constants from S.C.F. Functions. *Journal of Chemical Physics, 47*(14), 1300–1307.
7. Gombash, P. (1951). *Atom statistical model and its application* (p. 398). Moscow: L.
8. Clementi, E. (1965) Tables of atomic functions. *Janos Bolyai Mathematical Society of Research and Development, 9*(2), 76, Supplement.
9. Korablev, G. A. & Zaikov, G. E. (2009). Spatial-Energy Parameter as a Materialised Analog of Wafe Function. *Progress on Chemistry and Biochemistry* (pp. 355–376). New York: Nova Science Publishers, Inc.

10. Korablev, G. A. & Zaikov, G. E. (2006). Energy of chemical bond and spatial-energy principles of hybridization of atom orbitalls. *Journal of Applied Polymer Science, 101*(3), 2101–2107.
11. Korablev, G. A. & Zaikov, G. E. (2009). Formation of carbon nanostructures and spatial-energy criterion of stabilization. *Mechanics of Composite Materials and Structures, RAS, 15*(1), 106–118.
12. Batsanov, S. S. (2009). *Structural chemistry (facts and dependencies)*. Moscow: Michigan State University Press.
13. Volkenshtein (1988). *Biophysics* (p. 598). Moscow: Nauka.
14. Govindzhi. (Ed.). (1987). *Photosynthesis* (vol.1, p. 728, vol.2, p. 460). Moscow: Mir.
15. Clayton, R. (1984). Photosynthesis. *Physical mechanisms and chemical models* (p. 350). Moscow: Mir.
16. Schmidt, M. W., Baldridge, K. K., Boatz, J. A., et al. (1993). General atomic and molecular electronic structure system. *Journal of Computational Chemistry, 14*, 1347–1363.
17. Khokhriakov, N. V. & Kodolov, V. I. (2009). Influence of active nanoparticles on the structure of polar liquids. *Chemical Physics and Mesoscopy, 11*(3), 388–402.
18. Feniouk, B. A. (1998). Study of the conjugation mechanism of ATP synthesis and ATP proton transport. *Biology and Natural Science*, 108.
19. Feniouk, B. A., Junge, W., & Mulkidjanian, A. (1998). Tracking of proton flow across the active ATP-synthase of Rhodobacter capsulatus in response to a series of light flashes. *EBEC Reports, 10,* 112.

CHAPTER 6

ELECTRONIC EXCITATION ENERGY TRANSFER BETWEEN MOLECULES OF CARBOCYANINE DYES IN COMPLEXES WITH DNA: DISTRIBUTION OF DISTANCES IN DONOR–ACCEPTOR PAIRS

P. G. PRONKIN and A. S. TATIKOLOV

CONTENTS

6.1 INTRODUCTION

Electronic excitation energy transfer (EEET) between molecules of carbocyanine dyes, which form non covalent complexes with DNA, has been studied by steady-state and time-resolved technique. Three oxacarbocyanine dyes have been used as electronic excitation energy donors, and 3,3'-diethylthiacarbocyanine iodide has served as an acceptor dye. The fluorescence decay kinetics of the donors and its quenching by the acceptor in the presence of DNA were measured by picosecond spectrofluorometer. Monte-Carlo simulation of the fluorescence decay kinetic has been performed. An analysis of the kinetic dependences permitted obtaining the data on distribution of the distances in donor–acceptor pairs upon EEET, which have been compared with the results of the steady-state fluorescent experiments. The effect of the acceptor concentration on the parameters of distribution of its molecules in the quenching microphase has been revealed.

Non radiative electronic excitation energy transfer (EEET) has been a subject matter of numerous studies, which is largely due to the appearance of new efficient fluorescent dyes. The studies of EEET plays an important role in investigation of the structure and functioning of biomacromolecules (DNA, proteins) [1, 2], because non radiative energy transfer is capable of providing researchers with valuable information on the structure of donor–acceptor pairs.

EEET experiments in dye-DNA system are of particular interest due to unique structure of biopolymer matrix [3-6]. For example, EEET between dye molecules covalently bound to DNA described in the work of Hannestad et al. [7]. In such a system, the chromophores are rigidly fixed with respect to one another in the biopolymer matrix. A coumarin dye (Pacific Blue) was used as the donor of electronic excitation energy, and an indocyanine dye (Cy3) served as the acceptor [4]. These dyes were bound to the ends of the fragments of DNA double helices (20 and 50 base pairs). The electronic excitation energy was transferred along the biopolymer helix from the donor to the acceptor by means of mediator molecules (Oxazole Yellow dye) non covalently bound to DNA. The use of the donor and the acceptor bound covalently to the ends of the biopolymer helix determined the directed character of the EEET process. This permits ensuring favorable mutual orientation of the dipoles and, hence, high efficiency of EEET, and on the other hand, eliminating the uncertainty related to determination of the distance in the donor–acceptor pairs. Hence, the DNA helix acted as the peculiar "nanosized light guide".

It is known that cationic thia- and oxacarbocyanine dyes in aqueous solutions are capable of forming non covalent complexes with DNA [8, 9]. In a complex with the biopolymer, molecules of dyes are located close to one another, their fluorescent properties change drastically; in particular, the fluorescence quantum yield increases several times, which is a favorable factor for EEET detection [10].

The results of steady-state fluorescent measurements provide averaged information on the spatial location of donor and acceptor fluorophores, whereas the analysis of the kinetic data permits obtaining information on distribution of the distances between the donor and the acceptor of electronic excitation energy [2, 4-6]. Studying the fluorescence kinetics upon EEET is a unique instrument capable of providing researchers with information on structural and dynamic properties of systems under study [2], including the systems containing dyes and DNA [1, 8–11].

In this work, EEET from a donor dye to an acceptor dye and fluorescence quenching of the donor dye by the acceptor in a complex with DNA were studied. The energy transfer kinetics (the fluorescence decay kinetics of the donors in the presence of the acceptor) was measured. To estimate the EEET constants, Monte-Carlo simulation of the fluorescence decay kinetic was performed. The analysis of the kinetic data gave distributions of the donor–acceptor distances, which were compared with the results of the steady-state fluorescent experiments [10, 11].

The oxacarbocyanine dyes: 3,3′-dimethyl-9-ethyloxacarbocyanine iodide (D1), 3,3′,9-triethyl-5,5′-dimethyloxacarbocyanine iodide (D2), and 3,3′,9-triethyl-6,6′-dimethoxyoxacarbocyanine iodide (D3) were used as the dyes–donors of electronic excitation energy, and 3,3′-diethylthiacarbocyanine (A) served as the acceptor dye.

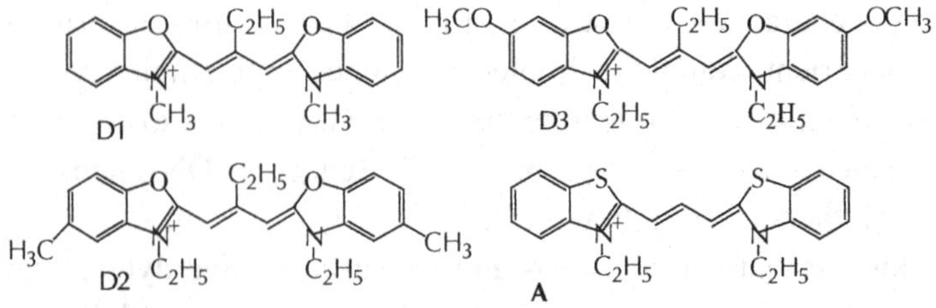

6.2 MATERIALS AND METHODS

The absorption spectra of the dyes were measured with a Shimadzu UV-3101 PC spectrophotometer (Japan), and the fluorescence and fluorescence excitation spectra were recorded with Flyuorat-02-Panorama (Russia) and Shimadzu RF-5301 PC (Japan).

The fluorescence quantum yields (Φ_f) of the donors in solutions in the absence and in the presence of DNA were determined using the standard reference method, where the standard was fluorescein in an aqueous NaOH solution ($\Phi_f = 0.85$) [12]. To determine the fluorescence quantum yields of the acceptor (A), we used the data reported in [8].

The dyes provided by the NIIKHIMFOTOPROEKT Research Center and commercial chicken DNA (Reanal, Hungary) were used. The pH7 phosphate buffer (at a concentration of 20 mmol l^{-1}) in distilled water was used as a solvent. The DNA concentration in the phosphate buffer solution was determined using the absorption coefficient of a base pair $\varepsilon = 13{,}200$ l mol^{-1} cm^{-1} at a wavelength of 250nm [13].

The efficiency (r) of EEET between molecules of the dyes (donor and acceptor) was determined from the spectral and fluorescent data using Eq. (1):

$$r = \frac{I_{ex.D} / I_{ex.A}}{Abs_D / Abs_A}, \tag{1}$$

where, $I_{ex.D}$, $I_{ex.A}$, Abs_D, and Abs_A are the intensities the donor (D) and acceptor (A) bands in the fluorescence excitation (I_{ex}) and absorption (Abs) spectra, respectively [14]. For the dyes studied, r was determined at different DNA concentrations. The critical radius of energy transfer (R_0) was determined from the Forster theory [14]:

$$R_0 = 0.2108(\kappa^2 \Phi_{f0} n^{-4} \int F(\lambda)\varepsilon(\lambda)\lambda^4 d\lambda)^{1/6}, \quad (2)$$

where κ^2 is the orientation factor, which depends on mutual orientation of the transition moments of the donor and the acceptor; Φ_{f0} is the fluorescence quantum yield for the donor in the absence of the acceptor; n is the refractive index of the medium; $F(\lambda)$ is the normalized fluorescence spectrum of the donor, $\varepsilon(\lambda)$ is the molar absorption coefficient of the acceptor, $\text{lmol}^{-1}\ \text{cm}^{-1}$; and λ is the wavelength, nm. The value of the calculated refractive index for the dye–DNA system $n = 1.4$ [10].

The distance (R) between the donor and the acceptor was determined from Eq. (3) [14]:

$$r = R_0^6 / (R_0^6 + R^6). \quad (3)$$

In the calculations of R_0 and R by Eqs. (2) and (3) the value of the orientation factor $\kappa^2 = 2/3$ was used, which corresponded to the random orientation of donor and acceptor molecules owing to rotational diffusion.

The dynamic quenching of donor fluorescence upon introduction of an acceptor into the system, which is due to random collisions between the fluorophore and the quencher, is described by the Stern–Volmer equation [5]:

$$I_0 / I = 1 + k_q \tau_0 [Q] = 1 + K_{dyn}[Q], \quad (4)$$

where I_0 and I are the fluorescence intensities (quantum yields) in the absence and in the presence of the quencher, respectively; k_q is the bimolecular quenching rate constant; I_0 is the fluorescence decay time in the absence of the quencher; $[Q]$ is the quencher concentration; and $K_{dyn} = k_q \tau_0$ is the Stern–Volmer quenching constant. In the case of dynamic fluorescence quenching, the experimental dependence of I_0/I on $[Q]$ is linear [14].

To allow the deactivation processes caused by static quenching, the additional exponential factor $e^{v[Q]}$ is introduced into the Stern–Volmer equation [5]:

$$I_0 / I = (1 + k_q \tau_0 [Q]) e^{v[Q]} = (1 + K_{dyn} [Q]) e^{v[Q]}, \qquad (5)$$

where v is the volume of the sphere of energy donor quenching by an acceptor (with the radius R_0), which does not depend on donor and acceptor concentrations in the whole volume of the solution (the Perrin equation).

The fluorescence decay kinetics were measured with a Fluo-time 200 (PicoQuant, GmbH, Germany) picosecond spectrofluorometer with semiconductor laser excitation at a wavelength of 470 nm (pulse duration of about 100ps). Standard quartz cells (1cm) were used in the measurements. The measurements were conducted at room temperature (20 ± 2°C). The fluorescence decay kinetics of dyes D1, D2, and D3 were measured at a constant DNA concentration $c_{DNA} = 1.25 \times 10^{-5}$mol l^{-1} ($\lambda_{reg} = 520$, 530, and 535nm, respectively). Solutions of the dyes were prepared immediately before the measurements.

The fluorescence decay kinetics of D1–D3 in the presence of A was simulated by the Monte-Carlo method [15, 16] (Eqs. (6), (7)) and also analyzed in terms of Eqs. (9), (10) using the computational methods [2, 4, 6].

The EEET Monte-Carlo model takes into account the following conditions: random distribution of donor and acceptor positions (the value of the orientation factor $\kappa^2 = 2/3$); the donor molecules are excited by an extremely short pulse (delta pulse); the donor photoexcitation is non recurrent process; donor fluorescence and EEET are the competitive processes.

The numerical modeling of EEET kinetics has been fulfilled by Monte-Carlo technique for comparatively large number of photons (~1.2×10^5 units). The inverse functions approach has been used to calculate the fluorescence emission time of a donor ($\Delta\tau_D$, Eq. (6)) and EEET time from a donor to an acceptor ($\Delta\tau_A$, Eq. (7)) [15, 16].

$$\Delta\tau_D = -\tau_D \ln x_1 \qquad (6)$$

$$\Delta\tau_A = \Delta\tau_{ET} - \tau_A \ln x = -\frac{1}{k_{ET}} \ln x_2 - \tau_A \ln x \qquad (7)$$

where, k_{ET} is the EEET rate constant, τ_D, τ_A are the fluorescence decay times of the donor and acceptor fluorophores.

The EEET rate constant can be estimated from the steady-state fluorescence data (Eq. (8)):

$$k_{ET} = \frac{1}{\tau_D}\left(\frac{R_0}{R}\right)^6 \tag{8}$$

where R_0 and R are the critical radius of energy transfer, and distance (R) between the donor and the acceptor obtained using Eqs. (1)–(3).

In Eqs. (6) and (7) x, x_1, x_2 are the random variables, which have equally probable distributions [16]. The simulation results in the frequency bar chart.

The fluorescence decay kinetics of D1–D3 in the presence of A was analyzed in terms of Eqs. (9) and (10) using the computational methods. According to the published data [2, 4, 6], the fluorescence decay kinetics of the donors is determined by the expression of the form (9):

$$I_{D-A}(t) = I_0 \int_0^R P(R) \exp\left[-\frac{t}{\tau_0}\left(1+\left(\frac{R_0}{R_{D-A}}\right)^6\right)\right] dR, \tag{9}$$

where, I_0 is the initial amplitude of the kinetic dependence of dye fluorescence (I_{D-A} at $t = 0$), τ_0 is the fluorescence decay time in the absence of the quencher, $P(R_{D-A})$ is the distribution probability density for the distances between the donor and quencher (A) molecules (in the vicinity of the acceptor).

The distribution of the random function $P(R_{D-A})$ was given by the gamma-distribution ($P(R_{D-A}) \sim \Gamma\ (k,\theta)$) with the probability density (10):

$$P(R_{D-A}) \sim \Gamma(\mathrm{k},\ \theta) = x^{k-1}\frac{e^{-x/\theta}}{\theta^k \Gamma(k)}, \quad x \geq 0 \tag{10}$$

6.3 RESULTS AND DISCUSSION

The effect of the ds-DNA complexation on the spectral and fluorescent properties of dyes D1–D3, and A were studied in sufficient detail earlier [8, 9].

It will be remarked that, for all selected donor-acceptor pairs, the electronic excitation energy transfer (EEET) does not occur in a homogeneous solution (that is, in the absence of DNA) because of the long distance between donor and acceptor molecules (about 700 Å at the chosen concentrations of the dyes).

Fluorescence of the acceptor dye (λ_{ex} = 490nm, λ_{fl} = 569nm) observed in such cases is due to direct excitation of A to the edge of the absorption band, rather than energy transfer from donor to acceptor [10]. In the fluorescence excitation spectra of the D–A system (c_{DNA} = 0; spectra not shown), only one band is present, corresponding to fluorescence excitation of the acceptor dye (at λ_{ex} = 552nm). The fluorescence excitation band of dyes D1–D3 (at λ_{ex} ~ 485–510nm) is lacking in the spectra.

At low DNA concentrations (less than 5×10^{-5}mol l^{-1}) in the solution, a marked decrease in the absorbance is observed for dyes D1–D3, which is accompanied by some broadening of the bands (for example, the absorption spectra of D2 and A are shown in inset in Figure 6.1). Dye–DNA interaction leads to a long-wavelength shift of the maximums of the dyes absorption bands. The effects observed are traditionally explained by changes in the dye molecules environment (microsolvation) upon interaction with DNA and by the formation and decomposition of non structured aggregates in the presence of DNA.

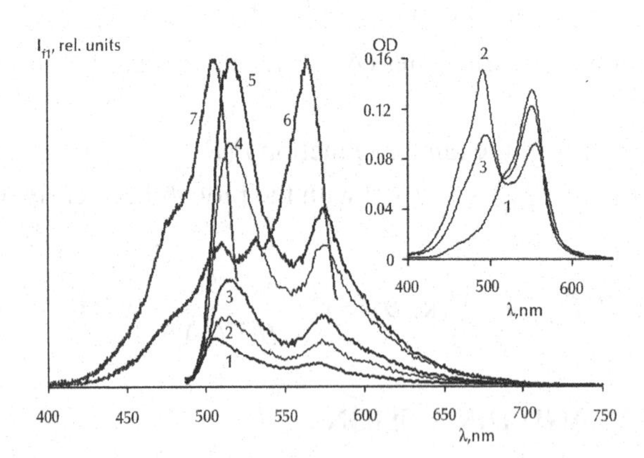

FIGURE 6.1 Fluorescence spectra of mixtures of dyes D2 and A in phosphate buffer solution (pH7) at c_{DNA} = (*1*) 0, (*2*) 9.5×10^{-6}, (*3*) 2.39×10^{-5}, (*4*) 7.2×10^{-5}, and (*5*) 1.19×10^{-4}mol l^{-1} at λ_{ex} = 475nm, and fluorescence excitation spectra at λ_{fl} = 590 (*6*), 530 (*7*)nm at c_{DNA} = 7.2×10^{-5}mol l^{-1}. Inset: absorption spectra of A (*1*), D2–A (*2*) (at c_{DNA} = 0mol l^{-1}, c_{D2} = 2.02×10^{-6}mol l^{-1}, c_A = 1.9×10^{-6}mol l^{-1}), and D2–A at c_{DNA} = 1.19×10^{-4}mol l^{-1}.

In the dye-DNA complex, the fluorescence quantum yield of dyes increases several times (see Figure 6.1, curves 1–5). For instance, in the fluorescence spectra of D2–A (λex = 475nm, cDNA = 1.19×10^{-4} mol l–1) obtained in the presence of DNA, two peaks at 517 and 576nm are observed, which correspond to fluorescence of the donor dye and the acceptor dye, respectively (Figure 6.1, curve 5). In the fluorescence excitation spectra of the acceptor in the presence of DNA, the D1–D3 bands are observed, which is indicative of EEET. In particular, in the fluorescence excitation spectra the band of D2 bound to the biopolymer appears, which is due to EEET from excited D2 to A (Figure 6.1, curve 6, λreg = 590nm). The positions of the maximums in the fluorescence excitation spectra of the dyes in the presence of DNA agree well with those in the absorption spectra. With an increase in the DNA concentration in the solution, their maximums are shifted to the long-wavelength side (Δλ ~10nm). Since the fluorescence spectra of D1–D3 overlap well with the absorption spectrum of the acceptor, we may consider the energy transfer process to occur by the Forster inductive-resonance mechanism.

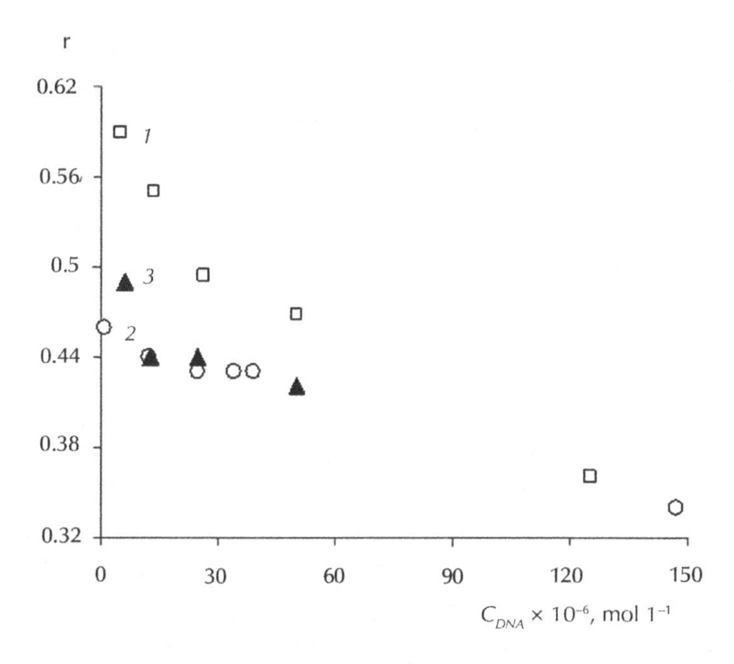

FIGURE 6.2 Data on the efficiency of the EEET process (r; Eq. (1)) for D1 (1), D2 (2), D3 (3) as a function of DNA concentration in a phosphate buffer solution (pH 7)

Figure 6.2 shows the dependences of the EEET efficiency (r) for the dyes D1–D3 calculated by Eq. (1) upon the DNA concentration increase in a phosphate buffer solution (pH 7). An increase in the DNA concentration leads to decrease in the values of the EEET efficiency: for D1 r decreases from 0.59 (at $c_{DNA} = 5.0 \times 10^{-6}$mol l^{-1}) to 0.36 (at $c_{DNA} = 1.25 \times 10^{-4}$mol l^{-1}); for the dye pair D3–A, the value of r decreases from 0.49 to 0.42 (with an increase in the biopolymer concentration in the solution from 6.25×10^{-6} to 5.0×10^{-5}mol l^{-1}). The value of r does not depend also on the registration wavelength of the excitation spectra. The data obtained rule out a substantial contribution in EEET of the trivial mechanism of excitation of the acceptor by reabsorption of donor fluorescence. In the case of reabsorption of donor fluorescence by the acceptor, this would lead to higher light energy of donor emission absorbed by dye A. In that case, with growing DNA concentration, the band belonging to donor dyes would increase in the fluorescence excitation spectrum of acceptor, that is, the efficiency r would increase, which is not observed in the experiment [10, 11].

The calculations of the critical radius of energy transfer (R_0) and the distance between the donor and the acceptor (R) were performed for different DNA concentrations using Eqs. (2) and (4). The distances R between the donor and the acceptors were found to be similar (35–45Å) for all donor–acceptor pairs. The values of the critical radius of energy transfer were found to be rather high for the D3–A pair (37–47Å upon growing the DNA concentration from 6.25×10^{-6} to 5.0×10^{-5}mol l^{-1}); with allowance for the EEET efficiency, the distances between the donor and the acceptor (R) in this system are 37–50Å. For the D1–A pair, comparable values of the critical radius of energy transfer (35 and 37Å at $c_{DNA} = 5.0 \times 10^{-6}$ and 1.25×10^{-5}mol l^{-1}, respectively) were obtained [11].

The obtained values of the critical EEET radii and the distances between the donor and the acceptor were found to be close to the data obtained earlier [10] and correspond to the values determined for other donor–acceptor pairs of dyes non covalently bound to DNA [5, 17].

Quenching of fluorescence of D1–D3 by molecules of A was studied at different DNA concentrations by the spectral and kinetic methods. In the experiment, an increase in the acceptor concentration (c_A increased from 0 to 1.6×10^{-5}mol l^{-1}) led to a marked drop in the intensity of the fluorescence bands of donors (data on quenching of D3 fluorescence shown in Figure 6.3 (a)). The experimental dependence of I_0/I on [A] exhibits a pronounced upward deviation from the linear plot (Figure 3 (b), curve *1*).

The fitting of the experimental data to Eq. (4) without allowance for concentrating of the dyes in the biopolymer microphase gives inadequate results [10], since in this case the calculations give for the bimolecular quenching rate constant k_q the values 3–4 orders of magnitude greater than the diffusion quenching rate constant (for aqueous solutions, k_{diff} ~10^{10}l mol^{-1} s^{-1}), which has no physical meaning.

The upward deviation of the Stern–Volmer plot is observed in the case when donor fluorescence can be quenched via both collisions with an acceptor quencher (dynamic quenching) and static quenching, which occurs practically instantaneously and takes place if a quencher molecule is located in the quenching sphere of the fluorophore (Eq. (5)).

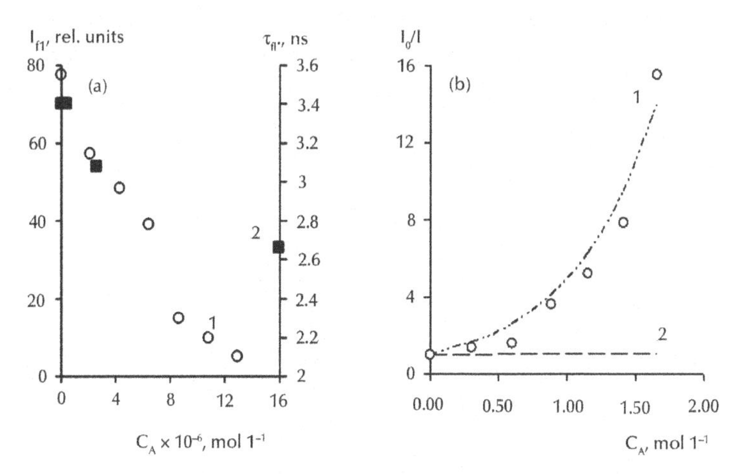

FIGURE 6.3 (a) Experimental data on quenching of D3 fluorescence as a function of A concentration: (*1*) fluorescence intensity at λ_{fl} = 529nm, and (*2*) fluorescence decay time (λ_{ex} = 470nm, λ_{reg} = 535nm).
(b) Data on quenching of D3 fluorescence in the coordinates of the Stern–Volmer plot (I_0/I versus the concentration of quencher A) at c_{DNA} = 2.5 × 10^{-5}mol l^{-1} calculation (*1*) by Eq. (5), and (*2*) by Eq. (4). In the calculations, we allowed for concentrating of the acceptor in the DNA microphase

As a result of the interaction of the dyes with DNA, the actual concentration of quencher A in the vicinity of molecules of donor fluorophores turns out to be higher than that averaged over the whole solution. The microphase model [18, 19] permitted the authors to allow for the concentrating effect upon complexation with DNA and to estimate the radius of the sphere of donor fluorescence quenching by the acceptor (R_q ~45Å [11]) from the assumption that the value of the bimolecular quenching rate constant (k_q) corresponds to the diffusion constant (k_{dif} ~10^{10}l mol^{-1} s^{-1}).

In the absence of the quencher, the fluorescence decay kinetics for dyes D1–D3 in complexes with DNA can be described by exponential dependences, with the fluorescence decay times for D1, D2, and D3 being 3.8, 3.86, and 3.4ns, respectively.

Figure 6.4 (a) shows in the semilogarithmic coordinates the fluorescence decay kinetics for dyes D1 and A (curves *1*, *2*) at c_{DNA} = 1.25 × 10^{-5}mol l^{-1}.

Frequency bar chart of photon emission decay of D1 and A (Figure 6.4 (b), curves *1, 2*) was calculated by the Monte-Carlo technique. The calculation parameters r = 0.36, R_0 = 37Å, R = 41Å were obtained from the EEET steady-state fluorescence experiments. The D1 and A fluorescence decay times (τ_D, τ_A) were determined from kinetics measurements. Thus the EEET rate constant (k_{ET}) was equal to 1.42 × 10^8s^{-1}.

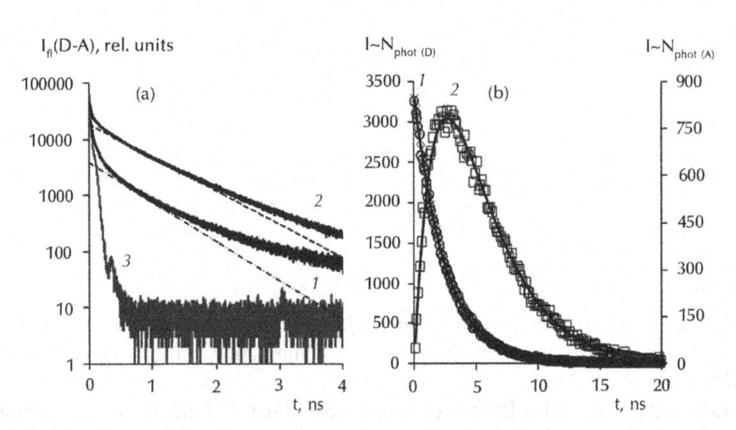

FIGURE 6.4 (a) Fluorescence decay kinetics for dyes D1 (*1*) at λ_{reg} = 530nm and A (*2*) at λ_{reg} = 620nm (c_{DNA} = 1.25 × 10^{-5}mol l^{-1}; c_{D1} = 1.6 × 10^{-6}mol l^{-1}, c_A = 3.09 × 10^{-5}mol l^{-1}. The dashed line are the calculations of fluorescence decay kinetics when τ_D(ET) = 2.5ns and τ_A(ET) = 2.9ns. The fluorescence decay time τ_D(ET) and τ_A(ET) were obtained by the Monte-Carlo calculation. (*3*) Instrumental function (in the semilogarithmic coordinates). **(b)** Frequency bar chart of photon emission decay of D1 (1) and A (2), calculated by the Monte-Carlo technique (Eqs. (6)–(8))

According to the numerical modeling results, EEET should lead to a substantial decrease in the fluorescence decay times of D1 (τ_D(ET) = 2.5ns). This decrease is considered to be the result of dynamic quenching of fluorescence upon EEET. However, the experimental fluorescence decay kinetics for D1–D3 in the presence of A are accelerated negligibly. For dye D1 τ_{D1} = 3.33ns at c_A = 3.09 × 10^{-5} mol l^{-1}, for D2 τ_{D2} = 3.23ns at c_A = 1.31 × 10^{-5} mol l^{-1}, and for D3 τ_{D3} = 2.66ns at c_A = 1.59 × 10^{-5}mol l^{-1} (Figure 3 (b), curve *2*). The frequency bar chart of photon emission decay for A (Figure 6.4 (b), curve *2*) has the growing part with $\tau \sim \tau_D$(ET) = 2.5ns and the decay part (τ_A(ET) = 2.9ns), which does not correspond to the experimental data (Figure 6.4 (a)). The absence of the growing part in the experimental kinetics of acceptor emission can be explained by direct photoexcitation of acceptor molecules to the short-wavelength edge of the absorption band (it was observed in steady-state fluorescence experiments at c_{DNA} = 0 mol l^{-1}).

Consequently, the comparison of the experimental data on fluorescence quenching with the numerical modeling results allow us to conclude that the main contribution to the quenching of D1–D3 by the acceptor in the presence of DNA is made by the mechanism of static quenching, because even at sufficiently high quencher concentrations (c_A = 1 × 10^{-5}mol l^{-1}), the observed drop of the fluorescence intensity of the donors (Figure 6.3 (a)) is accompanied by the insignificant decrease in the fluorescence decay time [11]. In the presence of the quencher, the fluorescence decay kinetics for D1 differ noticeably from the exponentials (as well as for D2 and D3), which cannot be explained using the Perrin model [5], according to which fluorescence is emitted by the donor molecules located beyond the quenching sphere of the acceptor, and the fluorescence decay kinetics is exponential with $\tau \sim \tau_0$.

The non exponential fluorescence decay kinetics is explained by the presence of distribution of the *D–A* distances in the donor–acceptor pairs, that is, it is necessary to allow for the distribution probability density for the distances between the donor and the acceptor, $P(R_{D-A})$ [2]. Non radiative energy transfer in the pairs with shorter distances (and more favorable mutual orientations) will occur much faster (because the EEET efficiency for these pairs is higher). This will lead to inhomogeneous distribution of the EEET rates for excited molecules of the donor dyes in the dye–DNA system at the expense of faster deactivation of donor molecules in such pairs.

With reference to the kinetic data, this is reflected in stronger non exponential character of the fluorescence decay kinetics of the donor (Figure 6.4 (a), curve *1*) with growing the acceptor concentration. The fluorescence decay kinetics of the donor can be described by the integral Eq. (11) [2, 4, 6]. We will believe here that averaging of mutual orientation of molecules of the dyes will have time to occur in the solution, that is, $\kappa^2 = 2/3$ (which also follows from the results of the stationary measurements [10]).

The choice of the distribution type in this case is a separate problem, because the characteristics of the distribution of distances in donor–acceptor pairs are not known beforehand. Practically, continuous distributions are used (for example, the Gaussian distribution [4]). In this work, as a probability distribution for interpretation of the experimental data, we chose the more universal gamma-distribution Eq. (12), which at large k approaches the normal distribution. The results of the calculations showed the presence of the concentration dependence (Figure 6.5) for the distribution of dye A ($P(R_{D-A})$).

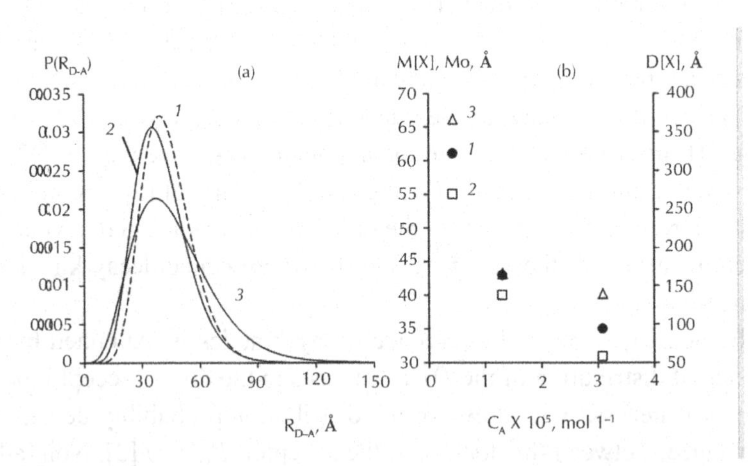

FIGURE 6.5 (a) Gamma-distribution of the probability density $P(R_{D-A})$ for dye D1 (1), D2 (2), D3 (3) at $c_{DNA} = 1.25 \times 10^{-5}$mol l^{-1}. The curves correspond to comparable acceptor concentrations (for D1 and D2 $c_A = 1.31 \times 10^{-5}$mol l^{-1}, and for D3 at $c_A = 1.59 \times 10^{-5}$mol l^{-1}. **(b)** Parameters of the distribution probability density for the donor–acceptor distance $P(R_{D-A})$ for dye D1: the expected value (*M[X]*, (*1*)), the mode (*Mo*, (*2*)), and the dispersion (*D[X]*, (*3*)), at different volume concentrations of the acceptor ($c_{DNA} = 1.25 \times 10^{-5}$mol l^{-1})

The low probability density for the location of the acceptor in the immediate vicinity of the donor (0–20 Å) is explained in terms of Coulomb repulsion of the cations and corresponds to the geometrical size of the dyes [4, 5]. A practically ten-fold increase in the volume concentration of the dye (from 4.25×10^{-6} to 3.09×10^{-5} mol l^{-1}) leads to a shift of the maximum of the probability density curve toward shorter R_{D-A}, which corresponds to shortening the average distances in the D1–A pairs due to the concentrating effect. Figure 6.5 (b) shows that an increase in the volume concentration of the acceptor, along with decreasing the expected value $M[X]$, is also accompanied by a decrease in the dispersion $D[X]$ of distribution of the distances R_{D-A} (which corresponds to narrowing the respective curves in Figure 5 (a)). For dyes D2 and D3, an increase in the acceptor concentration leads to the analogous effect: a three-fold increase in the volume concentration of the acceptor (from 4.25×10^{-6} to 1.31×10^{-5} mol l^{-1}) leads to a decrease in the dispersion of $P(R_{D-A})$ by a factor of 1.6 and is accompanied by a decrease in the values characterizing the average distances in the D2–A pairs, that is, the expected value and the mode. The change in the dispersion can be explained by the decrease in the Debye radius due to the local (in the biopolymer microphase) increase in the ionic strength of the solution with growing the concentration of acceptor dye molecules in the DNA microphase. It can be seen that the expected value $M[X]$ of $P(R_{D-A})$ is much the same for the three compounds and satisfactorily corresponds to the energy transfer distance R (~30–40Å) under these conditions [10, 11].

In this work, the spectral-fluorescence data and the fluorescence decay kinetics for the donors in the presence of the acceptor were analyzed in detail. It was shown, with the use of Monte-Carlo calculations, that upon quenching of fluorescence of the donor dyes by the acceptor dye, the dynamic (diffusion) quenching does not play a significant role, and the quenching occurs mainly by the static mechanism, which causes an upward deviation and concavity of the experimental Stern–Volmer plots. The analysis of the kinetic data permitted us to determine the characteristics of the distribution of acceptor molecules with respect to the donor in the DNA microphase; the concentration effect of the acceptor on the parameters of the distribution of its molecules was found. The kinetic measurements of the fluorescence decay were shown to permit obtaining the parameters of the gamma-distribution of the distances (R_{D-A}) even at small deviations of the fluorescence decay kinetics of the donor from the exponential dependence (at low quencher concentrations).

The values of R_{DA} apparently depend on the ratio of the portion of dye molecules bound in a complex to the number of DNA base pairs. From the parameters of the B-DNA helix, we may take this ratio to be ~1:12–1:14, that is, 0.71–0.83 of the acceptor molecule per the DNA turn. These results are satisfactorily correspond to the data obtained not only for the dyes of the similar structure (for example, for Cyan 2 this ratio is 1:11–1:12 [20]), but also for a number of thiazole orange derivatives [21].

KEYWORDS

- **Carbocyanine dyes**
- **Electronic excitation energy transfer**
- **Carbocyanine dyes**
- **Monte-Carlo simulation**
- **Stern–Volmer plot**

REFERENCES

1. Lilley, D. M. J., & Wilson, T. J. (2000). Fluorescence resonance energy transfer as a structural tool for nucleic acids. *Current Opinion in Chemicall Biology,* 4, 507.
2. Klostermeier, D., & Millar, D. P. (2002). Time-resolved fluorescence resonance energy transfer: A versatile tool for the analysis of nucleic acids. *Biopolymers,* 61(3), 159.
3. Kolpashchikov, D. M. (2010). Binary probes for nucleic acid analysis. *Chemical Reviews,* 110(8), 4709.
4. Sindbert, S., Kalinin, S., Nguyen, H., Kienzler, A., Clima, L., Bannwarth, W., Appel, B., Miller S., & Seidel, C. A. M. (2011). Accurate distance determination of nucleic acids via Forster resonance energy transfer: Implications of dye linker length and rigidity. *Journal of American Chemical. Society,* 133(8), 2463.
5. Kakiuchi, T., Ito, F., & Nagamura (2008). Time-resolved studies of energy transfer from meso-tetrakis(N-methylpyridinium-4-yl)- porphyrin to 3,3′-Diethyl-2,2′-thiatricarbocyanine iodide along deoxyribonucleic acid chain. *The Journal of Physical Chemistry. B,* 112, 3931.

6. Ranjit, S., Gurunathan, K., & Levitus, M. (2009). Photophysics of backbone fluorescent DNA modifications: reducing uncertainties in FRET. *Journal of Physical Chemistry.,* 113 (22), 7861.
7. Hannestad, J. K., Sandin, P., & Albinsson, B. (2008). Self-assembled DNA photonic wire for long-range energy transfer. *Journal of American Chemical. Society,* 130(47), 15889.
8. Anikovsky, M. Yu., Tatikolov, A. S., & Kuzmin, V.A. (2002). Fluorescent properties of some thia- and oxacarbocyanine dyes in the presence of DNA. *High Energy Chemistry,* 36(3), 179.
9. Pronkin, P. G., Tatikolov, A. S., Anikovsky, M. Yu., & Kuzmin, V. A. (2005). The study of cis–trans equilibrium and complexation with DNA of meso-substituted carbocyanine dyes. *High Energy Chemistry,* 39(4), 237.
10. Pronkin, P. G., & Tatikolov, A. S. (2009). Electronic excitation energy transfer between molecules of carbocyanine dyes in complexes with DNA. High Energy Chemistry,43(6), 471.
11. Pronkin, P. G. & Tatikolov, A. S. (2011). Study of electronic excitation energy transfer between cyanine dye molecules noncovalently bound to DNA. *High Energy Chemistry,* 45 (2), 140.
12. Magde, D., Wong, R., & Seybold, P. G. (2002). Fluorescence quantum yields and their relation to lifetimes of rhodamine 6g and fluorescein in nine solvents: improved absolute standards for quantum yields. *Photochemistry Photobiology,* 75(4), 327.
13. Baguley, B. C. & Falkenhang, E. M. (1978). The interaction of ethidium with synthetic double-stranded polynucleotides at low ionic strength. *Nucleic Acids Research,* 5, 161.
14. Lakowicz, J. R. (1983). *Principles of fluorescence spectroscopy.* New York: Plenum.
15. Bielajew, A. F. (2001). *Fundamentals of the Monte-Carlo method for neutral and charged particle transport.* Michigan: University of Michigan.
16. Berney. C.& Danuser, G. (2003). FRET or No FRET: A quantitative comparison. *Journal of. Biophysics,,* 84, 3992.
17. Ito, F., Kakiuchi, T., & Nagamura, T. (2007)). Excitation energy migration of acridine orange intercalated into deoxyribonucleic acid thin films. *Journal of Physical Chemistry C,* 111, 6983.
18. Berezin, I. V., Martinek, K., & Yatsimirskii, A. K. (1973). Physicochemical foundations of micellar catalysis. *Russian Chemical. Reviews,* 42 (10), 787–802.
19. Kuzmin, M. G. & Zaitsev, N. K. (1987). *The Interface structure and electrochemical processes at the boundary between two immiscible liquids.* Berlin: Springer.
20. Yarmoluk, S. M., Lukashov, S. S., Losytskyy, M. Yu., Akerman, B., & Kornyushyna, O. S. (2002). Interaction of cyanine dyes with nucleic acids: XXVI. Intercalation of the trimethine cyanine dye Cyan 2 into double-stranded DNA: study by spectral luminescence methods. *Spectrochimica. Acta, Part A,* 58, 3223.
21. Ogulchansky, T. Yu., Yashchuk, V. M., Yarmoluk, S. M., & Losytskyy, M. Yu. (2000)). Interaction of cyanine dyes with nucleic acids. 14. Spectral peculiarities of several monomethyne benzothiazole cyanine dyes and their interaction with DNA. *Biopolymers and Cell,* 16, 345.

CHAPTER 7

QUANTUM-CHEMICAL MODELING: PART I

AV. A. BABKIN, D. S. ZAKHAROV, and G. E. ZAIKOV

CONTENTS

7.1 QUANTUM-CHEMICAL CALCULATION OF SOME MOLECULES AROMATIC OLEFINS BY METHOD MNDO

7.1.1 INTRODUCTION

Quantum-chemical calculation of molecules of α-cyclopropyl-p-izopropylstyrene, α -cyclopropyl-2,4-dimethylstyrene, α -cyclopropyl-p-ftorstyrene was done by method MNDO. Optimized by all parameters geometric and electronic structures of these compounds were received. The universal factor of acidity was calculated (pKa=32). Molecules of α-cyclopropyl-p-izopropylstyrene, α-cyclopropyl-2,4-dimethylstyrene, and α-cyclopropyl-p-ftorstyrene pertain to class of very weak H-acids (pKa>14).

The aim of this work is a study of electronic structure of molecules α-cyclopropyl-p-izopropylstyrene, α-cyclopropyl-p-izopropylstyrene, α-cyclopropyl-2,4-dimethylstyrene, α-cyclopropyl-p-ftorstyrene [1], and theoretical estimation of its acid power by quantum-chemical method MNDO. The calculation was done with optimization of all parameters by standard gradient method built-in in PC GAMESS [2]. The calculation was executed in approach the insulated molecule in gas phase. Program MacMolPlt was used for visual presentation of the model of the molecule. [3].

7.1.2 MATERIALS AND METHODS

Geometric and electronic structures, general and electronic energies of molecules α-cyclopropyl-p-izopropylstyrene, α-cyclopropyl-2,4-dimethylstyrene, and α-cyclopropyl-p-ftorstyrene were received by method MNDO and are shown in Figure 7.1, 7.2, and 7.3 and in Table 7.1, 7.2, and 7.3, respectively. The universal factor of acidity was calculated by formula: pKa = 49.4-134.61*qmaxH+[4], which used with success, for example in [5–35] (where, qmaxH+ – a maximum positive charge on atom of the hydrogen (by Milliken [1]) R=0.97, R– a coefficient of correlations, qmaxH+=+0,06, +0,06, +0,08, respectively. pKa=30-33.

7.1.3 CONCLUSION

Quantum-chemical calculation of molecules α-cyclopropyl-p-izopropylstyrene, α-cyclopropyl-2, 4-dimethylstyrene, and α-cyclopropyl-p-ftorstyrene by MNDO method was executed for the first time. An optimized geometric and electronic structures of these compound were received. Acid power of molecules α-cyclopropyl-p-izopropylstyrene, α-cyclopropyl-2,4-dimethylstyrene, and α-cyclopropyl-p-ftorstyrene was theoretically evaluated (pKa = 30–33). These compound pertain to class of very weak H-acids (pKa > 14).

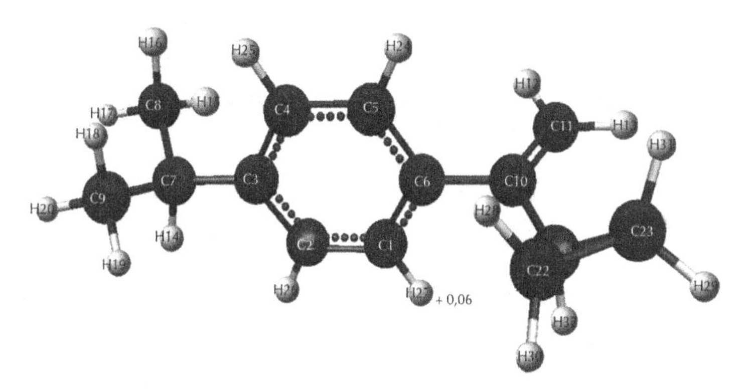

FIGURE 7.1 Geometric and electronic molecular structure of α-cyclopropyl-p-izopropylstyrene. (E0= -196,875kDg/mol, Eel= -1215375kDg/mol)

TABLE 7.1 Optimized bond lengths, valence corners and charges on atoms of the molecule of α-cyclopropyl-p-izopropylstyrene.

Bond lengths	R,A	Valence corners	Grad	Atomw	Charge (by Milliken)
C(1)-C(6)	1.40	C(6)-C(1)-C(2)	121	C(1)	-0.04
C(2)-C(1)	1.39	C(1)-C(2)-C(3)	121	C(2)	-0.05
C(3)-C(2)	1.40	C(2)-C(3)-C(4)	119	C(3)	-0.07
C(3)-C(7)	1.50	C(3)-C(4)-C(5)	121	C(4)	-0.05
C(4)-C(3)	1.40	C(4)-C(5)-C(6)	121	C(5)	-0.05
C(5)-C(4)	1.39	C(2)-C(3)-C(7)	120	C(6)	-0.03

TABLE 7.1 *(Continued)*

Bond lengths	R,A	Valence corners	Grad	Atomw	Charge (by Milliken)
C(6)-C(5)	1.40	C(3)-C(7)-C(8)	111	C(7)	-0.02
C(6)-C(10)	1.47	C(3)-C(7)-C(9)	111	C(8)	+0.04
C(7)-C(8)	1.52	C(8)-C(7)-C(9)	110	C(9)	+0.04
C(7)-C(9)	1.52	C(1)-C(6)-C(10)	121	C(10)	-0.06
C(10)-C(11)	1.34	C(6)-C(10)-C(11)	122	C(11)	-0.03
C(10)-C(21)	1.47	C(10)-C(11)-H(12)	122	H(12)	+0.04
H(12)-C(11)	1.10	C(10)-C(11)-H(13)	122	H(13)	+0.04
H(13)-C(11)	1.10	C(3)-C(7)-H(14)	108	H(14)	+0.01
H(14)-C(7)	1.13	C(7)-C(8)-H(15)	110	H(15)	0.00
H(15)-C(8)	1.12	C(7)-C(8)-H(16)	111	H(16)	0.00
H(16)-C(8)	1.12	C(7)-C(8)-H(17)	110	H(17)	-0.01
H(17)-C(8)	1.12	C(7)-C(9)-H(18)	111	H(18)	0.00
H(18)-C(9)	1.12	C(7)-C(9)-H(19)	110	H(19)	0.00
H(19)-C(9)	1.12	C(7)-C(9)-H(20)	110	H(20)	-0.01
H(20)-C(9)	1.12	C(6)-C(10)-C(21)	116	C(21)	-0.07
C(21)-C(22)	1.51	C(11)-C(10)-C(21)	123	C(22)	-0.05
C(22)-C(23)	1.50	C(22)-C(23)-C(21)	60	C(23)	-0.06
C(23)-C(21)	1.51	C(10)-C(21)-C(22)	121	H(24)	+0.06
H(24)-C(5)	1.10	C(21)-C(23)-C(22)	60	H(25)	+0.06
H(25)-C(4)	1.10	C(21)-C(22)-C(23)	60	H(26)	+0.06
H(26)-C(2)	1.10	C(22)-C(21)-C(23)	60	**H(27)**	**+0.06**
H(27)-C(1)	1.10	C(4)-C(5)-H(24)	120	H(28)	+0.04
H(28)-C(22)	1.10	C(3)-C(4)-H(25)	120	H(29)	+0.04
H(29)-C(23)	1.10	C(1)-C(2)-H(26)	120	H(30)	+0.04
H(30)-C(22)	1.10	C(2)-C(1)-H(27)	120	H(31)	+0.04
H(31)-C(23)	1.10	C(21)-C(22)-H(28)	119	H(32)	+0.05
H(32)-C(21)	1.11	C(21)-C(23)-H(29)	118		

TABLE 7.1 *(Continued)*

Bond lengths	R,A	Valence corners	Grad	Atomw	Charge (by Milliken)
		C(21)-C(22)-H(30)	119		
		C(21)-C(23)-H(31)	120		
		C(10)-C(21)-H(32)	111		

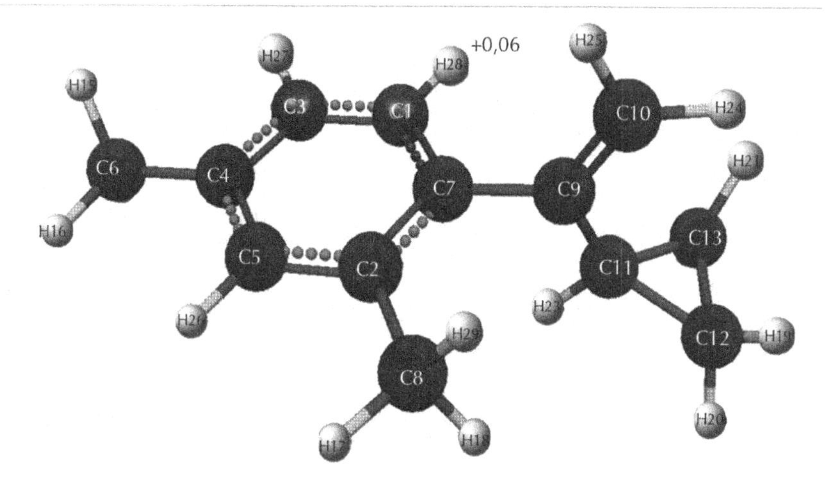

FIGURE 7.2 Geometric and electronic molecular structure of α-cyclopropyl-2,4-dimethylstyrene (E0= -181,125 kDg/mol, Eel= -1,084,125 kDg/mol)

TABLE 7.2 Optimized bond lengths, valence corners and charges on atoms of the molecule of α-cyclopropyl-2,4-dimethylstyrene

Bond lengths	R,A	Valence corners	Grad	Atom	Charge (by Milliken)
C(1)-C(7)	1.42	C(1)-C(7)-C(2)	119	C(1)	-0.0508
C(2)-C(5)	1.42	C(7)-C(1)-C(3)	121	C(2)	-0.0835
C(3)-C(1)	1.40	C(1)-C(3)-C(4)	121	C(3)	-0.0412
C(4)-C(3)	1.41	C(2)-C(5)-C(4)	123	C(4)	-0.1009
C(5)-C(4)	1.41	C(3)-C(4)-C(5)	118	C(5)	-0.0280
C(6)-C(4)	1.51	C(3)-C(4)-C(6)	121	C(6)	0.0814
C(7)-C(2)	1.42	C(5)-C(2)-C(7)	119	C(7)	-0.0187

TABLE 7.2 *(Continued)*

Bond lengths	R,A	Valence corners	Grad	Atom	Charge (by Milliken)
C(7)-C(9)	1.50	C(5)-C(2)-C(8)	119	C(8)	0.0807
C(10)-C(9)	1.35	C(1)-C(7)-C(9)	118	C(9)	-0.0545
C(11)-C(9)	1.50	C(7)-C(9)-C(10)	120	C(10)	-0.0416
C(12)-C(11)	1.54	C(7)-C(9)-C(11)	115	C(11)	-0.0617
C(13)-C(12)	1.52	C(9)-C(11)-C(12)	125	C(12)	-0.0561
C(13)-C(11)	1.54	C(9)-C(11)-C(13)	125	C(13)	-0.0568
H(14)-C(6)	1.11	C(4)-C(6)-H(14)	111	H(14)	-0.0028
H(15)-C(6)	1.11	C(4)-C(6)-H(15)	111	H(15)	-0.0027
H(16)-C(6)	1.11	C(4)-C(6)-H(16)	113	H(16)	-0.0050
H(17)-C(8)	1.11	C(2)-C(8)-H(17)	112	H(17)	-0.0072
H(18)-C(8)	1.11	C(2)-C(8)-H(18)	111	H(18)	-0.0002
H(19)-C(12)	1.10	C(11)-C(12)-H(19)	121	H(19)	0.0389
H(20)-C(12)	1.10	C(11)-C(12)-H(20)	118	H(20)	0.0368
H(21)-C(13)	1.10	C(11)-C(13)-H(21)	121	H(21)	0.0387
H(22)-C(13)	1.10	C(11)-C(13)-H(22)	118	H(22)	0.0370
H(23)-C(11)	1.10	C(9)-C(11)-H(23)	111	H(23)	0.0451
H(24)-C(10)	1.09	C(9)-C(10)-H(24)	124	H(24)	0.0394
H(25)-C(10)	1.09	C(9)-C(10)-H(25)	123	H(25)	0.0425
H(26)-C(5)	1.09	C(2)-C(5)-H(26)	119	H(26)	0.0550
H(27)-C(3)	1.09	C(1)-C(3)-H(27)	119	H(27)	0.0581
H(28)-C(1)	1.09	C(3)-C(1)-H(28)	119	**H(28)**	**0.0600**
H(29)-C(8)	1.11	C(2)-C(8)-H(29)	111	H(29)	-0.0019

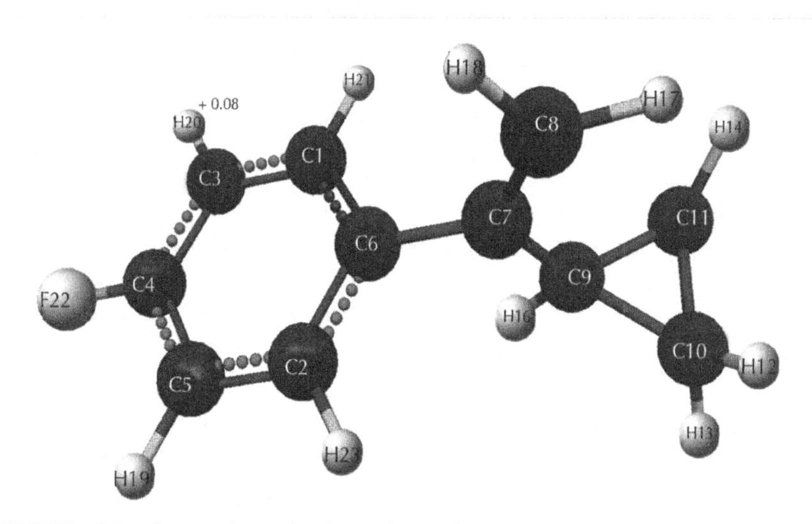

FIGURE 7.3 Geometric and electronic molecular structure of α-cyclopropyl-p-ftorstyrene (E0= -196,875 kDg/mol, Eel= -1,155,000 kDg/mol)

TABLE 7.3 Optimized bond lengths, valence corners and charges on atoms of the molecule of α-cyclopropyl-p-ftorstyrene.

Bond lengths	R,A	Valence corners	Grad	Atom	Charge (by Milliken)
C(1)-C(3)	1.40	C(1)-C(6)-C(2)	119	C(1)	-0.02
C(2)-C(6)	1.41	C(4)-C(5)-C(2)	120	C(2)	-0.02
C(3)-C(4)	1.42	C(5)-C(4)-C(3)	120	C(3)	-0.09
C(4)-C(5)	1.42	C(1)-C(3)-C(4)	120	C(4)	+0.15
C(5)-C(2)	1.40	C(6)-C(2)-C(5)	121	C(5)	-0.09
C(6)-C(7)	1.49	C(3)-C(1)-C(6)	121	C(6)	-0.06
C(7)-C(9)	1.50	C(1)-C(6)-C(7)	121	C(7)	-0.06
C(8)-C(7)	1.35	C(6)-C(7)-C(8)	120	C(8)	-0.04
C(9)-C(10)	1.54	C(6)-C(7)-C(9)	115	C(9)	-0.07
C(10)-C(11)	1.52	C(7)-C(9)-C(10)	124	C(10)	-0.06

TABLE 7.3 *(Continued)*

Bond lengths	R,A	Valence corners	Grad	Atom	Charge (by Milliken)
C(11)-C(9)	1.54	C(7)-C(9)-C(11)	125	C(11)	-0.06
H(12)-C(10)	1.10	C(9)-C(10)-H(12)	121	H(12)	+0.04
H(13)-C(10)	1.10	C(9)-C(10)-H(13)	118	H(13)	+0.04
H(14)-C(11)	1.10	C(9)-C(11)-H(14)	121	H(14)	+0.04
H(15)-C(11)	1.10	C(9)-C(11)-H(15)	118	H(15)	+0.04
H(16)-C(9)	1.10	C(7)-C(9)-H(16)	111	H(16)	+0.04
H(17)-C(8)	1.09	C(7)-C(8)-H(17)	124	H(17)	+0.04
H(18)-C(8)	1.09	C(7)-C(8)-H(18)	123	H(18)	+0.04
H(19)-C(5)	1.09	C(2)-C(5)-H(19)	120	H(19)	+0.08
H(20)-C(3)	1.09	C(1)-C(3)-H(20)	120	**H(20)**	**+0.08**
H(21)-C(1)	1.09	C(3)-C(1)-H(21)	119	H(21)	+0.07
F(22)-C(4)	1.33	C(3)-C(4)-F(22)	120	F(22)	-0.18
H(23)-C(2)	1.09	C(5)-C(2)-H(23)	119	H(23)	+0.07

TABLE 7.4 General and energies (E0), maximum positive charge on atom of the hydrogen (qmaxH+), universal factor of acidity (pKa).

Molecules of aromatic olefins	E0	qmaxH+	pKa
á-cyclopropyl-p-izopropylstyrene	-196,875	+0.06	33
á -cyclopropyl-2,4-dimethylstyrene	-181,125	+0.06	33
á -cyclopropyl-p-ftorstyrene	-196,875	+0.08	30

KEYWORDS

- **Aromatic olefins**
- **Bond lengths**
- **Geometric molecular structure**
- **Quantum-chemical calculation**
- **Valence corners**

REFERENCES

1. Kennedy, J. (1978). Cation polymerization of olefins. *The world* (p. 430). Moscow: John Wiley & Sons.
2. Shmidt, M. W., Baldrosge, K. K., Elbert, J. A., Gordon, M. S., Enseh, J. H., Koseki, S., Matsvnaga, N., Nguyen, K. A., Su, S. J. et al. (1993). *Journal of computational chemistry, 14,* 1347–1363.
3. Babkin, V. A., Fedunov, R. G., Minsker, K. S. et al. (2002) *Oxidation communication, №1, 25,* 21–47.
4. Bode, B. M. & Gordon, M. S. (1998). *Journal of molecular graphics and modelling, 16*, 133–138.
5. Babkin, V. A., Dmitriev, V. Yu., & Zaikov, G. E. (2010). Quantum chemical calculation of molecule hexene-1 by method MNDO. In *Quantum chemical calculation of unique molecular system* (Vol. I, pp. 93–95). Volgograd: Volgograd State University.
6. Babkin, V. A., Dmitriev, V. Yu., & Zaikov, G. E. (2010). Quantum chemical calculation of molecule heptene-1 by method MNDO. In *Quantum chemical calculation of unique molecular system* (Vol. I, pp. 95–97). Volgograd: Volgograd State University.
7. Babkin, V. A., Dmitriev, V. Yu., & Zaikov, G. E. (2010). Quantum chemical calculation of molecule decene-1 by method MNDO. In *Quantum chemical calculation of unique molecular system* (Vol. I, pp. 97–99). Volgograd: Volgograd State University.
8. Babkin, V. A., Dmitriev, V. Yu., & Zaikov, G. E. (2010). Quantum chemical calculation of molecule nonene-1 by method MNDO. In *Quantum chemical calculation of unique molecular system* (Vol. I, pp. 99–102). Volgograd: Volgograd State University.
9. Babkin, V. A. & Andreev, D. S. (2010). Quantum chemical calculation of molecule isobutylene by method MNDO. In *Quantum chemical calculation of unique molecular system* (Vol. I, pp. 176–177). Volgograd: Volgograd State University.
10. Babkin, V. A. & Andreev, D. S. (2010). Quantum chemical calculation of molecule 2-methylbutene-1 by method MNDO. In *Quantum chemical calculation of unique molecular system* (Vol. I, pp. 177–179). Volgograd: Volgograd State University.
11. Babkin, V. A. & Andreev, D. S. (2010). Quantum chemical calculation of molecule 2-methylbutene-2 by method MNDO. In *Quantum chemical calculation of unique molecular system* (Vol. I, pp. 179–180). Volgograd: Volgograd State University.
12. Babkin, V. A. & Andreev, D. S. (2010). Quantum chemical calculation of molecule 2-methylpentene-1 by method MNDO. In *Quantum chemical calculation of unique molecular system* (Vol. I, pp. 181–182). Volgograd: Volgograd State University.
13. Babkin, V. A., Dmitriev, V. Yu., & Zaikov, G. E. (2010). Quantum chemical calculation of molecule butene-1 by method MNDO. In *Quantum chemical calculation of unique molecular system* (Vol. I, pp. 89–90). Volgograd: Volgograd State University.
14. Babkin, V. A., Dmitriev, V. Yu., & Zaikov, G. E. (2010). Quantum chemical calculation of molecule hexene-1 by method MNDO. In *Quantum chemical calculation of unique molecular system* (Vol. I, pp. 93–95). Volgograd: Volgograd State University.
15. Babkin, V. A., Dmitriev, V. Yu., & Zaikov, G. E. (2010). Quantum chemical calculation of molecule octene-1 by method MNDO. In *Quantum chemical calculation of unique molecular system* (Vol. I, pp. 103–105). Volgograd: Volgograd State University.

16. Babkin, V. A., Dmitriev, V. Yu., & Zaikov, G. E. (2010). Quantum chemical calculation of molecule pentene-1 by method MNDO. In *Quantum chemical calculation of unique molecular system* (Vol. I, pp. 105–107). Volgograd: Volgograd State University.
17. Babkin, V. A., Dmitriev, V. Yu., & Zaikov, G. E. (2010) Quantum chemical calculation of molecule propene-1 by method MNDO. In *Quantum chemical calculation of unique molecular system* (Vol. I, pp. 107–108). Volgograd: Volgograd State University.
18. Babkin, V. A., Dmitriev, V. Yu., & Zaikov, G. E. (2010). Quantum chemical calculation of molecule ethylene-1 by method MNDO. In *Quantum chemical calculation of unique molecular system* (Vol. I, pp. 108–109). Volgograd: Volgograd State University.
19. Babkin, V. A. & Andreev, D. S. (2010). Quantum chemical calculation of molecule butadien-1,3 by method MNDO. In *Quantum chemical calculation of unique molecular system* (Vol. I, pp. 235–236). Volgograd: Volgograd State University.
20. Babkin, V. A. & Andreev, D. S. (2010). Quantum chemical calculation of molecule 2-methylbutadien-1,3 by method MNDO. In *Quantum chemical calculation of unique molecular system* (Vol. I, pp. 236–238). Volgograd: Volgograd State University.
21. Babkin, V. A. & Andreev, D. S. (2010). Quantum chemical calculation of molecule 2,3-dimethylbutadien-1,3 by method MNDO. In *Quantum chemical calculation of unique molecular system* (Vol. I, pp. 238–239). Volgograd: Volgograd State University.
22. Babkin, V. A. & Andreev, D. S. (2010). Quantum chemical calculation of molecule pentadien-1,3 by method MNDO. In *Quantum chemical calculation of unique molecular system* (Vol. I, pp. 240–241). Volgograd: Volgograd State University.
23. Babkin, V. A. & Andreev, D. S. (2010). Quantum chemical calculation of molecule trans-trans-hexadien-2,4 by method MNDO. In *Quantum chemical calculation of unique molecular system* (Vol. I, pp. 241–243). Volgograd: Volgograd State University.
24. Babkin, V. A. & Andreev, D. S. (2010). Quantum chemical calculation of molecule cis-trans-hexadien-2,4 by method MNDO. In *Quantum chemical calculation of unique molecular system* (Vol. I, pp. 243–245). Volgograd: Volgograd State University.
25. Babkin, V. A. & Andreev, D. S. (2010). Quantum chemical calculationof molecule cis-cis-hexadien-2,4 by method MNDO. In *Quantum chemical calculation of unique molecular system* (Vol. I, pp. 245–246). Volgograd: Volgograd State University.
26. Babkin, V. A. & Andreev, D. S. (2010). Quantum chemical calculation of molecule trans-2-methylpentadien-1,3 by method MNDO. In *Quantum chemical calculation of unique molecular system* (Vol. I, pp. 247–248). Volgograd: Volgograd State University.
27. Babkin, V. A. & Andreev, D. S. (2010). Quantum chemical calculation of molecule trans-3-methylpentadien-1,3 by method MNDO. In *Quantum chemical calculation of unique molecular system* (Vol. I, pp. 249–250). Volgograd: Volgograd State University.
28. Babkin, V. A. & Andreev, D. S. (2010). Quantum chemical calculation of molecule cis-3-methylpentadien-1,3 by method MNDO. In *Quantum chemical calculation of unique molecular system* (Vol. I, pp. 251–252). Volgograd: Volgograd State University.

29. Babkin, V. A. & Andreev, D. S. (2010). Quantum chemical calculation of molecule 4-methylpentadien-1,3 by method MNDO. In *Quantum chemical calculation of unique molecular system* (Vol. I, pp. 252–254). Volgograd: Volgograd State University.
30. Babkin, V. A. & Andreev, D. S. (2010). Quantum chemical calculation of molecule cis-3-methylpentadien-1,3 by method MNDO. In *Quantum chemical calculation of unique molecular system* (Vol. I, pp. 254–256). Volgograd: Volgograd State University.
31. Babkin, V. A. & Andreev, D. S. (2010). Quantum chemical calculation of molecule 1,1,4,4-tetramethylbutadien-1,3 by method MNDO. In *Quantum chemical calculation of unique molecular system* (Vol. I, pp. 256–258). Volgograd: Volgograd State University.
32. Babkin, V. A. & Andreev, D. S. (2010). Quantum chemical calculation of molecule 2-phenylbutadien-1,3 by method MNDO. In *Quantum chemical calculation of unique molecular system* (Vol. I, pp. 260–262). Volgograd: Volgograd State University.
33. Babkin, V. A. & Andreev, D. S. (2010). Quantum chemical calculation of molecule 1-phenyl-4-methylbutadien-1,3 by method MNDO. In *Quantum chemical calculation of unique molecular system* (Vol. I, pp. 262–264). Volgograd: Volgograd State University.
34. Babkin, V. A. & Andreev, D. S. (2010). Quantum chemical calculation of molecule chloropren by method MNDO. In *Quantum chemical calculation of unique molecular system* (Vol. I, pp. 264–265). Volgograd: Volgograd State University.
35. Babkin, V. A. & Andreev, D. S. (2010). Quantum chemical calculation of molecule trans-hexathrien-1,3,5 by method MNDO. In Quantum chemical calculation of unique molecular system (Vol. I, pp. 266–267). Volgograd: Volgograd State University.

CHAPTER 8

QUANTUM-CHEMICAL MODELING: PART II

A. A. TUROVSKY, A. R. KYTSYA and L. I. BAZYLYAK

CONTENTS

8.1 QUANTUM–CHEMICAL MODELING OF THE KINETICS AND CHEMICAL MECHANISM OF THE METHYL ACETATE HYDROLYSIS IN REATIONS OF THE ENZYMATIC AND HOMOGENEOUS CATALYSIS BY CHEMO-TRYPSIN AND SERINE

8.1.1 INTRODUCTION

It has been done the quantum–chemical modeling of the methyl acetate hydrolysis reaction in the processes of enzymatic and homogeneous catalysis by chemotrypsin and serine respectively, it was shown that the structure of an active center of the chemotyrpsin taken in references is proved by our calculations. The kinetic parameters for the reactions of acyl compounds formation and also of their hydrolysis for both types of the catalysis were estimated. It was shown that the formation of acylchemotrypsin proceeds via a series of the intermediate complexes, caused by the conformation transition, which, in a great measure, caused by the hydrogen bonds of the chemotyrpsin and depend on the journey of serine active group hydrogen atom. It was established, that the stabilization of the activated complexes of the acidulating and deacidulating reactions in enzymatic and homogeneous catalysis in a great measure depends on a value of the charges of the reactive center that is indicated on the values of the kinetic parameters of such reactions. It has been calculated the thermodynamics of the elementary stages for the acidulating and deacidulating reactions.

Especially great successes under investigation of the driving forces of the enzymatic catalysis were achieved in a case of the chemotyrpsin. Chemotrypsin—this is endopeptidase, which cleaves the peptide bonds into peptides. The most important information about the structure of the chemotyrpsin's molecule has been obtained with the use of the roentgen investigations [1-4]. It was found, that all charged groups into the molecule of the ferment are directed sideways to the aqueous solution (with the except for the three, which have the special functions into a mechanism action of the active center). The successes in kinetic investigations in most cases were caused by the works of M. Bergmann, D. Frugonn, and H. Neyrag, who determined that the chemotyrpsin can hydrolyze also the simple low-molecular products (amides, esters).

The force of the chemotyrpsin catalytic action under the esters hydrolysis approximately in 10^6 times exceeds the catalytic action both of OH^-, and H_3O^+.

Hydrolysis of the substrate (amides, esters) on the active center of the chemotyrpsin proceeds in some stages. The first stage of the enzymatic process includes the sorption (so-called formation of the *Michaels's* complex *ES*). The next stages include the chemical transformation of the sorbed molecule with the formation of the intermediate compound of the acylferment EA in accordance with the following kinetic scheme:

$$E + S \underset{k_{-1}}{\overset{k_1}{\Leftrightarrow}} ES \underset{\downarrow P_1}{\overset{k_2}{\rightarrow}} EA \underset{\uparrow H_2O}{\overset{k_3}{\rightarrow}} E + P_2 \qquad (1)$$

$$K_S = \frac{k_{-1}}{k_1}$$

P_1 and P_2 are products of the hydrolysis.

The equilibrium position is determined only by the non valence interactions with the protein of side chemically inert fragments of the substrate's molecule. The intermediate product represents by itself the acyl–ferment, which is unstable compound (lifetime ~ 0.01 *s*. [5]).

Finally, the enzymatic hydrolysis can be presented in accordance with the following scheme:

$$\underset{(E)}{EH} + \underset{(S)}{RC(O)OCH_3} \overset{K_S}{\Leftrightarrow} \underset{(E \cdot S)}{HE \cdot RC(O)OCH_3} \qquad (2)$$

$$E \cdot S \overset{k_2}{\rightarrow} \underset{(EA)}{E - C(O)R} + \underset{(P_1)}{CH_3OH} \qquad (3)$$

$$EA \underset{H_2O}{\overset{k_3}{\rightarrow}} E + RCOOH \qquad (4)$$

However, the stages (3) and (4) are not elementary [6–9] and include the quick (and equilibrium) formation of the intermediate position, which corresponds to a new conformational state of the ferment.

Under hydrolysis of the molecule of a substrate which is sorbing on an active center, the *OH*-group of the serine stands out as the attacking nucleophile [6, 10-12]. It is assumed that the high activity of the serine related with its surroundings into the active center. Along with the serine, the imidazole group of *His* also takes part in its activity [6, 10, 11, 13]. At this, the Nitrogen atom of the Histidine forms the hydrogen bond with Oxygen of the serine hydroxyl. Accordingly to *Blow* [14], the second hydrogen bond exists between the atoms nitrogen and histidine and carbonyl group of the remains of *Asp*, which is located into the depths of the ferment globule. The system of the hydrogen bonds leads to increase of the negative charge on the OH-group of the serine that promotes to its nucleophilicity.

On the acidulating stage the nucleophilic attack of the carboxyl carbon of the substrate by generalized nucleophile of active center: *Ser*, *His*, *Asp* proceeds. As a result of the active center acidulating, the turn of the *Ser* remains around the bonds $C_\alpha - C_\beta$ which is accompanying by the displacement of the oxygen atom on ~ 2.5Å, takes place. At this, imidazole group of the *His* displaces sideways to the solvent [3]. As a result, the imidazole group of the *His* is included into a free ferment (and into the *Michaels's* complex) in hydrogen bond. *Ser* in acylferment contributes one's own atom N for the formation of the hydrogen bond with water. As a result, the activated molecule of water has the ability to effectively attack the carbonyl carbon of the substrate on the deacidulating stage. At this, it is formed the product of the hydrolysis and it is regenerated a free ferment. The above said chemical mechanism of the hydrolytic action of chemotyrpsin is described in such a way in references.

We were interested into the analysis of such approach to the enzymatic catalysis with the use of the quantum–chemical calculations.

8.1.2 EXPERIMENTAL TECHNIQUES

The aim of the presented work was the modeling of the kinetics and chemical mechanism of the enzymatic catalysis process for reaction of hydrolysis of methyl acetate by chemotyrpsin with the use of the quantum–chemical method.

Firstly, it was interesting to consider the stage-by-stage modeling of the process in accordance with the well-known schemes accepted in references and, as far as possible, to append them.

Secondly, it was interesting to estimate the kinetic parameters of the enzymatic catalysis reaction.

Thirdly, it was interesting to compare the kinetics of processes of enzymatic and homogeneous catalysis of methyl acetate.

For quantum–chemical calculations the semiempirical method PM6 was used.

Modeling objects:

– methyl acetate;

– generally accepted active form of chemotyrpsin containing of three fragments (Figure. 8.1). In Figure 8.1 also the geometry and electronic characteristics of chemotyrpsin is presented.

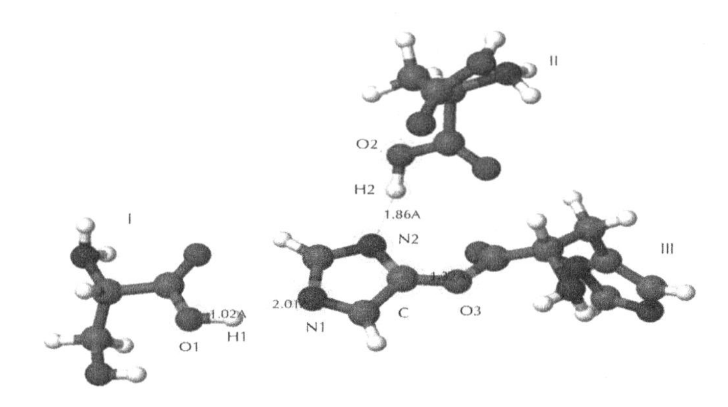

FIGURE 8.1 Geometry and electronic characteristics of the chemotyrpsin molecule: I Ser; II Asp; III His.

Ionization potential 9.605 eV; electron affinity: −2.601, angle O_2–N_2–O_3 104.8°

Atoms	H_1	O_1	N_1	H_2	O_2	N_2	C	O_3
Charge	+0.385	−0.546	−0.363	+0.403	−0.515	−0.474	+0.338	−0.429

TABLE 8.2 *(Continued)*

Bonds	O_1–H_1	N_1–H_1	O_2–H_2	N_2–H_2	C–O_3
Distance (Å)	1.02	2.01	1.05	1.86	1.32

8.1.3 RESULTS AND DISCUSSION

It can be seen from the optimal geometry of the active fragment of chemotyrpsin (Figure 8.1), that the hydrogen bonds correspondingly to the distances exist between the atoms H_1 and N_1, and also between atoms H_2 and N_2. Probably, the second bond is stronger (the distance is shorter). Let us note, that the electron affinity in the chemotyrpsin fragment is enough high. During the fitting of methyl acetate molecule to chemotyrpsin along the coordinate C_1–O_1 (Figure 8.2a) at a distance of 1.60Å, change of the complex's geometry and elongation of hydrogen bond O_1–H–N_1 takes place, that is caused by the conformational transformations of chemotyrpsin molecule. Such conformational transformation proceeds with enough great energy charge (see peak no. 1 in Figure 8.3).

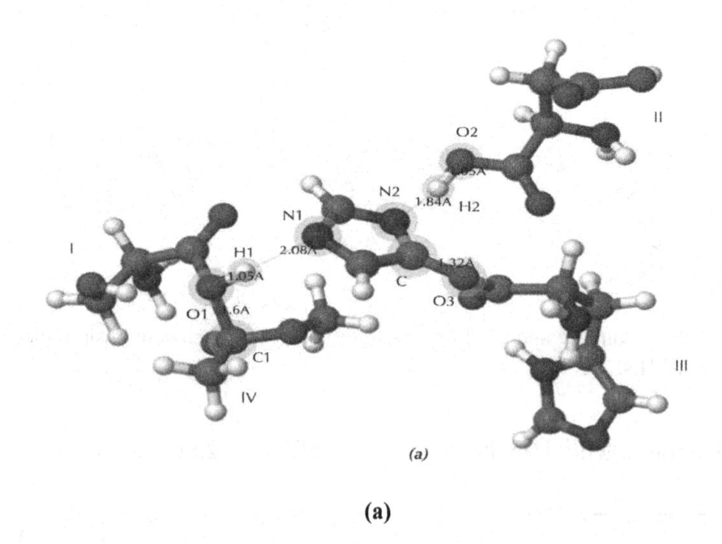

(a)

FIGURE 8.2 *(Continued)*

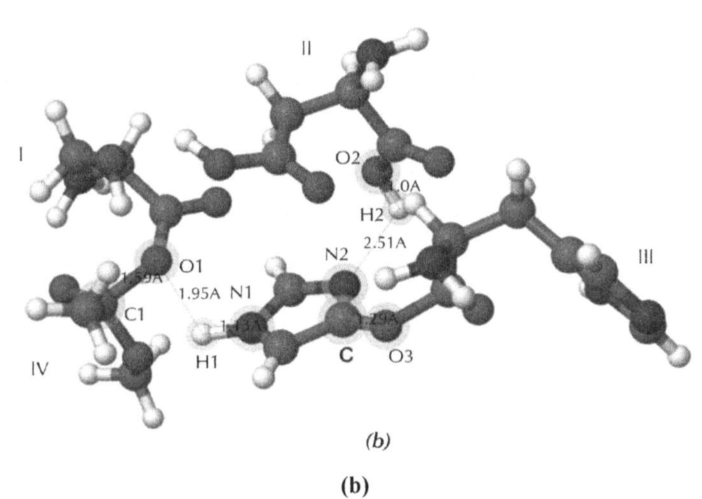

(b)

FIGURE 8.2 Change of the geometry of complex at point 1 (correspondingly to Figure 8.3). **I** Ser, **II** Asp, **III** His, **IV** methyl acetate. (a) – Distance C_1–O_1: 1.60Å, angle O_2–N_2–O_3: 104.70. (b) – Distance C_1–O_1: 1.59Å, angle O_2–N_2–O_3: 122.90

Distance C_1–O_1: 1.60 Å, angle O_2–N_2–O_3: 104.7^0

Atoms	H_1	O_1	N_1	C_1	H_2	O_2	N_2	C	O_3
Charge	+0.399	−0.531	−0.358	+0.758	+0.401	−0.519	−0.476	+0.339	−0.432

Bonds	O_1–H_1	N_1–H_1	O_2–H_2	N_2–H_2	C–O_3
Distance (Å)	1.05	2.08	1.05	1.84	1.32

Distance C_1–O_1: 1.59Å, angle O_2–N_2–O_3: 122.9^0

Atoms	H_1	O_1	N_1	C_1	H_2	O_2	N_2	C	O_3
Charge	+0.405	−0.706	−0.212	+0.811	+0.360	−0.500	−0.457	+0.423	−0.420

Bonds	O_1–H_1	N–H_1	O_2–H_2	N_2–H_2	C–O_3
Distance (Å)	1.95	1.13	1.00	2.51	1.29

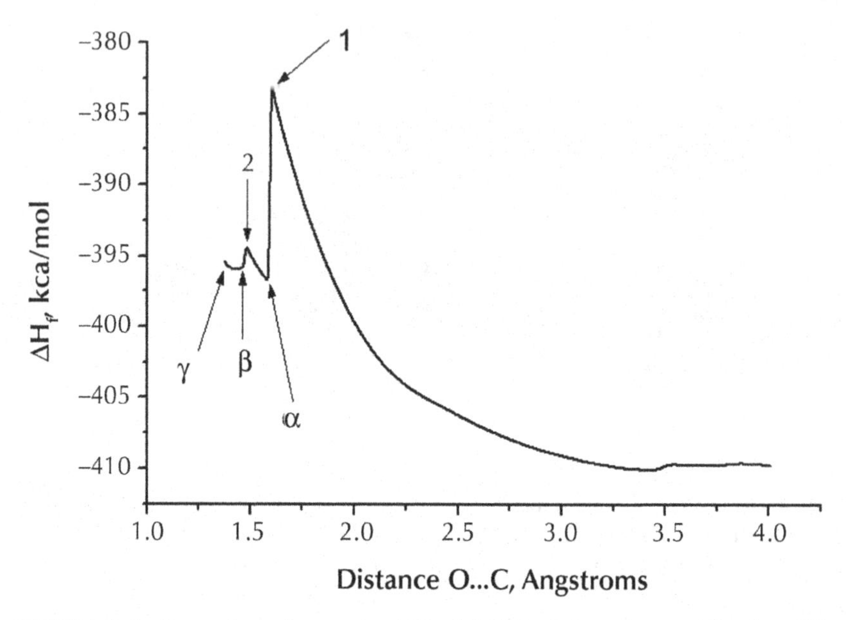

FIGURE 8.3 Formation of complex [Kh–CH_3CO]. *1* complex coordination [Kh–CH_3COOCH_3] (1.60Å), transfer of hydrogen atom from serine to nitrogen atom; *2* detachment of methanol (1.48Å)

It can be seen from Figure 8.2a, that at the distance C_1–O_1 (1.60Å), the charges on H_1 and O_1 are increased; the charge on N_1 is decreased. The change of angle O_2–N_2–O_3 is insignificant. The interatomic distance O_1–H_1 is also enhanced. Insignificant change of a distance C_1–O_1 leads to considerable changes in geometry and values of charges on atoms of the chemotyrpsin fragment. At distance C_1–O_1 (1.59Å), as a result of new conformational changes, a new complex (α) is formed (Figure 8.2b and 8.3 (α)). From Figure 8.2b, a great increase of charge on the reactive center O_1 can be seen. It also increased charge on C_1 of methyl acetate. A charge on N_1 is considerably decreased as a result of great decrease in distance N_1–H_1. The charges on N_2 and H_2 are correspondingly decreased at the expense of the hydrogen bond distance increasing as a result of change of interatomic distances and angles of the surrounding atoms. An essential change in the dihedral angle O_2–N_2–O_3 is observed, that leads to more stable conformation of chemotyrpsin (Figure 8.3 *(α)*).

Due to the fitting of atom C_1 to $O_{1,}$ the atom O_4 (Figure 8.3 (peak no. 2) and 8.4a) gets near to the atom H1; and at a distance of 1.48Å, the transition of atom H_1 to O_4 and detachment of methanol molecule from the complex take place. Such process is illustrated by peak no. 2 in Figure 8.3.

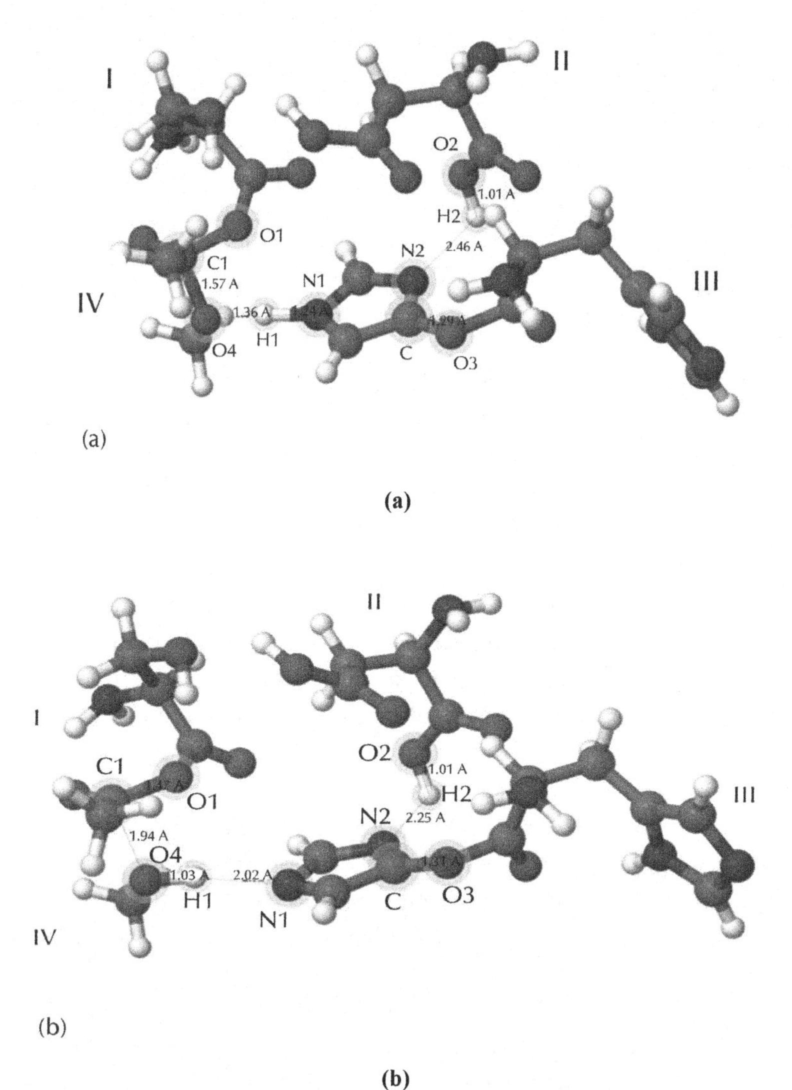

FIGURE 8.4 The change of the geometry of complex at point 2. (a) – **I** Ser, **II** Asp, **III** His, **IV** methyl acetate; (b) – **I** Ser + CH_3CO, **II** Asp, **III** His, IV methanol

Distance C_1–O_1: 1.48Å, angle O_2–N_2–O_3: 123.3°

Atoms	H_1	O_1	N_1	C_1	O_4	H_2	O_2	C	O_3
Charge	+0.420	−0.648	−0.280	+0.801	−0.550	+0.363	−0.502	+0.399	−0.425

Bonds	O_4–H_1	C_1–O_4	N_2–H_1	O_2–H_2	N_2–H_2	C–O_3
Distance (Å)	1.36	1.57	1.24	1.01	2.46	1.29

Distance C_1–O_1: 1.47Å, angle O_2–N_2–O_3: 117.8°

Atoms	H_1	O_1	N_1	C_1	O_4	H_2	O_2	C	O_3	N_2
Charge	+0.376	−0.623	−0.374	+0.794	−0.523	+0.378	−0.513	+0.349	−0.434	−0.481

Bonds	O_1–H_1	C_1–O_4	N_1–H_1	O_2–H_2	N_2–H_2	C–O_3
Distance (Å)	1.03	1.94	2.02	1.01	2.25	1.31

For complex (Figure 8.4a) which itself represents the activated complex for the formation of intermediate unstable compound, the charges on atoms H_1 and N_1 are increased most essentially. The charges on C_1 and O_1 are decreased. As a result of addition of hydrogen atom to O_4 of methyl acetate (Figure 8.4a), the distances of N_1−H_1 are increased and of N_2−H_2 are decreased correspondingly.

In complex (β) (Figure 8.3) at distance C_1−O_1 = 1.47Å, the distances of O_4−H_1 and N_2−H_2 are considerably decreased and the interatomic distances of C_1−O_4, N_1−H_1, and C−O_3 are increased as a result of essential changes of the general geometry of complex. At this, the charges on N_4, N_2, and H_2 atoms are increased and on H_1, O_1, C_1, and O_4 atoms are decreased. The angle O_2−N_2−O_3 is essentially changed.

The complex (β) (Figure 8.3) transfers into the intermediate compound via the conformation (γ) (Figure 8.3). The formation of the intermediate compound [Kh–CH_3CO] (Figure 8.5) leads to the decrease of charges on atoms C_1, O_1, N_1, N_2, and O_3 and to the increase of charges on atoms N_2, O_2, H_2, and O_1.

Interatomic distance C_1−O_1 for intermediate compound consists of 1.39Å. The distance N_2–H_2 is considerably increased. The angle O_2–N_2–O_3 is greatly changed. The electron affinity is decreased in comparison with the starting active center of the chemotyrpsin.

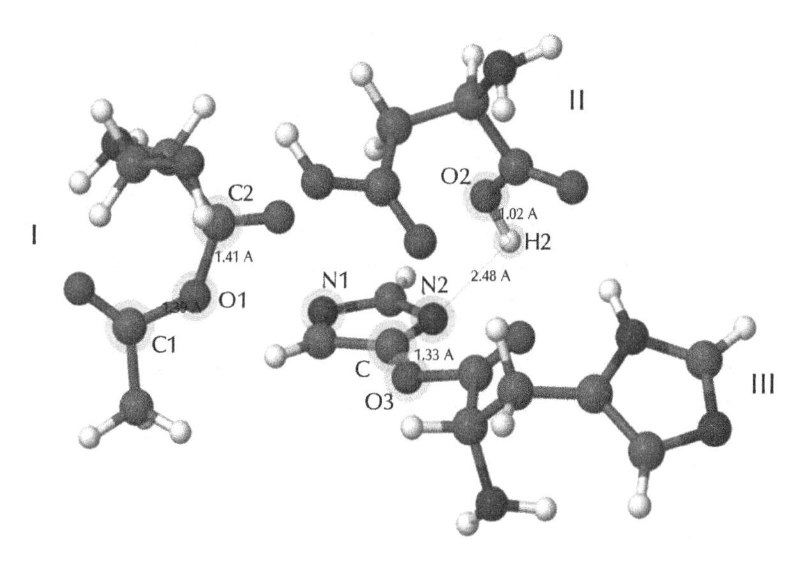

FIGURE 8.5 Geometry and electronic characteristics of complex [Kh–CH_3CO]. I Ser + CH_3CO, **II** *Asp*, **III** *His*

Angle O_2–N_2–O_3 is 83.2^0, ionization potential = 9.050 eV, electron affinity = –2.120 eV

Atoms	C_2	O_1	N_1	C_1	H_2	O_2	C	O_3	N_2
Charge	+0.595	−0.559	−0.324	+0.717	+0.388	−0.520	+0.302	−0.412	−0.419

Bonds	O_1–C_1	C_2–O_1	O_2–H_2	N_2–H_2	C–O_3
Distance (Å)	1.39	1.41	1.02	2.48	1.33

Let us consider the thermodynamics of the enzymatic catalysis by chemotyrpsin for the reaction of C−O bond breakdown in methyl acetate. Below, there is la scheme of a process, thermodynamics of which calculated without taking into account the interactions of complexes with the methanol (1) and acetic acid (2):

Kh + CH_3COOCH_3 → [Kh−CH_3CO] + CH_3OH (1)

(DH_p = +12.9 kilocalorie/mole; ΔS_p = +6.0 calorie/mole·K; DF_p = +14.7 kilocalorie/mole)

[Kh−CH_3CO] + H_2O → Kh + CH_3COOH (2)

(DH_p = −10.6 kilocalorie/mole; ΔS_p = +10.1 calorie/mole·K; DF_p = −13.6 kilocalorie/mole)

Here: Kh–chemotrypsin; Kh–CH_3CO–acylchemotrypsin

ΣDF = + 1,1 kilocalorie/mole; K = exp(−DF/RT) = 0.2.

Thermodynamic characteristics of the starting substances of the reaction products and of the intermediate compound (*acylchemotrypsin*) are represented in Table 8.1. As we can see from the calculations, the acylchemotrypsin formation process is endothermic. Such intermediate compounds are not thermodynamically stable and easily transforms into the final products of the deacidulating reaction.

TABLE 8.1 Thermodynamic characteristics of the reagents and of the products of the methyl acetate hydrolysis reaction.

No.	Formula/substance	DH (kilocalorie/mole)	ΔS (calorie/mole·K)	DF (kilocalorie/mole)
1	Chemotrypsin	−306.6	238.9	−377.8
2	CH3C(O)OCH3	−97.5	75.8	−120.1
3	H2O	−54.3	45.0	−67.7
4	CH3OH	−48.3	55.9	−65.0
5	CH3COOH	−101.2	69.0	−121.8
6	[Kh–CH3CO]	−342.9	252.8	−418.2

Let's consider the kinetics of the acidulating and deacidulating reactions of the fermentative part of chemotyrpsin. Kinetic data for the acylferment formation reaction and for the stages of its deacidulating are represented in Table 7.2.

TABLE 8.2 Kinetic parameters of the enzymatic catalysis of the acylferment formation

Stage	Point of calculation	DH (kilocalorie/mole)	ΔS (calorie/mole·K)	DH≠ (kilocalorie/mole)	ΔS≠ (calorie/mole·K)
	Starting substances	−410.1	275.0		
1	Activated complex	−394.9	250.0	15.2	24.5
	Products of the reaction	398.9	262.8		

TABLE 8.2 *(Continued)*

Stage	Point of calculation	DH (kilocalorie/mole)	ΔS (calorie/mole·K)	DH≠ (kilocalorie/mole)	ΔS≠ (calorie/mole·K)
	Starting substances	406.4	212.8		
2	Activated complex	395.2	250.4	11.2	37.6
	Products of the reaction	433.2	242.3		

Figure 8.3 presented the potential curve of reaction of acylferment formation. It can be seen from the presented curve that, as a result of the interaction of chemotyrpsin and methyl acetate, a series of intermediate complexes are formed, which are caused by the conformational changes of the ferment under the action of the methyl acetate. Energy of the process of the substrate adsorption on the ferment is sufficient for the formation of Michael's complex (conformation on Figure 8.3 (1)). Complex (1), which is characterized by enough high free energy, is easily transformed into the complex (α). At this, the transition from the complex (1) to the complex (α) is the exothermal process with the heat effect ~10 kilocalorie/mole. The complex (α) is easily transformed into the activated complex (2) (see Figure 8.3, peak no. 2) with the activation energy ~2 kilocalorie/mole. Activated complex (2) transfers into the conformation (β), and (β) is transformed into the conformation (γ), which with a little barrier is transformed into the intermediate compound—acylferment. Activation enthalpies of the acylferment formation are overestimated as a result of the inaccuracy of the semiempirical calculation methods. However, kinetic characteristics of the process may be trusted qualitatively (Table 8.2). Free activation energy of the acidulating reaction consists of 7.8 kilocalorie/mole, and of the deacidulating reaction this value consists of DF≠ = 4.0 kilocalorie/mole, in other words the deacidulating reaction rate is considerably great. The acidulating reaction is the limiting stage of the methyl acetate hydrolysis process.

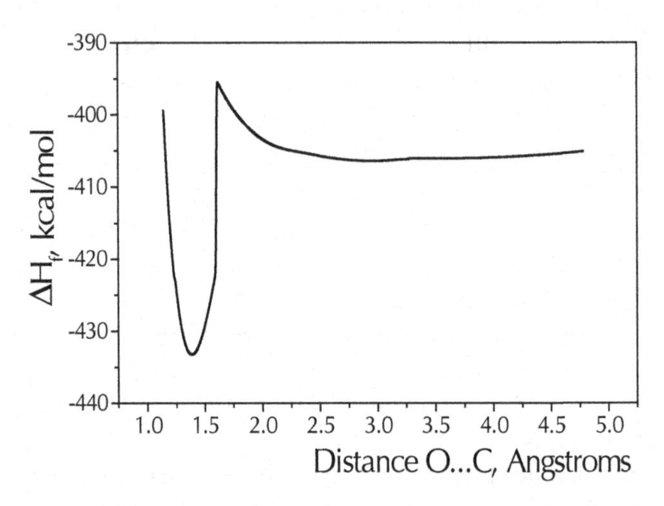

FIGURE 8.6 Potential curve of the deacidulating for chemotyrpsin complex

It can be seen from Figure 7.6, that the activated complex of the intermediate product deacidulating reaction is formed at the interatomic distance of C–O (1.75 Å). The deacidulating process is exothermal and proceeds with a heat effect 35 kilocalorie/mole. Probably, the deacidulating process is one stage, in other words it can be considered as elementary one. A great difference in activation entropies for acidulating and deacidulating reactions is explained by more distanded structure of the activated complex for the deacidulating reaction.

Generally, it is necessary to note that a great contribution into the values of the rate constants in chemotyrpsin acidulating and deacidulating reactions is characterized by entropic factors; this is connected with a great contribution of the activation oscillating constituents of entropy, which are caused by the low frequency vibrations of bonds in $\Delta S^{\neq}$.

It was interesting for comparison *to study the modeling of the reaction of the methyl acetate acidic homogeneous catalysis.*

The objects of investigation:

– methyl acetate;

– serine.

It was assumed that, the process of homogeneous catalysis proceeds also via the stage of acidulating and deacidulating. Calculations of the thermodynamic parameters of reagents, intermediate products, and the final products are represented in Table 8.3.

TABLE 8.3 Thermodynamic characteristics of the reagents

No.	Formula/substance	DH (kilocalorie/mole)	ΔS (calorie/mole·K)	DF (kilocalorie/mole)
1	Ser–H	−138.7	87.8	−164.9
2	$CH_3C(O)OCH_3$	−97.5	75.8	−120.1
3	H_2O	−54.3	45.0	−67.7
4	CH_3OH	−48.3	55.9	−65.0
5	CH_3COOH	−101.2	69.0	−121.8
6	$CH_3CO–O–Ser$	−175.6	108.9	−208.0

Scheme of process:

CH_3COOCH_3 + Ser–H → $CH_3CO–O–Ser$ + CH_3OH –$^{(+ H2O)}$ → CH_3COOH + Ser–H

CH_3COOCH_3 + Ser–H → $CH_3CO–O–Ser$ + CH_3OH (1)

(ΔH_p = +12.3 kilocalorie/mole, ΔS_p = +1.2 calorie/mole·K, ΔF_p = +11.9 kilocalorie/mole)

$CH_3CO–O–Ser$ + H_2O → CH_3COOH + Ser–H (2)

(ΔH_p = −10.0 kilocalorie/mole, ΔS_p = +2.9 calorie/mole·K, ΔF_p = −10.8 kilocalorie/mole)

$$\Sigma\Delta F = +1.1 \text{ kilocalorie/mole}, \ K = \exp\left(-\frac{\Delta F}{RT}\right) = 0.16.$$

It can be seen from the calculations, that the stage of the intermediate product formation is endothermic and its formation is not thermodynamically stable. Probably, the time of the intermediate product existing is very little. The hydrolysis reaction of the intermediate compound is thermodynamically efficient. Generally, free energy of the hydrolysis process consists of +1.1 kilocalorie/mole. Approximately within the limits of experimental error, the methyl acetate hydrolysis reaction is thermodynamically allowed.

Let us consider the kinetics of the catalysis process. In Figure 8.7, there are interatomic distances and electronic characteristics of the serine.

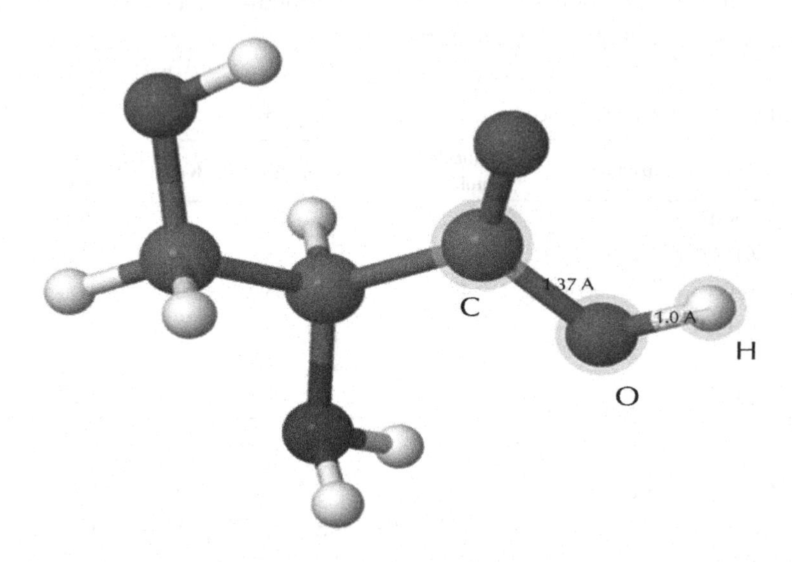

FIGURE 8.7 Geometry and electronic characteristics of the serine

Angle C–O–H: 113.0°, ionization potential: 10.232 eV, electron affinity: –0.066 eV

Atoms	H	O	C	Bonds	O–H	C–O
Charge	+0.348	−0.536	+0.561	Distance (Å)	1.00	1.37

Ionization potential is little higher than in the case of the chemotyrpsin. Electron affinity is sufficient low.

In activated complex of the acylserine (see Figure 8.8), a great change of the charges on reactive center C−O in comparison with the starting system is observed.

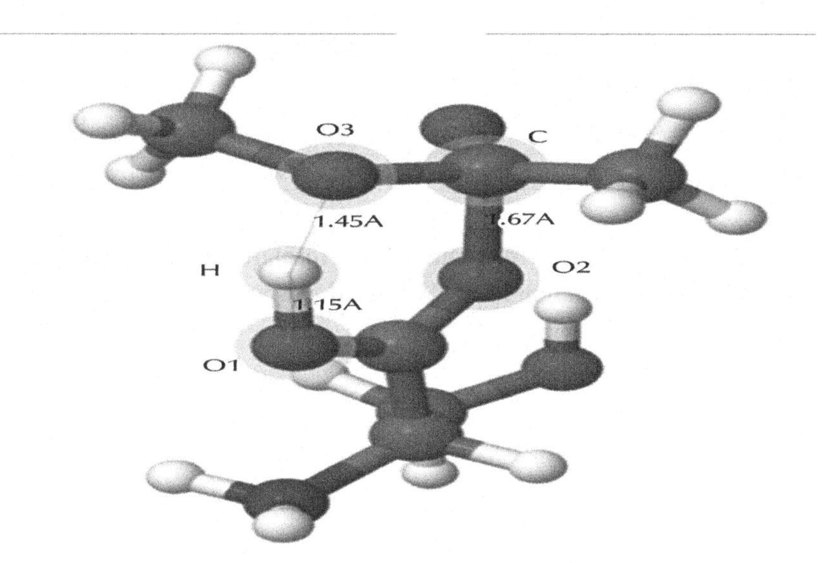

FIGURE 8.8 Geometry and the values of charges for activated complex of the acylserine formation.

The final intermediate product—acylserine—is formed at a distance of ~1.5Å (C–O) (Figure 8.9)

Atom	H	O_1	O_2	C	O_3	Bonds	$C–O_2$	$O_1–H$	$H–O_3$
Charge	+0.418	−0.537	−0.570	+0.771	−0.585	Distance (Å)	1.67	1.15	1.45

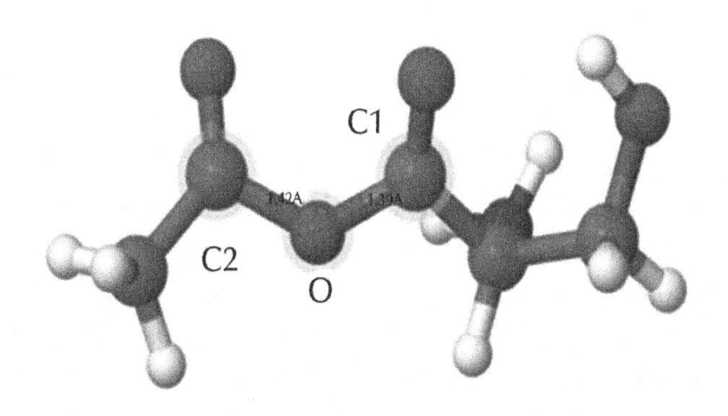

FIGURE 8.9 Geometry and electronic characteristics of the complex $CH_3CO–O–Ser$ (acidulating reaction)

Angle C–O–C: 126.1⁰, ionization potential: 10.012 eV, electron affinity: −0.401eV

Atoms	C1	O	C_2	Bonds	C_1–O_1	O–C_2
Charge	+0.584	−0.570	+0.703	Distance (Å)	1.390	1.42

The process of the intermediate product formation is endothermic (Figure 8.10) with the reaction heat ~11 kilocalorie/mole.

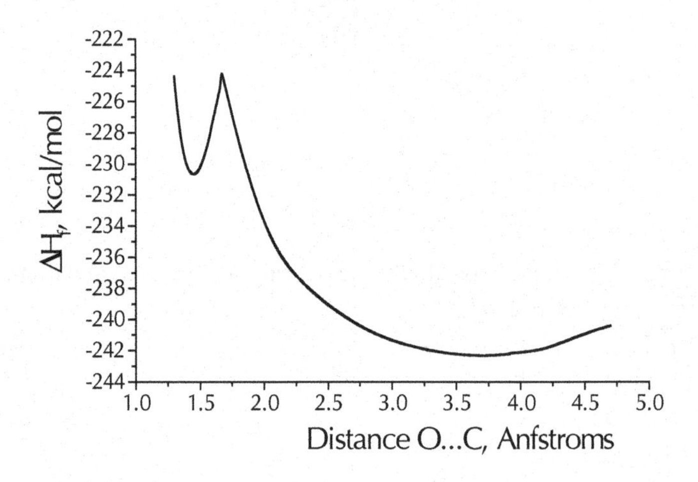

FIGURE 8.10 Potential curve of the intermediate product acylserine CH_3CO–O–Ser formation

Transformation of the activated complex into the intermediate product proceeds exothermally with heat ~6 kilocalorie/mole.

Activated complex of the hydrolysis reaction (Figure 8.11) of the intermediate product is formed at the internuclear distance ~1.6Å (C−O). Reaction of the hydrolysis is exothermal and proceeds with heat ~6.5 kilocalorie/mole.

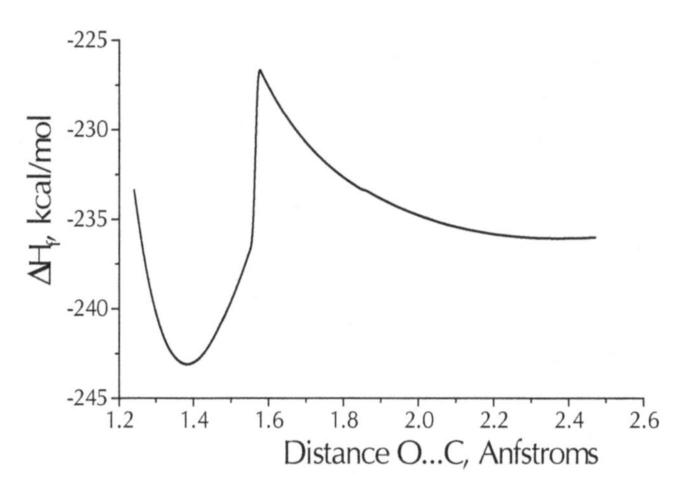

FIGURE 8.11 Potential curve of the deacidulating CH_3CO–O–Ser reaction

Kinetic parameters of the reaction of intermediate product formation (acidulating) and final products (hydrolysis) are represented in Table 8.4. The potential curves of the stages are represented on Figure 8.10 and 8.11 correspondingly above.

TABLE 8.4 Kinetic parameters of the acidulating and deacidulating reactions of serin

Stage	Point of calculation	ΔH (kilo-calorie/mole)	ΔS (calorie/mole·K)	EA (kilocalorie/mole)	ΔS# (calorie/mole·K)
	Starting substances	–242.3	137.1		
1	Activated complex	–224.4	106.1	17.9	31.0
	Products of reaction	–230.7	136.6		
	Starting substances	–236.0	122.7		
2	Activated complex	–226.3	114.4	9.7	8.3
	Products of reaction	–243.1	103.7		

Heat of activation of acidulating reaction is considerably greater, than the $\Delta H^{\neq}$ of the deacidulating reaction. However, the activation entropy of first reaction is considerably greater.

Free activation energy of the acidulating reaction consists of −0.7 kilocalorie/mole. Free activation energy of the deacidulating reaction at T = 298K consists of +4.7 kilocalorie/mole. In other words, the second stage is the limiting one, whereas for the enzymatic reaction the limiting stage is the acidulating stage.

More electron affinity of the active fragment of chemotrypsin (−2.601 eV) explains the acceleration of the enzymatic catalysis in comparison with the homogeneous catalysis by serine, electron affinity of which is sufficiently low (−0.0066 eV).

If to assume, that the reactive centers C_1–O_1 for chemotrypsin and serine are characterized by approximately the same resonance integrals and overlapping integrals, then the orbital energy of these both reactions will be determined accordingly to the formula $E_{orbit} \approx \frac{const}{I-E}$ [15], where I is the ionization potential (donor), and E is electron affinity (acceptor). For the methyl acetate I ≈ 11.66eV; E of chemotrypsin consists of −2.66eV, whereas for serine this value is equal to −0.0066eV. Correspondingly, the orbital energies of interaction for the reaction of intermadiate acylcompounds formation will be consist of $\frac{const}{14.32}$ and $\frac{const}{11.67}$ respectively; in other words, the orbital energy of acylchemotrypsin formation will be 1.22 times lesser in comparison with the intermediate energy of acylcompound formation in the case of homogeneous catalysis by serin.

It is necessary to note, that the electron structure of the reagents has an essential influence on the hydrolysis reaction of C–O bond in methyl acetate both via the enzymatic and via the homogeneous (acidic) catalysis. For example, the values of charges on the active atoms of the oxygen of hydroxyl groups of the serinic fragment of chemotrypsin and free serine itself are respectively equal to −0.546 i −0.536, that in some way leads to the higher nucleophilicity of chemotrypsin.

During the enzymatic catalysis, the electronic characteristics of the reagents are greatly changed, that leads to the essential changes of the nucleophilicity of the reactive center of chemotrypsin and substrate (methyl acetate). If to assume the respective charges as the characteristics of such reactive centers, then the following picture will be obtained. The values of the atoms charges along the reaction way $C1^{+}$–$O1^{-}$ (equilibrium state) is changed from +0.666, −0.559 to +0.801, −0.648 into the activated complex.

For the reaction of homogeneous catalysis of methyl acetate by serine into the equilibrium state, the charges on the reactive centers C and O are equal to +0.666 and –0.536 respectively, and for the activated complex +0.771 and –0.537.

It is following from the all above said, that the nucleophilicity of the reactive center (oxygen atom of serine) of chemotrypsin is more than that of serine. Generally, it can be concluded that the stabilization of the activated complex at the expense of the charges of reactive center C–O will be higher in comparison with the homogeneous catalysis by serine molecule. According to this, heat of the activated complex formation will also be less in the case of enzymatic catalysis that leads to the less activation heat of the acylferment formation.

Let us note, that the activated complex of the acylferment in a case of the enzymatic catalysis is stabilized also at the expense of the formation of hydrogen bond of *His* and atom of oxygen of methyl acetate (finally, atom of hydrogen transfers on the oxygen with the formation of methanol).

Let us consider rather more detailed reaction of chemotrypsin deacidulating. The Carbon atom of C1 (*Fig.* 12) is attacked by the atom of oxygen O3 of water molecule with the simultaneous attack of hydrogen of water the nitrogen atom of the imidazole group of the *His*. Geometry and the charges distribution into the activated complex of acylchemotrypsin are represented on Figure 8.12.

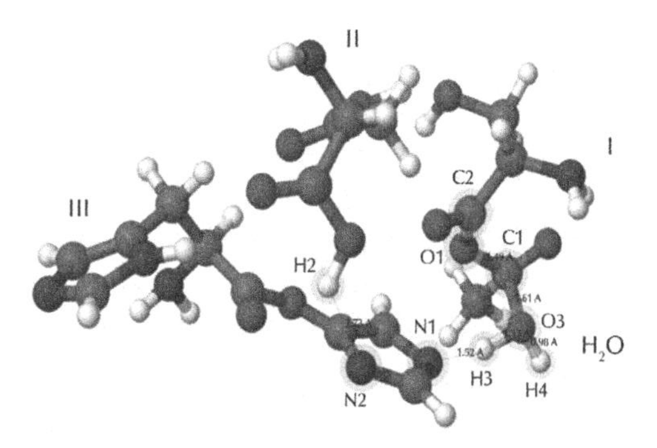

FIGURE 8.12 Geometry of activated complex and the charges of some atoms [Kh–CH_3CO]–H_2O. I Ser + CH_3CO, **II** Asp, and **III** His

Atoms	C_2	O_1	N_1	C_1	H_2	H_3	H_4	O_3	N_2
Charge	+0.633	−0.651	−0.420	+0.794	+0.374	+0,412	+0.336	−0.554	−0.401

	O_1–C_1	C_2–O_1	O_3–C_1	O_3–H_3	O_3–H_4	N_1–H_3	N_2–H_2
Distance (Å)	1.49	1.37	1.61	1.13	0.98	1.52	2.73

The values of charges on the reactive center for activated complex of the acylchemotrypsin C_1–O_3 are respectively equal to +794 i –0.554, whereas for the activated complex of the acylserine the values of such charges are equal to C_2 (+0.757)–O_2 (−0.549) (see Figure 8.13).

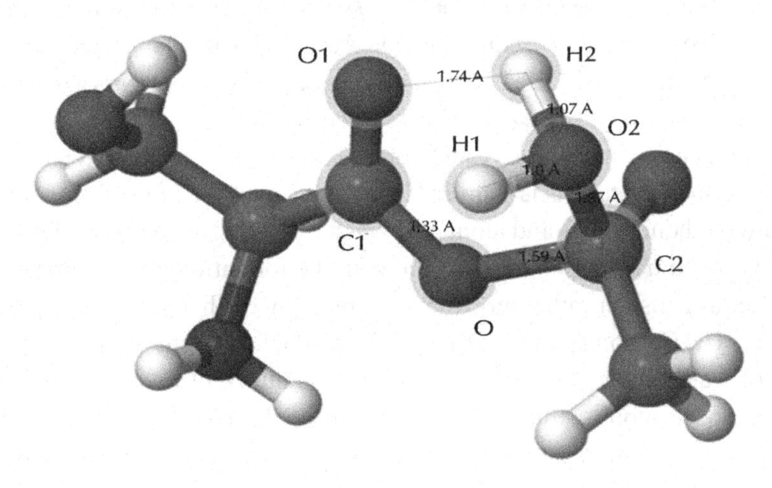

FIGURE 8.13 Geometry of the activated complex [CH_3CO–O–Ser]–H_2O

Atoms	C1	O	C_2	O_1	O_2	H_1	H_2
Charge	+0.634	−0.636	+0.757	−0.645	−0.459	+0.346	+0.404

Bonds	C_1–O	O–C_2	O_2–C_2	O_2–H_1	O_2–H_2	O_1–H_2
Distance (Å)	1.33	1.58	1.57	1.00	1.07	1.74

H_2O	H_3O^+
Charge on O: –0.6188	Charge on O: –0.122
Charge on H: +0.3094	Charge on H: +0.374
Distance O–H: 0.95Å	Distance O–H: 1.04Å
Ionization potential: 11.906eV	Ionization potential: 22.869eV
Electron affinity: 4.068eV	Electron affinity: –6.346eV

From the comparison of both reactions of acylchemotrypsin and acylserine hydrolysis it can be concluded that most stabilization of the activated complex is observed in a case of the acylchemotrypsin deacidulating stage. Due to such higher stabilization of the activated complex of the deacidulating reaction, less activation energy for the enzymatic reaction is observed.

In a case of the acylcompounds hydrolysis, a molecule of water plays the role of nucleophile. Electronic characteristics of water molecule are represented in legend of the Figure 8.13. Taking into account that the ionization potential of the acylchemotrypsin is equal to 9.05 eV, and of the acylserine ~10eV (the difference is 23 kilocalorie/mole), it can be concluded that the deacidulating reaction will proceed with less activation energy for acylchemotrypsin. It possible that the deacidulating reactions will took place with the participation of hydronium ion which is characterized with a high electron affinity.

Also it was considered, the reaction of the methyl acetate hydrolysis in the presence of sulphuric acid. It was shown that the first acidulating stage proceeds with $DH_{\text{реакц.}} = +1.4$ kilocalorie/mole, $\Delta S_{\text{реакц.}} = -2.9$ calorie/mole·K, $DF_{\text{реакц.}} = +2.3$ kilocalorie/mole. The second deacidulating stage is characterized by the following thermodynamic parameters: $DH_{\text{реакц.}} = +0.9$ kilocalorie/mole, $\Delta S_{\text{реакц.}} = 7{,}0$ calorie/mole·K, $DF_{\text{реакц.}} = -1.2$ kilocalorie/mole. Total reaction: $\Sigma DF_{\text{реакц.}} = +1.1$ kilocalorie/mole; $K = \exp\left(-\frac{\Delta F}{RT}\right) = 0.16$.

Activation parameters of the acidulating reaction are: $DH^{\neq} = 34.2$ kilocalorie/mole, $\Delta S^{\neq} = 10.1$ calorie/mole·K, $DF^{\neq} = \sim 31$ kilocalorie/mole.

Activation parameters of deacidulating reaction are: $DH^{\neq} = 10.3$ kilocalorie/mole, $\Delta S^{\neq} = 8.3$ calorie/mole·K, $DF^{\neq} = 7.8$ kilocalorie/mole.

At the comparison of the methyl acetate hydrolysis kinetics in the presence of the sulphuric acid, it is necessary to note, that the deacidulating process proceeds with the same kinetic parameters as in a case of serin. However, the activation energy for H_2SO_4 is practically in two times higher, and the activation entropy is too lesser.

It is seen from the value $DF^{\neq}$ that the limiting stage in the methyl acetate hydrolysis with the sulphuric acid reaction is the stage of the acylproduct formation, whereas in a case of the serine opposite situation is observed.

Probably, it is not unambiguously stated, that for all the reactions of ethers hydrolysis by different substrates, the reaction of the intermediate product (namely, acylferment) formation is the limiting stage.

Quantum–chemical calculations show that, really into the activated state of the ferment, there are hydrogen bonds at the expense of which the nucleophilicity of the reactive atom of oxygen of serine is increased. These bonds substantially determine the chemical mechanism of the acylferment formation and have an influence on the values of their kinetic parameters for the enzymatic catalysis and on the geometry of conformers on the reaction way.

8.1.4 CONCLUSION

The following conclusions can be done from the all above said:

A reaction of the intermediate product acylchemotrypsin formation is not elementary (as it is described in [16]), but probably consists of the following stages:

1) Hydrogen bond (serinic group—imidazole group) O–H...N is somewhat elongated; the *Michael's* complex is endothermicly formed;
2) Exothermal transition of hydrogen atom to nitrogen atom (imidazol group of *His*) (conformer α);
3) Formation of activated complex of acidulating reaction (2);
4) Transformation of the activated complex into acylferment complex (conformer β), which transfers into the conformer γ, and finally into the intermediate product (acylchemotrypsin).

In other words, the reaction of the chemotyrpsin acidulation can be presented according to the following scheme:

ferment + *methyl acetate* ↔ *Michael's complex* ↔ *conformer α* ↔ *conformer β* ↔ *conformer γ* ↔ *acylferment.*

Thereby, a reaction of the acylchemotrypsin formation consists of a series of conformational transformations of reagents, themselves on the reaction way.

However, we have some doubts as to generally accepted positions of the *Michael's* complex formation; such doubts cannot be solved with the use of the semi-empirical calculations. It was accepted that the energy of the substrate adsorption process on the active center of ferment is enough for the *Michael's* complex formation, which requires the high energy of the formation. Ambiguously, the physical adsorption can provide such energy.

As the practice of quantum–chemical calculations for such systems show the interaction in them proceeds with enough low activation energies, however also with insufficient heat effects, which cannot able initiate the *Michael's* complex formation.

Chemical adsorption or some other approach can lead to such complex formation. For example, the substrate is adsorbed on active part and is overbalanced.

Each ferment has the essential protein part besides the reactive centers. Free energy of the ferment is into equilibrium non working (non activated) state and is relatively stable; that is, the ferment is characterized with minimal free energy. At adsorption of the substrate on the active center, the conformation of ferment is greatly changed as a result of interaction. The result of this are the changes of non valence and valence bonds into molecules that lead to the increase of free energy of a system, which is distributed between the bonds of system, including the reactive center. Such (excited) system of ferment leads to the short-term chemical act with the formation of products plus free ferment, which has the initial conformation which is again excited by the substrate.

Theoretically, such hypotheses can be in some ways confirmed by quantum–chemical calculations based on the excitation theory.

Maybe, positive result about the distribution of the energy of an excited system can be obtained according to PPKM theory [17]. In accordance with this theory, the transformation of complex ferment–substrate, as it is known, connected with the energy redistribution between different freedom degrees. However, zero energy of the excited complex and the activated complex, and also the potential energy of bonds which form the active complex, are not ordered to such distribution. Residual energies of the excited and activated complex are called disconnected energies. Different freedom degrees play a different role in energy change. Adiabatic freedom degrees practically do not transfer the energies of bonds, which are broken as a result of a low probability of energy transfer from the external freedom to the internal degrees, which correspond to three rotary and three transitional movements of a system as a whole.

Internal degrees of freedom will be consisting of the active and non active degrees of freedom. Active degrees of freedom are the degrees without limitations of the energy of bonds which are formed or broken; and the non active degrees of freedom can transfer energy of the bonds only in the case when the active complex is activated complex.

Thereby, the reaction of ferment initiated by substrate can be schematically represented in accordance with the following scheme:

ferment (domain) + H_2O → solvate–ferment + substrate → excited ferment–substrate (new conformation) → ferment–substrate with distributed energy on degrees of freedom into *Michael's* complex → activated complex → products + ferment.

In references, the most essential reasons of the enzymatic reaction acceleration in comparison with the homogeneous catalytic processes are as follows:

1) sorption interaction with protein of the side substrate groups accelerate the reaction by 10^7 times;
2) polyfunctional catalysis (general acidic–base catalysis) can lead to the reaction acceleration by 10 times;
3) micro effects of medium of the active center can change the rate of catalysis by dozens orders.

It is possible, that the very important role into the processes of the enzymatic catalysis play the micro effects of medium of the active center. The rate of catalysis in such cases can be changed to dozens orders. However, unfortunately now nothing has done in the presented direction.

KEYWORDS

- **Acidulating reaction**
- **Acylchemotrypsin**
- **Acylferment**
- **Imidazol group**
- **Michael's complex**

REFERENCES

1. Blow, D. M. & Steitz, T. A. (1970). Annual Review of Biochemistry, 39, 63.
2. Matthews, B. W. et al. (1967). Nature, 214, 652.
3. Henderson, R. J. (1970). Journal of Molecular Biology, 54, 341.
4. Birktoft, J. J. et al. (1970). Philosophical Transactions of the Royal Society (London), B 257, 67.
5. Miller, Ch. G. & Bender, M. L. (1968). Journal of the American Chemical Society, 90, 6850.
6. Bernhard, S. (1971). Structure and function of the enzymes. Moscow: Myr (in Russian).
7. Hess, G. P. et al. (1970). Philosophical Transactions of the Royal Society (London), B 257, 89.
8. Bernhard, S. A. & Gutfreund, H. (1970). Philosophical Transactions of the Royal Society (London), B 257, 105
9. Himoe, A., Brandt, K., & Hess, G. P. (1971). Journal of Molecular Biology, 55, 215.
10. Mosolov, V. V. (1971). Proteolytic enzymes. Moscow: Nauka (in Russian).
11. Cunningham, L. (1965). Comprehensive Biochemistry, 16, 85.
12. Bender, M. L. & Kezdy, F. J. (1964). Journal of the American Chemical Society, 86, 3704.
13. Berezin, I. V. & Martynek, K. (1972) In Colln.: structure and functions of enzymes. Moscow: Moscow State University Edition (in Russian).
14. Blow, D. M., Birktoft, J. J., & Hartley, B. S. (1969). Nature, 221, 337.
15. Klopmann, G. (1977). Reactive ability and the reaction ways (p. 383). Moscow: Myr (in Russian).
16. Berezin, I. V. & Martynek, K. (1978). The principles of the physical chemistry of the enzymatic catalysis (p. 350). Moscow: Vysshaya Shkola (in Russian).
17. Stiepukhovich, A. D. & Ulitsky, V. A. (1975). Kinetics and thermodynamics of cracking radical reactions (p. 255). Moscow: Khimiya (in Russian).

CHAPTER 9

THE INFLUENCE OF MELAFEN: PLANT GROWTH REGULATOR TO SOME METABOLIC PATHWAYS OF ANIMAL CELLS

O. M. ALEKSEEVA

CONTENTS

9.1 INTRODUCTION

The effects of plant growth regulator, Melafen, to some animal cell metabolic pathways were investigated at our work. Earlier literature data recognized that the fluctuations of animal cellular volume are correlated with some Ca^{2+}-dependent cellular metabolic pathways. The cellular volume was registered by spectral method—the first light scattering of diluted cellular suspensions that were incubated under the Melafen aqua solutions at wide concentration region (10^{-21}–10^{-3}M). The Ehrlich ascetic carcinoma (EAC) cells, that have the metabotropic purinoreceptors P2Y at its surface, showed two maximums at light scattering kinetic curve that correlated with two maximums of volumes increasing after the one addition of ATP. The leukocytes, which have the Ca^{2+}-channel-former purinoreceptors P2X at its surface, showed that the lag-phase at light scattering kinetic curve before the cellular volume increasing after the ATP addition was in dependence of Melafen concentration. The thymocytes, which have the Ca^{2+} channel-former purinoreceptors P2Z (and other too) at its surface, showed two maximums at light scattering kinetic curve after the one addition of ATP also. But the second answer that correlated with the CRAC activation was in bimodal dependence of Melafen concentration. We conclude that the three main types of purinoreceptors: metabotropic P2Y and two Ca^{2+}-channel-formers P2X and P2Z were under the direct Melafen actions. Whereas CRAC-CIF-activated Ca^{2+}-store regulated Ca^{2+} channel was under the indirect Melafen actions. Thus, the first three points of purine-dependent Ca^{2+} transduction pathways—three types of receptors are under the direct Melafen influence.

The main aim of investigation was the finding the quintessential specific targets for Melafen, plant growth regulator, actions at the animal cells, which have the different origin.

This investigation deals with the melamine salt of bis (oximethyl) phosphinic acid—Melafen, that was synthesized by works of laboratory of A. E. Arbuzov Institution of organic and physical chemistry [1]. We tested the influence of this hydrophilic substance at the wide concentration range (10^{-21}–10^{-3}M) on the functional properties of animal cells.

The animal cells must have a number of the variable targets for hydrophilic substance action.

If deal with the hydrophobic substance, one must note certain points. The hydrophobe substances go into the membrane directly or to the hydrophobe pockets into the protein molecules, and are incorporated there, and then the substances are dissipated among hydrophobic phase, or form its own phase in the cellular compartments [2]. Its action has a negligible specificity without certain targets almost always. The hydrophobe substances have the possibility to show its specificity at protein hydrophobe pocket only. It can influence to the micro viscosity of lipid and protein membrane components [3]. Its actions to the protein structure and functions may be mediated by hydrophobic phase changing or immediate by hydrophobic targets bindings.

On the contrary, the hydrophilic substance, as Melafen may have the contacts with any charged or polar surfaces. Thus, the molecules of Melafen may have their interactions with number of non specific targets. But the incorporation to the membrane and dissipate among the lipid molecules does not happen. The targets are unknown, but the aftermath's actions of Melafen were known. In this chapter, we will describe the overall results of Melafen actions.

From literature data, it is known that the hydrophilic Melafen change the fatty acid composition and lipid and protein microviscosity of cellular microsome and mitochondrial membranes of vegetable [2]. Low concentrations (from 4×10^{-12} to 2×10^{-7}M) of Melafen changed the structural characteristics of plant and animal cell membranes. Melafen changes the microviscosity of free bilayer lipids and annular lipids bounded with protein clusters with different effective concentrations for plant and animal membranes. Melafen decreased the level of lipid peroxide oxidation (LPO) in biological membranes under bed environment conditions also. The Melafen concentrations that affect to the microviscosity of free and annular lipids decreases the intensity of LPO processes too [2]. On a foundation of the research conducted, authors assumed that high physiological activity of Melafen is linked to its actions to the physical and chemical state of biological membranes resulting in change of lipid-protein interaction, influencing the activity by membrane-associated enzymes and channels. Melafen increases the effectiveness of energy metabolism of plant cells [3]. The exclusive influence of Melafen was shown on electron transport in respiratory chain of mitochondria [4], and stress-resistance of vegetables and cereal corn in bed environment conditions as result [5]. The seed treatment by the Melafen or Pyrafen (Melafen analog) reduces the intensity of lipid peroxidation and strengthens the mitochondria energy of 6-day pea seedlings, which have undergone stress in conditions of moistening shortage and moderate cooling [6]. Melafen changes the fatty acid composition of mitochondria greatly in the presence of its effective concentrations [7]. These are the reasons for increase in crop yield. The influences to the animal microsomes and mitochondrial were also showed [4–6].

Our laboratory investigated the possibility of Melafen influence to the first targets at animal objects: the protein in blood-vascular system, cellular membranes and its components. We found that Melafen under large concentrations changes greatly (may be loosed) the quarter structure of bovine serum albumin (BSA)—the main soluble protein in blood-vascular system. The intrinsic fluorescence of two tryptophanils that contained BSA molecule was quenched because the tryptophanils became access to the main quencher—H_2O [11].

But to the integral membrane-bounded proteins, Melafen influences were negligible even under the large concentrations. Thus the organization of protein microdomains of erythrocyte ghost did not change, that we registered with aid of differential scanning microcalorymetry (DSK). The thermo-induced parameters of usual cytoskeleton protein components of cellular membrane: spektrin, ancyrin, actin, demantin, fragments of ion-channels, and other stood unchanged under the Melafen presence [12].

However, the thermo-induced parameters of lipid microdomains at membranes multilammelar liposomes formed by individual neutral saturated phospholipids dimyristoilphosphatidylcholine (DMPC) were changed greatly under the Melafen influence at the wide concentration diapason. Having the use of the specific method of membrane extraction from steady state, the different rates of heat supplied to the cell with our liposomes suspension, we received the glaring picture of dependence of thermally induced parameters—enthalpies, maximum temperatures, thermally induced transition, and cooperatives transition, under Melafen concentrations in the range 10^{-17}–10^{-3}M. The main extreme was under the 10^{-14}–10^{-8}M of Melafen for all rates (1, 0.5, 0.25, and 0.125°C/min) of heat supplied to the cell with our liposomes suspensions [13].

But the reciprocal location and density of packaging of membranes at multilammelar liposomes remain unchanged under the Melafen influence at the wide concentration diapason. In this case, the membranes of multilammelar liposomes were formed by egg lecithin that is the mixture of natural saturated and no saturated, neutral, and charged phospholipids [14]. In this type of membrane, where the bilayer structure is reinforced by balancing on charge and by the location of length of fatty acids tails at nature phospholipids mix, the membrane thickness from data of x-ray diffraction method remain unchanged under the Melafen presence [15]. Concerning the microdomains organizations in such mixture we say nothing. It is impossible, as the application of differential scanning calorimetry for lipids mixture have been hampered.

The structure (and functions and fate may be) of native cells were under Melafen influenced too. The small doses 10^{-11}–10^{-5}M of Melafen influence *in vivo* on the morphology of the erythrocytes in mice. We obtained the decrease of height, area, and volume of the AFM image of red blood cells, that registered by AFM (atomic force microscopy) [16].

The low doses 10^{-12}–10^{-5}M of Melafen influence to the fate of animal tumor cells *in vitro*. The content of protein "labels of apoptosis": protein-regulator p53 (increase) and antiapoptosis protein Bcl-2 (decrease), that were showed by immunoblotting methods at the EAC cells. This fact indicated that the apoptosis was developed within 1.5 hr of Melafen action: whereas similar doses of Melafen *in vivo* suppressed the growth of Luis carcinoma [17].

Thus, Melafen looses the structure of the soluble protein, changes the microdomaine organization of DMPC membrane and does not change the structural organizations of ghost and egg lecithin membranes. But Melafen globally change the erythrocyte morphology and delayed the rate of growth of solid Luis carcinoma.

Taking into account the data obtained by the A. E. Arbuzov Institution of organic and physical chemistry about formation by the Melafen in aqueous solutions of supramolecular structures involving of water molecules [18], it can be assumed that such structures can change the microdomains organization in attackable delicately organized and labile structures. These assumptions may be real only for interactions between biological objects with Melafen in aqua solutions, but in the cellular interior there is not sufficient amount of free water molecules for such supramolecular structures formation. It is hard to believe that the linkage with Melafen will be stronger than with nature cellular chelating agents of water. To withdraw the water molecules from the coats of cell components in the presence of Melafen is not an easy task, on the contrary, to structure the bulk water and the near membrane water layers is quite possible.

However, in present chapter we emphasize the attention toward the Melafen aqua solutions actions at the animal cellular level *in vitro*.

That is why we tried to clear up the most possible specific targets for Melafen on animal's cells surface. Several types of easily emitted cells of different etiology or different origin were having picked up. The main task of the work was to investigate the Melafen action on three types of cellular plasmalemmal receptors—purinoreceptors P2Y, P2X, P2Z, which were present in three types of cells: EAK cells, thymocytes, and lymphocytes.

Melafen is a melamine salt of bis (hydroximethyl) phosphinic acid. It is a hydrophilic polyfunction substance with multitargets for its actions (Figure 9.1). There are the phosphoric, hydroximethyl groups, and nitrogen contained structures at Melafen molecule, potentially pointed to the certain targets at the biological cells. The pretreatment of crop seeds by aqueous solutions of Melafen at concentration 10^{-11}–10^{-9}M increased the yield of plant production by 11% or more due to the increasing of plants stress tolerance under the bed environment [10]. But the increasing concentration of Melafen to 10^{-7}M and higher inhibits the processes of plants growth completely [19]. Therefore, our studies were carried out at a wide range of concentrations (10^{-21}–10^{-2}M). The main purpose was to determine how the aqueous solutions of Melafen in a wide range of concentrations influence the function of animal cells *in vitro* only.

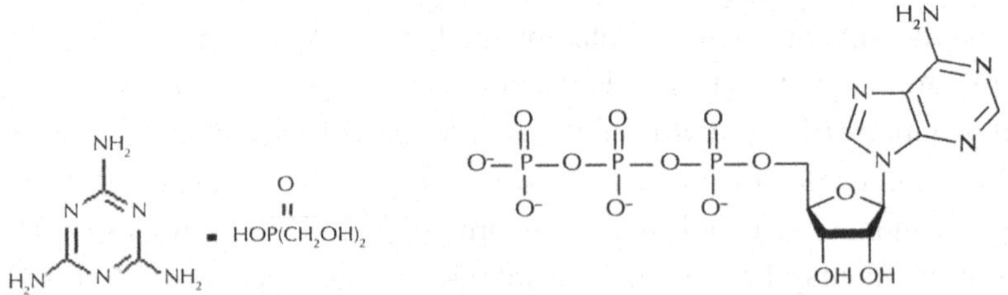

FIGURE 9.1 Structural formula of Melafen and ATP

It can be assumed that some fragmental structural similarity of Melafen-organophosphorus plant growth regulator, and ATP molecules (Figure 9.1 and 9.2) may define the binding with similar active sites. As result we may observe the both substances actions on purine receptors. But the vectors of consequences of Melafen and ATP molecules actions are opposed to each other. Also we must note the global activating influence of ATP applications to the Ca^{2+} signal transductions at the EAK cells at 7–8 days of development that was described by Zamai [20] and Zinchenko [21]. The main metabolic pathway points are indicated [22] and in this case, we deal with the testing of Melafen actions to the ATP-binding purinoreceptors at the base of some similarity of ATP and Melafen molecular construction.

9.2 MATERIALS AND METHODS

Materials: ATP ("Bochringer" Germany); HENKS (138mM NaCl; 5.4mM KCl; 1.2mM $CaCl_2$; 0.4mM KH_2PO_4; 0.8mM $MgSO_4$; 0.3mM Na_2HPO_4; 5.6mM D-glucose; 10mM HEPES pH 7.2), cytrat Na ("PAN EKO" Russia); DMPC dimyristoilphosphatidylcholine ("Sigma"); NaCl, KCl, $CaCl_2$, KH_2PO_4, $MgSO_4$, Na_2HPO_4, HEPES ("Sigma), and A23187 (Sigma).

Methods: The EAK cells received by the following methods [21, 23].

The EAK was induced in pubescent white male mice by NMRI introduction intraperitoneally on 10^6 cells of diploid strain of ascetic Ehrlich carcinoma. The EAK cells were isolated from mice on 7–8 days after the transplantation.

Thymocytes were separated by pulping through the Capron net from thymus of the white Wister rat. The cells were washed out three times by centrifugation at 800rpm for 10min in HENKS medium at 4°C. The pellet was resuspended in HENKS medium of concentration about 1–5 x 10^8cells/ml.

The pooled fraction of lymphocytes and platelets were received after spontaneous deposition of erythrocytes from the blood of white Wister rat in the presence of medium of HENKS (1:1) and 5% citrate Na. The cell viability was estimated on cells overtone 0.04% trypan blue and compiled in all experiences not less than 95%.

The registration of light diffusion by dilute suspension of EAK cell, of thymocytes, of lymphocytes and platelets, was held by the method Cornet [23], modified by Zinchenko [21] at a right angle under the wavelength 510nm with aid of fluorescent spectrophotometer "MPF-44B" Perkin-Elmer.

9.3 RESULTS AND DISCUSSION

The use of light scattering method allowed us to investigate the overall cellular answers under the Melafen additions without any artificial messengers. Thanks to presented method we tested the action of Melafen to the animal cells and its components under a wide range of concentrations.

The three cellular objects were used as the dilute cells cellular suspension: ascetic Ehrlich carcinoma (EAC) cells—transformed cells with uncontrolled growth, and normal cells thymocytes and lymphocytes (the white fraction of blood without the platelets and erythrocytes).

First—EAC cells as a good model of cells with the complete cellular transduction system: the active P2Y purino receptors are presented at the cellular plasmalemma surface at 7–8 days of carcinoma growth [20]. Thus, the ATP or ADP additions initiated the Ca^{2+} signal transduction (Figure 9.2). ATP is the first messenger that deals with the extracellular signal transductions pathways. ATP is released to the extracellular space. Atleast two subtypes of receptors for extracellular ATP are currently known: the G-protein coupled P2Y receptors, which are metabotrophic receptors, and the ATP-gated cation channels classified as P2X receptors (and its subspecy—P2Z receptors). EAC cells gave a typical cellular response to a signal (ATP-addition). It sent the signal from the cell surface to the inside of endoplasmic reticulum (ER) $InsP_3$-receptor, and backward to the CRAC at the cell surface.

In Figure 9.2, we show our cumulative scheme of purine-dependent Ca^{2+} transduction at cells. The metabotropic purinoreceptors P2Y throughout the G-proteins activate PLC that produces $InsP_3$, then $InsP_3$ bounds with $InsP_3$-receptors, that release Ca^{2+} from ER and retrograde signal CIF (related to ER Ca^{2+}-store depletion) go to the cellular surface, and the large flow of Ca^{2+} introduced to the cell throughout the Ca^{2+} activated channels (CRAC) [24].

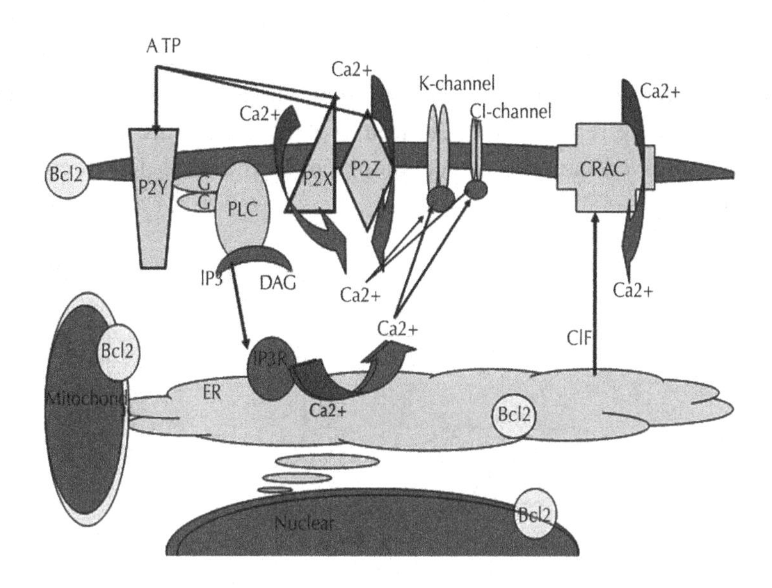

FIGURE 9.2 The total scheme of P2Y-, P2X-, and P2Z-related systems of signal transduction. (Scheme was modified from Alekseeva O.M., 2010)

So, after ATP addition, two large increases in intracellular Ca^{2+} concentrations were occured. As a result Ca^{2+}-dependent K^+ и Cl^- channels are activated. The volume of cells changed because these channels regulate the cell volume. The compensators H_2O flows occurred to the cellular interiors and cellular volumes are increased greatly and bitterly. Then, Ca^{2+} concentrations are decreased quickly, Ca^{2+} is attenuated by mitochondria/ER or it is removed from the cell. After that, Ca^{2+} concentration in the cell increases again smoothly. Ca^{2+} dependent K^+ and Cl^- channels are activated again. The cellular volume increases slowly too. Thus, the light scattering of dilute suspension of EAC cells were changed too twice. There are two large increases in intracellular Ca^{2+} concentrations, as the result, we obtain two maximums at light scattering kinetic curves.

We recorded the right angle light scattering of dilute suspension of EAC cells with Perkin–Elmer-44B spectrophotometer at a wavelength of 510nm. The control samples showed the bimodal cellular responses (Figure 9.3). Urgently after the first addition of ATP to the cell suspension, the big maximum appeared quickly on the kinetic curve (first peak). Then it bust momentary. After that, low plate is appeared and then the slow rise up is developed in plate, that is equal or above the reference levels (second peak). When we repeated the addition of ATP, the picture was repeated. Melafen inhibited the response development by the doze-dependent manner (Figure 9.3).

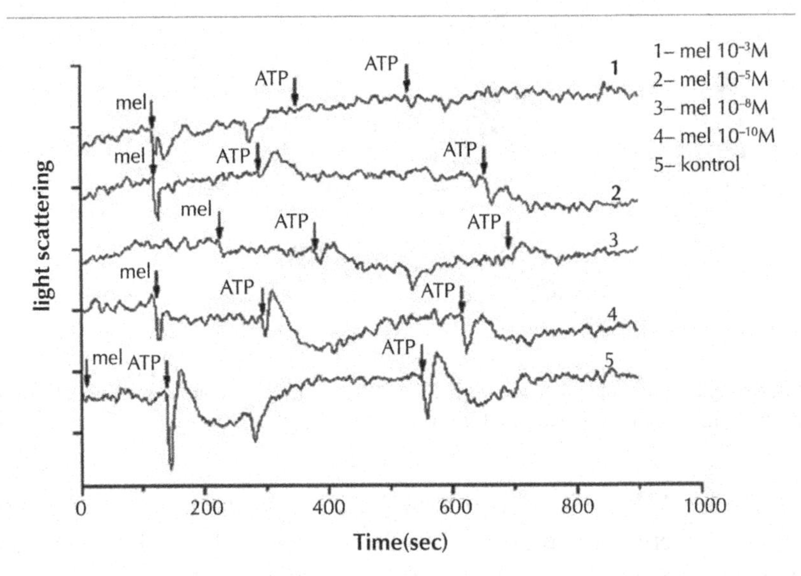

FIGURE 9.3 The influences of Melafen to the first and second cellular answer of EAC cells. Kinetic curves of light scattering of the cell suspension under the wide concentrations region of Melafen

But Melafen has a bidirectional effect on EAC cells (Table 9.1). We found that at super low concentrations, it (10^{-12} and 10^{-13}M) stimulated the signal transduction increasing the Ca^{2+} releasing from intracellular Ca^{2+} store (the first peak). But under the bigger concentrations, its actions changed their vector and the depressing of the overall cell responses were began. In case, when its concentrations were increased, the amplitudes of first and second extremes were decreased. The second cell response—the Ca^{2+} entering through the plasma membrane (second peak) was not activated by Melafen (under the anywhere concentrations), and it shown the bigger sensitivity to its action. It was inhibited by Melafen at smaller concentrations. Thus, Melafen concentration 10^{-7}M decreased the first extreme by 50% and the second extreme by 70%. Melafen concentration 10^{-4}M really eliminated the overall cell response fully; it inhibited the purine-dependent Ca^{2+} transduction—both peaks. Hence, the carcinoma cells that characterized by uncontrolled cell division can be depressed by such doses of Melafen which are harmless for erythrocytes that we obtained earlier [16].

TABLE 9.1 The Melafen influences to EAC cell bimodal responses (peak first and second) under the ATP adding

Sample + Melafen	Peak 1 (rel. un.)	Δ (%)	Peak 2	Δ (%)
Control	27.5 ± 0.1	-	20 ± 0.1	-
+10^{-13}M	35 ± 0.1	+27 ± 0.01	19 ± 0.1	-5 ± 0.01
+10^{-12}M	33 ± 0.1	+20 ± 0.01	19 ± 0.1	-5 ± 0.01
+10^{-11}M	27.5 ± 0.1	0	13 ± 0.1	-35 ± 0.01
+10^{-10}M	25 ± 0.1	-9 ± 0.01	13 ± 0.1	-35 ± 0.01
+10^{-9}M	22 ± 0.1	-20 ± 0.01	10 ± 0.1	-50 ± 0.01
+10^{-8}M	16 ± 0.1	-42 ± 0.01	7 ± 0.1	-65 ± 0.01
+10^{-7}M	13 ± 0.1	-53 ± 0.01	6 ± 0.1	-70 ± 0.01
+10^{-6}M	12 ± 0.1	-56 ± 0.01	3 ± 0.1	-85 ± 0.01
+10^{-5}M	11 ± 0.1	-60 ± 0.01	2 ± 0.1	-90 ± 0.01
+10^{-4}M	7 ± 0.1	-74.5 ± 0.01	0	-100
+10^{-3}M	0	-100	0	-100

Thus, Melafen influence on two target surfaces of cells simultaneously—on purinoreceptors PY2 and on CRAC, considerably reducing their activity in plant's growth stimulated doses. Really, Melafen changes the functions of surface receptors and intracellular signal transduction.

It will be interesting to test the Melafen influence on the overall cellular answer of the normal cells that may be activated by ATP additions. Because that we recorded of right angle of light scattering of dilute suspension thymocytes and leucocytes. These cells have ranking among P2Y receptors and have another types of purinoreceptors non methabotrophic channel formers P2X (at leukocytes) [25] and its modification P2Z (at thymocytes) [26]. The compositions of groups of P2 receptors at thymocytes are showing both P2Z and P2X receptor activation characteristics, and are depending on the stage of cellular growth. The additions of ATP to the thymocytes suspension caused the two-phase change of cell volume (Figure 9.4).

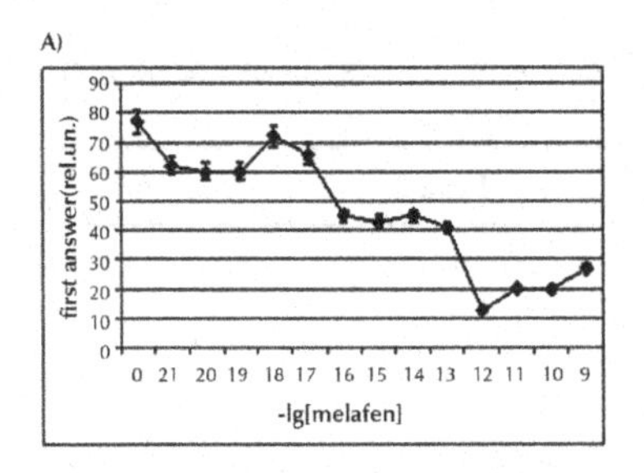

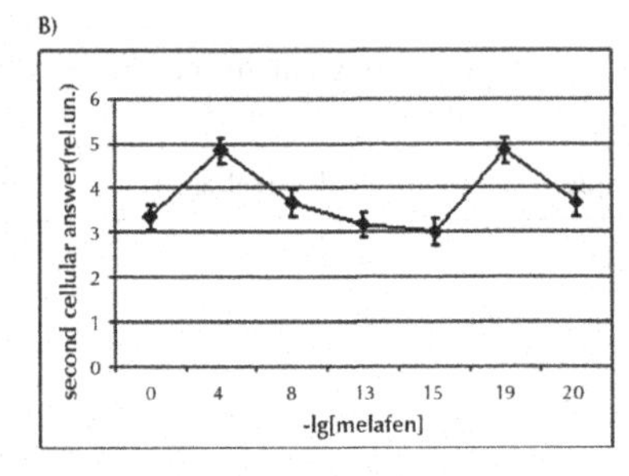

FIGURE 9.4 The influence of Melafen to the first (A) and second (B) cellular answers of cell suspension of rat thymocytes under the wide concentration regions of Melafen. The amplitude of light scattering risings after the ATP additions to the cell suspension in the dependence of Melafen concentrations

Melafen influences both phases. The question, which certain points of Ca^{2+} transduction in rat thymocytes are involved in Melafen effects, were open. We may conclude that P2X and P2Z Ca^{2+} channel formed receptors are susceptible to Melafen influence under a wide region of concentrations.

Additions of ATP to leucocytes suspension caused releasing of Ca^{2+} from intracellular stores through activating of metabotropic P2Y purine receptors (at the medium of measurement Ca^{2+} do not present). Thus, we eliminated the possibility of introducing the extracellular Ca^{2+} to cell interior. We used the measurement medium without the Ca^{2+} ions. In this case, the non metabotrophic P2X- and P2Z-channel formers were silent structure and as itself will lead the P2Y metabotrophic receptor. What activate the Ca^{2+} ions releasing from intracellular Ca^{2+} stores (endoplasmic reticulum)? The Melafen attendance shortens the time which the phase lag up to cellular answer development. The EAC cells behave analogously. It goes out faster to stable behavior in presence of Melafen. Impact of Melafen on the channel-former P2X is the clear coming up of leucocytes receptors (Figure 9.5).

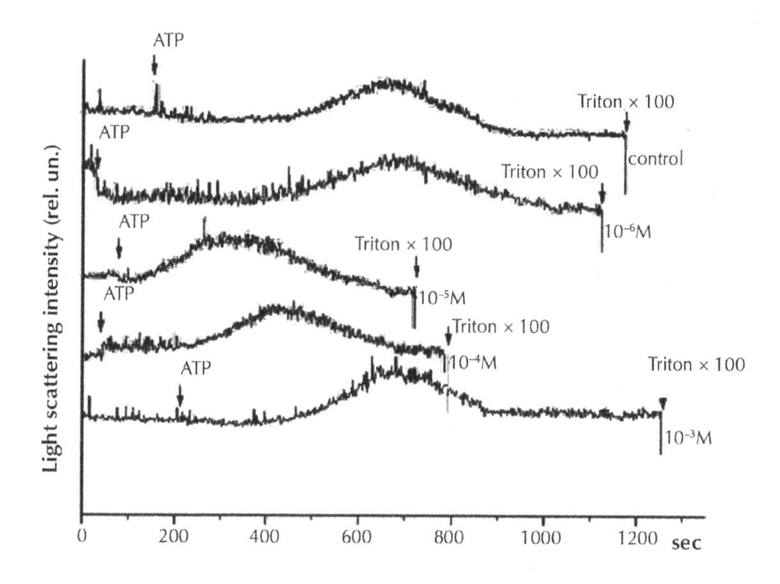

FIGURE 9.5 The influences of Melafen on the lag-phase of cellular answers of leucocytes. Kinetic curves of light scattering of the cell suspension under the wide concentrations region of Melafen

9.4 CONCLUSION

The Melafen actions to the three types of animal cells have the three global targets P2Y, P2X, P2Z—purinoreceptors. Plant growth regulator has the negative actions to the animal cells of different origin. And its actions were present under the low and ultra low concentrations.

KEYWORDS

- **Ca^{2+} transduction**
- **Lipid peroxide oxidation**
- **Melafen**
- **Microviscosity**
- **Purinoreceptors**

REFERENCES

1. Fattachov, S. G., Reznik, V. S., & Konovalov, A. I. (2002). In set of articles. Reports of 13th International Conference on chemistry of phosphorus compounds. Melamine salt of bis (hydroxymethyl) phosphinic acid. (Melaphene) as a new generation regulator of plant growth regulator, St. Petersburg (p. 80).
2. Alekseeva, O. M., Narimanova, R .A, Yagolnik, E. A., & Kim, Yu. A. (2011). Phenosan influence and its hybrid derivative on membrane components, Bashkir State University (pp. 15–19). All-Russian Conference Biostimulators in medicine and agriculture, Ufa, March 15, 2011.
3. Vekshina, O. M., Fatkullina, L. D, Kim, Yu. A., & Burlakova, E. B. (2007). The changes of structure and functions of erythrocyte membranes and cells of ascetic when action of hybrid antioxidant of rising generation ICHFAN-10. *Bulletin of Experimental Biology and Medicine. Ehrlich carcinoma, 4,* 402–406.
4. Zhigacheva, I. V., Fatkullina, L. D., Burlakova, E. B., Shugaev, A. G., Generozova, I. P., Fattakhov, S. G., & Konovalov, A. I. (2008). Effects of the organo-phosphorus compound Melaphen—plant growth regulator—on structural characteristics of membranes of plant and animal origin. *Biologicheskie Membrany, 25*(2), 128–134.
5. Zhigacheva, I. V., Fatkullina, L. D., Shugaev, A. G., Fattakhov, S. G., Reznik, V. S., & Konovalov, A. I. (2006). Melafen and energy status of cells of plant and animal origin. *Doklady Biochemistry and Biophysics, 409,* 200–202.
6. Zhigacheva, I. V., Evseenko, L. S., Burlakova, E. B., Fattakhov, S. G, & Konovalov, A. I. (2009). Influence of organophosphorus plant growth regulator on electron transport in respiratory chain of mitochondria. *Doklady Akademii Nauk, 427*(5), 693–695.
7. Zhigacheva, I. V., Fatkullina, L. D., Rusina, I. F., Shugaev, A. G., Generozova, I. P., Fattakhov, S. G, & Konovalo, A. I. (2007). Antistress properties of preparation Melaphen. *Doklady Akademii Nauk, 414*(2), 263–265.
8. Zhigacheva, I. V. et al. (2011). Fatty acids membrane composition of mitochondria of pea seedlings in conditions of moistening shortage and moderate cooling. *Doklady Akademii Nauk, 437*(4), 558–560.
9. Zhigacheva, I. V. et al. (2010). Insufficient moistening and Melafen change the fatty-acid membrane composition of mitochondria from pea seedlings. *Doklady Akademii Nauk, 432*(1), 124–126.
10. Kostin, V. I., Kostin, O. V., & Isaichev, V. A. (2006). Research results concerning the application of Melafen when cropping. Investigation state and utilizing prospect of growth regulator Melafen in agriculture and biotechnology (pp. 27–37). Kazan: Kazan University.
11. Alekseeva, O. M., Yagolnik, E. A., Kim, Yu. A., Albantova, A. A., Mil, E. M., Binyukov, V. I., Goloshchapov, A. N., & Burlakova, E. B. (2011). *Melafen action on some links of intracellsignalling of animal cells* (pp. 435–438). International Conference on receptors and the intracellular signalling, 24–26 May, Puschino.
12. Alekseeva, O. M., Krivandin, A. V., Shatalova, O. V., Kim, Yu. A., Burlakova, E. B., Goloshapov, A. N., & Fattakhov, S. G. (2010). *No lipid microdomains destruction, but stabilization by melafen treatment of dimyristoilphosphatidylcholine liposomes* (pp. 8–12). International Symposium "Biological Motility: from Fundamental Achievements to Nanotechnologies", Pushchino, Moscow, Russia, May 11–15, 2010.

13. Alekseeva, O. M., Shibryaeva, L. S., Krementsova, A. V., Yagolnik, E. A., Kim, Yu. A., Golochapov, A. N., Burlakova, E. B., Fattakhov, S. G., & Konovalov, A. I. (2011). The aqueous melafen solutions influence to the microdomains structure of lipid membranes at the wide concentration diapason. *Doklady Akademii Nauk, 439*(4), 548–550.
14. Tarakhovsky, Yu. S., Kuznetsova, S. M., Vasilyeva, N. A., Egorochkin, M. A., & Kim, Yu. A. (2008). Taxifolin interaction (digidroquercitine) with multilamellar liposomes from dimitristoyl phosphatidylcholine. *Biophysicist, 53*(1), 78–84.
15. Alekseeva, O. M., Krivandin, A. V., Shatalova, O. V. Rykov, V. A., Fattakhov, S.G., Burlakova,E. B., Konovalov, A. I. (2009). The Melafen-Lipid interrelationship determination in phospholipid membranes. *Doklady Akademii Nauk, 427*(6), 218–220.
16. Binyukov, V. I., Alekseeva, O. M., Mil, E. M., Albantova, A. A. Fattachov, S. G., Goloshchapov, A. N., Burlakova, E. B., & Konovalov, A. I. (2011). The investigation of melafen influence on the erythrocytes in vivo by AFM method. *Doklady Biochemistry and Biophysics, 441*, 245–247.
17. Alekseeva, O. M., Erokhin, V. N., Krementsova, A. V., Mil, E. M., Binyukov, V. I. Fattachov, S. G., Kim, Yu .A., Semenov, V. A. Goloshchapov, A. N., Burlakova, E. B., & Konovalov, A. I. (2010). The investigation of melafen low dozes influence to the animal malignant neoplasms *in vivo* and *in vitro*. *Doklady Akademii Nauk, 431*(3), 408–410.
18. Rizkina, I. S., Murtazina, L. I., Kiselyov, J. V., & Konovalov, A. I. (2009). Property of supramolecular nanoassociates formed in aqueous solutions low and ultra-low concentrations of biologically active substance. *DAN, 428*(4), 487–491.
19. Osipenkova, O. V., Ermokhina, O. V., Belkina, G. G., Oleskina, Yu. P., Fattakhov, S. G., & Yurina, N. P. (2008). Effect of Melafen on expression of *Elip1* and *Elip2* genes encoding chloroplast light-induced stress proteins in barley. *Practical Biochemistry and Microbiology, 44*(6), 701–708.
20. Zamai, A. C., et al. (2005) Conference "Reception and intracellular signalization", Puschino. The ATP influence to the tumor ascetic cells at different stages of cellular growth (pp. 48–51).
21. Zinchenko, V. P., Kasimov, V. A., Li, V. V., Kaimachnikov, N. P. (2005). Calmoduline inhibitor of R2457I induces the short-time entry Ca^{2+} and the pulsed secretion ATP in cells of ascetic Ehrlich carcinoma. *Biophysicist, 50*(6), 1055–1069.
22. Pedersen, S. F., Pedersen, S., Lambert, I. H., & Hoffmann, E. K. (1998). P2 receptor-mediated signal transduction in Ehrlich ascites tumor cells. *Biochimica et Biophysica Acta, 1374*(1–2), 94–106.
23. Cornet, M., Lambert, I. H., & Hoffman, E. K. (1993). Relation between cytoskeletal hypoosmotic treatment and volume regulation in Erlich ascites tumor cells. *Journal of Membrane Biology*, 131, 55–66.
24. Artalej, O. A. & Garcia-Sancho, J. (1988). Mobilization of intracellular calcium by extracellular ATP and by calcium ionophores in the Ehrlich ascities tumor cells. *Biochimica et Biophysica Acta, 941*, 48–54.
25. Di Virgilio, F. et al. (2001). Nucleotide receptors: An emerging family of regulatory molecules in blood cells. *Blood*, 97, 587–600.
26. Nagy, P. V., Fehér, T., Morga, S., & Matkó, J. (2000). Apoptosis of murine thymocytes induced by extracellular ATP is dose- and cytosolic pH-dependent. *Immunology Letters*, 72(1), 23–30.

CHAPTER 10

SOME NEW ASPECTS OF BIOCHEMICAL TREATMENT: PART I

S. A. BEKUSAROVA, N. A. BOME, L. I. WEISFELD,
F. T. TZOMATOVA, and G. V. LUSCHENKO

CONTENTS

10.1 ACTIVATING BY PARA-AMINOBENZOIC ACID OF SOWING PROPERTIES OF SEED OF WINTER GRAIN CROPS AND FORAGE CEREALS

10.1.1 INTRODUCTION

Para-aminobenzoic acid (PABA) by way of well-known activator of growth and development was applied for treatment of ears of winter wheat and winter barley. The varieties of these cultures are destined for North-Caucasian region. Also panicles of selection samples of millet—the varieties are introduced in North Ossetia—were subject to treatment by PABA. PABA was dissolved in acetic acid, but not in water, as usual. Advantages of that mode are described. These methodologies allow to accelerate a breeding process, when often is necessary to sow in year, when was gathered harvest.

In the process of selection for the speed-up study of material, it is necessary to get the sufficient amount of seed in the earliest possible dates. For the speed-up estimation of plant-breeding material, it is necessary to get the sufficient amount of seed in the earliest possible dates. To that end conducted sowing of inflorescences without threshing from the plants of cereal cultures selected for a selection. The method of sowing by ears was previously tested during 2 years in Tyumen research station of N.I. Vavilov Research Institute of Plant Industry on more than 100 samples of winter wheat from world collections. The productivity of winter form of cereals depends on a set of biotic (pathogenic microorganisms) and abiotic (temperature, amount of precipitation, etc.) factors. We obtained positive results for winter hardiness, resistance to snow mold, which main pathogen is *Microdochium nivale* (Fr.) Samuels and I.C. Hallett (*Fusarium nivale* Ces. ex Berl. and Voglino).

Already in 1940s, PABA was discovered by J. A. Rapoport as modifier of metabolic processes. He showed on the example of experimental object, *Drosophila*, that PABA evokes positive changes of non hereditary character (i.e., it is not a mutagen) in the organism development [1]. Rapoport [2] proposed a scheme of relations between genotype, ferments, and phenotype: "genes → their hetero catalysis (on the substrate of ribonucleic acid (RNA) molecule) → messenger RNA (mRNA) → mRNA catalysis (substrate of amino acids) → ferments → their catalysis (different substrates) → phenotype".

Works on the application of PABA in agriculture continue and deepen. The experiments performed in the Republic of North Ossetia [4–9] showed, that the addition of PABA to the nutritive substances (potassium humate, irlit, leskenit, corn extract, juice of ambrosia, melted snow water, and others) gives a positive effect. Treatment with PABA seeds of pea and honey plant sverbiga east (*Bunias orientalis*) before planting [4], extra nutrition of clover [5], seed of leguminoze grasses [6] stimulates the germination plants in a greater degree, then nutritive substances without PABA. A positive result was got at treatment of sprouts of potato by mixture of the melted water with juice of ambrosia, leskenit, and PABA [7]. This method resulted in the increase of harvest of potato and decline of disease of Fusarium. Additional PABA in nutrient medium for treatment seed of triticale [8] results in the increase of maintenance of protein in green mass. Protracted treatment by PABA of handles of dogwood [9] increased engraftment of grafts. It was used for the receipt of potato without viruses [10].

In the joint research of scientists of the Institute of Biochemical Physics RAS and the North Caucasus Research Institute of mountain and foothill agriculture, it was shown that the treatment by PABA of potatoes tubers with the subsequent enveloping in an ash [11] increase yield of potato. Treatment of seed and seedlings of vegetable cultures by solution of PABA in mixture with boric acid and permanganate potassium [12] improves resistance to diseases of young plantlets and increases harvest of carrot, beet, cucumbers, and tomatoes, seeds and seedlings of vegetables.

After treatment by PABA of binary mixture of seed of winter wheat and winter vetch [13], the productivity and quality of green feed increases at mowing of mixture in a period from the beginning of exit in a tube of wheat and beginning of budding of vetch to forming of grain of milky ripeness of wheat.

In the conditions of the northern forest-steppe of Tyumen region, where beside of fertile soils occur saline, jointed impact of salt solutions and PABA (0.01% solution) on the seeds of three barley varieties with low salinity resistance considerably increased the salinity resistance of germs independently on NaCl concentration [14]. The positive results were obtained under spraying of inflorescences of mother plants of barley by the solutions of PABA before realization of crossing. In a series of hybrid combinations, the exceeding over control of length and width of flag leaf, number of leaves by plant and plant height was observed [15]. Spraying of inflorescences of four amaranth samples by PABA solutions increased seed productivity, the concentration of 0.02% showed to be the most effective [15].

In the present study, it is researched about the effect of PABA on germination, germination energy and winter-hardiness of cereals, namely ears of winter cereals and panicles of millet species. In contrast with most of above-mentioned studies, where PABA was dissolved in the hot water following the method developed by Rapoport with collaborators [3], in the present study PABA was dissolved in an acetic acid. The combination of PABA and an acetic acid creates an acid media and allows preventing of a set of fungi illnesses simultaneously with the preservation of genotype of samples under multiplying.

Below, we present the results of experiments performed on the experimental base of the North Caucasus Research Institute of mountain and foothill agriculture where perennial crops of legumes herbs were precursor.

10.1.2 MATERIALS AND METHODS

Cereals are most widely applied in the agricultural production and selections were applied as material. The following cultures of yield of 2009 were tested: winter wheat varieties Ivina, Vassa, Bat'ko, Don 107, Kollega, and Kalym and winter barley variety Bastion—the varieties designed for North-Caucasian region. It was also been tested the following varieties introduced in North Ossetia: selection samples of millet: Japanese millet (*Echinochloa frumentacea*), panic (*Setaria italica Panicumitalicum*) and Italian millet (*Setaria italica*).

PABA represents a fine-grained powder; it easily and completely dissolves in 3% solution of an acetic acid without heating. One Tea spoon of dry PABA powder (10 g) was dissolved in a small volume (20–25 ml) of 3% solution of an acetic acid. Then obtained mixture was dissolved in 1liter of tap water under room temperature. Thus, the PABA concentration of 0.1% was obtained. In order to obtain the concentration of 0.2%, 20 g of dry powder was dissolved.

Ten samples were soaked for the inflorescences without trashing of cereals with mature seeds, namely ears of cereals and panicles of millet species, in the solutions of PABA (0.1 or 0.2%) for 2–2.5; 3–4, and 5–6 hrs. Then the inflorescences were planted in the in the soil in the open field. The distance of 20–25cm was maintained between the samples. Untreated inflorescences, soaked in water for 3–4 hrs, were applied as control (control I). When preparing control II PABA was dissolved in hot water and soaked for inflorescences for 2–2.5 hrs.

In tables the variants of experiments were marked: in variant "0" is presented Control of I, where inflorescences were soaked in water (without treatment by PABA); in variant "1" is presented Control of II, where inflorescences were soaked in PABA dissolved in hot water; variants 2–5 are a soakage of inflorescences in PABA dissolved in acetic acid of concentrations of 0.1–0.2% for 2–6 hrs.

Simultaneously, thrashed seeds for germination in a Petri dish were placed with PABA dissolved in hot water (control I) or in acetic acid (control II). For the control II the data averaged for PABA concentrations of 0.1 and 0.2% are presented because their difference was insignificant.

10.1.3 RESULTS

The PABA concentration of 0.1 or 0.2% in acetic acid was sufficient for the penetration of substance in the embryo when treating cereal inflorescences. High concentrations (9–11%) inhibit height and development of plants [10].

Germination of threshed seeds soaked in PABA dissolved in an acetic acid, appeared 3–4 days earlier than when soaked in water. On seedlings from seed germinated in an aqueous solution of PABA developed fungal microflora and they died. When dissolved in acetic acid PABA seedlings persisted for long time, they continued to grow.

Germinating capacity of seeds in inflorescences under soaking in PABA, dissolved in an acetic acid, considerably exceeded germinating capacity in control variants (Table 10.1). The effectiveness of treatment by PABA solutions depended on its duration and concentration of solutions. The same trend concerning germination energy was observed for all five cereals under study.

TABLE 10.1 Germinating capacity of seeds of winter grain crops and millet species under different ways of treatments of cereal inflorescences by para-aminobenzoic acid

Variant of experiment	Winter wheat		Winter barley		Japanese millet		Panic		Italian millet	
	%	+/%	%	+/%	%	+/%	%	+/%	%	+/%
0. Control I without treatment: soaking in water for 3–4 hr	62	0	75	0	67	0	72	0	68	0
1. Control II: soaking in PABA (0.1–0.2%) dissolved in hot water for 2–2.5 hr	74	12/16	84	9/11	76	9/12	82	10/1.2	72	4/6
2. Soaking in PABA (0.1%) dissolved in an ac. a. for 2–2.5 hr	82	20/24	93	18/19	87	20/23	92	20/22	84	16/19
3. Soaking in PABA (0.2%) dissolved in an ac. a. for 2–2.5 hr	86	24/28	95	20/21	92	25/27	92	20/22	87	19/24
4. Soaking in PABA (0.2%) dissolved in an ac. a. for 3–4 hr	96	34/35	98	23/23	95	28/29	95	23/23	92	24/25
5. Soaking in PABA (0.2%) dissolved in an ac. a. for 5–6 hr	82	20/24	86	11/13	90	23/26	90	18/20	86	18/21
LSD_{05}	4.5		2.4		2.8		2.1		2.2	4.5

LSD the least substantial difference; *ac. a.* acetic acid

The middle percentage of wintering as compared with control of I (water without PABA) increases with the increase of concentration of PABA and durations of treatment. However, at more long-continued soakage of plants—for 5–6 hrs (variant 5) increase in relation to both controls was less than at a soakage for 3–4hrs (variant 4). At a soakage in water solution of PABA of 0.1% increase was observed, but on more low level than at dissolution in acetic acid.

The energy of germination, presented in Table 10.2, exceeded control without treatment of PABA in all variants of dissolution of PABA dissolved in acetic acid and in a variant with treatment of PABA dissolved in hot water (control of II, variant 1), and was higher than in control without addition of PABA (zero variant) at all five investigated cultures. Exceeding of indexes above control of I in control of II was below than in variants 1–5 (Table 10.2).

TABLE 10.2 Germination energy of seeds of winter grain crops and millet species under different ways of treatments of cereal inflorescences by para-aminobenzoic acid

Variant of experiment	Winter wheat		Winter barley		Japanese millet		Panic		Italian millet	
	%	+/%	%	+/%	%	+/%	%	+/%	%	+/%
0. Control I without treatment: soaking in water for 3–4 hr	54	0	68	0	53	0	60	0	54	0
1. Control II: soaking in PABA (0.1–0.2%), dissolved in hot water 2–2.5 hr	65	11/17	73	5/7	62	9/15	73	12/16	63	8/13
2. Soaking in PABA (0.1%), dissolved in an ac. a. 2–2.5 hr	72	55/76	81	13/16	75	22/29	81	22/31	73	19/26
3. Soaking in PABA (0.2%), dissolved in an ac. a. 2–2.5 hr	80	27/34	84	16/19	79	26/33	87	26/30	76	22/29
4. Soaking in PABA (0.2%), dissolved in an ac. a. 3–4 hr	85	32/38	89	21/24	83	30/36	89	29/33	79	25/32
5. Soaking in PABA (0.2%), dissolved in an ac. a. 5–6 hr	78	25/32	80	11/14	78	26/33	84	24/29	75	21/28
LSD_{05}		3.4		2.1		2.6		2.9		2.5

Winter-hardiness of winter cereals (Table 10.3)—wheat and barley in all variants with treatment of PABA dissolved in an acetic acid (variants 2–5) and in control of II (dissolution is in hot water) was higher than in control of 1 (without PABA). As well as other experiments indexes at dissolution of PABA in hot water in control II were below than in other variants (3–5).

TABLE 10.3 Winter-hardiness (% of plants overwintered) of winter cereals under different ways of treatments of ears by para-aminobenzoic acid

Variant of experiment	Winter wheat		Winter barley	
	%	+/%	%	+/%
0. Control I without treatment: soaking in water for 3–4 hr	78	0	84	0
1. Control II: soaking in PABA (0.1–0.2%), dissolved in hot water 2–2.5 hr	82	4/4.8	88	4/4.8
2. Soaking in PABA (0.1%), dissolved in an ac. a. 2–2.5 hr	86	8/9.3	95	11/22.6
3. Soaking in PABA (0.2%), dissolved in an ac. a. 2–2.5 hr	90	12/13.3	97	13/13.4
4. Soaking in PABA (0.2%), dissolved in an ac. a. 3–4 hr	98	20/20.4	98	14/14.5
5. Soaking in PABA (0.2%), dissolved in an ac. a. 5–6 hr	88	10/11.4	92	8/8.7
LSD_{05}		3.1		0.8

The dynamics of activating of phenotype under act of PABA is evidently shown on a histogram (Figure 10.1), reflecting the middle indexes of levels of increases of energy of germination, germination and winter-hardiness. The tendency of growth of middle indexes is evidently.

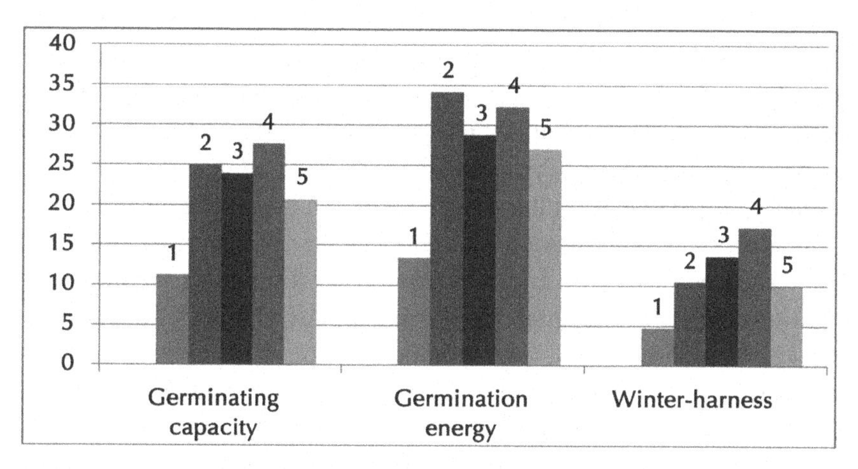

FIGURE 10.1 Comparison of averages (totally on all cultures, %) of germinated capacity, energies of germination of seed and winter-hardiness of plants at the different methods of treatment of ears of winter crops and panicles of kinds millet by para-aminobenzoic acid in the variants of experiments from 1 to 5 (see Tables 10.1–10.3) by comparison with control of I (without treatment, in tables—a zero variant). Ordinate axis—%

10.1.4 DISCUSSION

The higher indexes of germinating capacity, energy of germination at all cultures and winter-hardiness cold of winter crops are marked—winter wheat and barley in variants (2–5 in diagram) after treatment of PABA, dissolved in an acetic acid, as compared with control variants—without treatment of PABA (variant 0—control of I without PABA) and at dissolution of PABA in hot water (variant 1—control of II) in analogical variants (see Tables 10.1–10.3, Figure 10.1) were got. All indexes at a soakage in PABA in an acetic acid most long time—5–6 hrs (variant 5) as compared with analogical treatment for 3–4 hrs (variant 4) has a tendency to go down (see Tables 10.1–10.3, Figure 10.1). Maybe, in this case braking of development of plantules takes place [10].

This work executed in connection with the necessity of speed-up creation of new varieties steady to the ecologically diverse conditions and simultaneously responsive on the additional fertilizing. Work is important from the economic point of view. Sowing of inflorescences without threshing considerably simplifies work of breeder both in a scientific plan and in organizational: the methods of selection are simplified, reproduction of the best plants is accelerated for a selection and forming of new varieties, diminish expenses of labor. Usually at the unfavorable condones of growing, at introduction of plants, at the high doses of mutagens, or during distant hybridization of cultural plants with wild sorts fall down number in inflorescences, germination of seeds, viability of plants. At sowing of inflorescences—ears or panicles possibility to distinguish the most perspective families on the number of germinating seed in them appears with mature grain. From every separate inflorescence, it is possible to collect more seeds for the receiving of posterity, the terms of estimation of material and terms of reproduction of valuable selection samples grow short. A method especially touches the freshly material, when at preparation to sowing of winter crops in the year of harvest time not enough for threshing of ears and additional treatment of grain. Such quite often happens in those districts, where short vegetation period, for example, in the Tyumen area.

10.1.5 CONCLUSION

1. The method of acceleration of plant-breeding process offers at sprouting of mature seed in inflorescences of cereal cultures—ears of grain growing and panicles of millet. A method plugs the soakage of inflorescences in PABA (concentrations of 0.1 and 0.2%) dissolved in an acetic acid.
2. Dissolution of PABA in an acetic acid is more effective, than dissolution in hot water (control II). Activating of phenotype at soakage in PABA, dissolved in an acetic acid, in variants (3–5) was higher than in control II—at soakage in PABA dissolved in hot water.
3. Soakage of inflorescences during 3–4 hrs in PABA a dissolved in an acetic acid, gives the best result. More protracted soakage—5–6 hrs some reduces by all variants.
4. Concentrations of PABA applied to the treating of inflorescences were higher than in the cited works in which treated trashed seeds.

KEYWORDS

- **Acetic acid**
- **Hybrid combinations**
- **Inflorescences**
- **Para-aminobenzoic acid**
- **Winter-hardiness**

REFERENCES

1. Rapoport, I. A. (1948). *Phenogenetic analysis of dependent and independent differentiation.* Proceedings of the institute of cytology, histology, embryology of academy of sciences, 2(1), 3–128.
2. Rapoport, I. A. (1989). *The action of PABA in connection with the genetic structure. Chemical mutagens and para-aminobenzoic acid to increase the yield of crops* (pp. 3–37). Moscow: Science.
3. Rapoport, I. A. (1989). Editor of *chemical mutagens and para-aminobenzoic acid in enhancing the productivity of crops* (p. 253). Moscow: Science.
4. Bekuzarova, S. A., Abiyeva, T. S., & Tedeeva, A. A. (2006). *The method of pre-treatment of seeds*. Patent No. 2270548. Published 27. 02. 2006.
5. Bekuzarova, S. A., Farniyev, A. T., Basiyeva, T. B., Gaziyev, V. I., & Kaliceva, D. N. (2011). *The method of stimulation and development of clover plants*. Patent No. 2416186. Published on 20.04.2011.
6. Bekuzarova, S. A., Shtchedrina, D. I., Farnoyev, A. T., & Pliyev, M. A. (2006). *Method of additional fertilizing of leguminous grasses*. Patent No. 2282342. Published on 27.08.2006.
7. Ikaev, B. V., Marzoev, A. I., Bekuzarova, S. A., Basayev, I. B., Bolieva, Z. A, & Kizinov, F. I. (2010). *The method of treatment of pre-plant shoots of potato tubers*. Patent No. 2385558. Published on 10.04.2010.
8. Bekuzarova, S. A., Antonov, O. V., & Fedorov, A. K. (2003). *Method of increase of content of protein in green mass of winter triticale*. Patent No. 2212777. Published on 27.09.2003.
9. Cabolov, P. H., Bekuzarova, S. A., Tigiyeva, I. F., Tadtayeva, E. A., & Eiges, N. S. (2007). *The method of reproduction of dogwood drafts*. Patent No. 2294619. Published on 10.03.2007.
10. Shcherbinin, A. N. & Soldatova, T. B. (2004). *The nutrient medium for micropropagation of potato*. Patent No. 2228354. Published 10.05.2004.
11. Eiges, N. S., Weisfeld, L. I., Volchenko, G. A., Bekuzarova, S. A. (2007). *Method of pre-treatment of tubers of potatoes*. Patent No. 2202701. Published on 02.10.2007.

12. Eiges, N. S., Weissfeld, L. I., Volchenko, G. A., & Bekuzarova, S. A. (2003). *The method of pre-treatment of seeds and seedlings of vegetable crops*. Patent No. 2200392. Published 20. 03. 2003. Bull. No.6
13. Eiges, N. S., Weissfeld, L. I., Volchenko, G. A., Bekuzarova, S. A., Pliyev, M. A., & Hadarceva, M. V. (2008). *The method of receiving of feeds in green conveyer*. Patent No. 2330410. Published 10. 08. 2008. Bull. No. 22.
14. Bome, N. A. & Govorukhin, A. A. (1998). *The effectiveness of the influence of para-aminobenzoic acid on the ontogeny of plants under stress*. Bulletin of Tyumen State University. Tyumen: Tyumen State University. 2, 176–182.
15. Bome, N. A, Bome, A.Ja, & Belozerova, A. A. (2007). *Stability of crop plants to adverse environmental factors* (p. 192). Monograph. Tyumen: Tyumen State University.

CHAPTER 11

SOME NEW ASPECTS OF BIOCHEMICAL TREATMENT: PART II

LARISSA I. WEISFELD

CONTENTS

11.1 DAMAGE OF CHROMOSOMES AND DISRUPION OF MITOTIC ACTIVITY IN SEEDLINGS OF *CREPIS CAPILLARIS* BY ALKYLATING AGENT ANTINEOPLASTIC PREPARATION PHOSPHEMIDUM

11.1.1 INTRODUCTION

In this part of the chapter, we attempt to consider mechanism of origin of rearrangement of chromosome in the different phases of mitotic cycle. The appearance of rearrangements of chromosome is showed under the influence of antineoplastic medicament "phosphemid"—di- (ethylene imid)-pyrimidyl-2-amidophosphoric acid, named mutagen. Seeds *Crepis capillaris* and fibroblasts of human and mouse were being treated by this mutagen. In seedlings on metaphase plates rearrangements of chromosomes were analyzed. The rearrangements were analyzed at different terms after appearance of seedlings named "arisings". Also chromosomal rearrangements were examined at anaphases and telophases of fibroblasts.

Scientists attracted attention to the chemical compounds that cause heritable changes. It remains unsolved mechanism of their effects on the chromosome. The action of many compounds is similar with ionizing radiation; they are cause mutations of genes, disruptions of cell division, and rearrangement of chromosomes.

It is known that ethylene imine and its derivatives alkylate DNA and proteins (see review [1]). They induce mutations as it was shown in various model objects (Drosophila, higher plants, fungi, bacteria, viruses, and others) and breakage of the chromosome apparatus (see, for example, the review [2]). Chemical compounds that cause mutations and breakage of chromosomes usually are called "chemical mutagens".

In the history of the study of chemical mutagens a large role belongs to works of J. A. Rapoport [3]. He is discoverer of phenomenon of the chemical mutagenesis. He revealed out super-mutagens and discovered possibilities and their application in the breeding of crops and in other areas of agricultural production. J. A. Rapoport organized synthesis of super-mutagens. Since 1959, every year, he organized the All-Union conferences for scientists and breeders on chemical mutagenesis and its application in agriculture. These meetings served in those years a good genetic school, especially for young breeders, who were trained and the unscientific method of Lysenko. Based on the methodology developed by Rapoport and with the help of mutagens, which he distributed free of charge, breeders created the source material of crops and introduced new varieties [see the review 3]. A series of investigations on the application of the chemical mutagen ethylene imine and mechanism of its action in winter wheat was conducted and is ongoing, N. S. Eiges [see for example 6, 7], which is the follower of J. A. Rapoport.

The most effective and affordable way to analyze the mutagenesis is a cytogenetic method: the study of rearrangements (aberrations) of chromosomes observed in dividing cells (mitosis, karyokinesis), and disorders in passing of the mitotic cycle.

It was studied mechanism damage of the chromosomes and their relationship to the mitotic cycle and synthesis DNA at various biological objects—in the cells of the meristem of plants onions, Vicia faba, Crepis, and so on, embryonic cells of animals and human in vitro, at microbes, viruses, and other objects.

Ionizing radiation damages the chromosomes immediately after irradiation at all stages of the mitotic cycle—"undelayed" effect. In the cells that came into mitosis from G2 phase and S, arise aberrations of chromatid type (nature) in cells that come in mitosis from phase G1 (pre-synthesis) occur aberrations of chromosome type (double bridges, paired fragments at anaphase).

Under treatment by chemical mutagens of asynchronous cell cultures no rearrangements detect during the first 2–4 hrs (depending on the duration of G2 phase at different objects). This phenomenon is named "delayed" effect. Chromosomal rearrangements appear later—after the entry to mitosis of cells treated at the beginning or during DNA synthesis (phase S) or before phase S (named pre-synthetic phase G1). Chromosome rearrangements are usually analyzed in the ana-telophases or meta-phases. A large number of chromosomes in many objects complicate the identification of chromosome aberrations in metaphase plates. In this case, fragments of chromosomes or broken bridges cannot be identified. In this case, scientists analyze the anaphase and early telophases (ana-telophase method).

For the estimate of environmental contamination the method of analyzes of ana-telophases is sufficient. An example is the estimation of pollution in the industrial area of the city of Staryj Oskol on the meristem of birch [8].

In Moscow, 60–70 years extensive cytogenetic studies of chemical and radiation mutagenesis were carried by N. P. Dubinin, his colleagues and followers. The work was begun in the laboratory of radiation genetics of Institute of biophysics of Academy of Science the USSR and continued in Institute of general genetics. A large number of articles and monographs about cytogenetic effects of ionizing radiation or chemical compounds were published (see review [9]). N. P. Dubinin formulated the idea of the mechanism of action of chemical mutagens: mutagens cause potential changes in the chromosomes at all stages of the mitotic cycle, which are realized in a number of cell generations. He called it "chain processes" in mutagenesis [10–12].

At the same time period, staff of the laboratory under the direction of B. N. Sidorov and N. N. Sokolov (Andreev V. S., Generalova M. V., Grinih L. I., Durymanova S. E., Kagramanyan R. G., Protopopova E. M., Shevchenko V. V., and others) were conducted extensive work on the induction of chromosomal damage, their localization in the chromosomes, their association with the phases of the mitotic cycle and DNA synthesis after treatment of seedlings by ethylene imine, tio-TEF, radiation, and other mutagens on the model object Crepis capillaris. They studied chromosome aberrations in metaphase plates, which make it possible to take into account the polyploidy of metaphases.

In the early 1960s, working at N. P. Dubinin; The cytogenetic effect of alkylating agent phosphemidum (lat. synonym phosphemid, phosphasin)—di-(ethylene imid)-pyrimidyl-2-amidophosphoric acid was investigated (Figure 11.1).

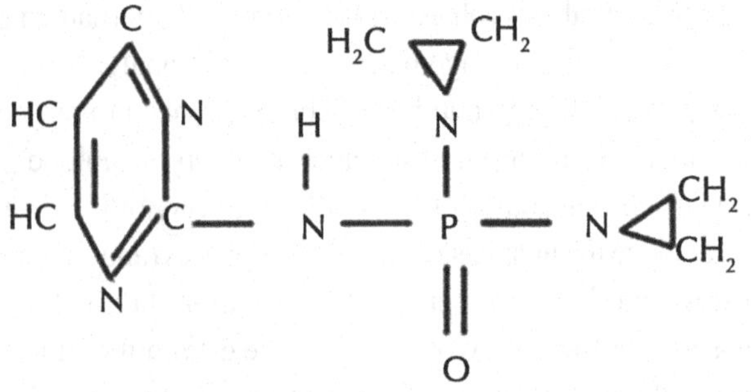

FIGURE 11.1 Phosphemidum (syn. phosphasin, phosphemid)

This compound is interesting because it consists of pyrimidine base and three molecules of ethylene imine. It was assumed that phosphemid will cause a lot of damages in DNA synthesis and thereby would inhibit tumor growth. The drug was synthesized in the laboratory of Chernov et al. [13], the All-Union Scientific Research Chemical-Pharmaceutical Institute (now the Center for Chemistry of drugs), and was referred to N. P. Dubinin for cytogenetic studies. The drug is a white crystalline powder, dissoluble in water and alcohol. Chernov et al. [13, 14] have shown that phosphemid (then he was named phosphasin) inhibits the growth of tumors in rats, mice, rabbits, but at the same time it causes leukopenia, leukocytosis, and suppressed erythropoiesis. But currently, the drug used in medicine for the treatment of certain tumors such as leukemia, lymphoma. In the early work with phosphemid (1963–1964gg), it was important to approach the mechanism of its antineoplastic action. It was assumed that due to pyrimidine bases and ethylene imine groups of drug will directly affect DNA during the synthesis and thereby destroys actively dividable tumor cells.

The work was carried out on primary culture of embryonic tissues—the mouse and human fibroblasts [15, 16]. We analyzed chromosomal aberrations in cells at the stages of anaphase or telophases (ana-telophases), after treatment culture by phosphemid in a concentration of $1 \cdot 10^{-4}$M.

It was shown that phosphemid delays entry of cells into mitosis and inhibits the mitotic activity of fibroblasts during the total period of culture growth. The waves of fall and of rise of mitotic activity generally are repeating waves of mitotic activity in control (without treatment), but at a lower level. The average frequency of mitosis made 54% from control. Lesions spindle was not observed, as a rule, but in individual cells were visible clumping of the chromosomes in the form of "stars". At later stages of fixing the number of nuclei was decreased by 3–4 times, which on indicates cell death or on the loss of their contact with surface of glass. At consideration the types of chromosomal rearrangements, it became clear that in the ana-telophases appear mainly chromatid type of rearrangements. At later stages of fixation, after 26 hrs growth of culture were observed double bridges (chromosomal type). They could have arisen as a result of doubling of the chromatid bridges in the second cycle (as a result their passing to one of poles during the first division) and as a result of the breaking-fusion of chromosomes before synthesis DNA (phase G_1) with the subsequent doubling during the synthesis of DNA, in accordance with the thesis of N. P. Dubinin.

It was necessary to set connection with the phases of mitotic cycle. To do this, it would be desirable to find an object that, firstly, would be synchronized (or at least partially synchronized) with the terms of the phases of the mitotic cycle, and secondly, it would be convenient for the analysis of chromosomes in metaphase. To do this, it is desirable to find an object which, firstly, would be synchronized (or at least partially be synchronized) with phases of the mitotic cycle, and secondly, was suitable for the analysis of chromosomes in metaphase.

In this regard *Crepis capillaris* (L.) Wallr. serves an ideal object. Mikhail S. Navashin studying chromosomes in the metaphase plates of seedlings published a number of works about taxonomy of genus *Crepis* [17]. It was showed emergence of a large number of chromosome rearrangements in aging seeds.

At metaphase plate *Cr. capillaris* clearly identified three pairs of homologous chromosomes (Figure 11.2). Analysis of rearrangements here is not in doubt.

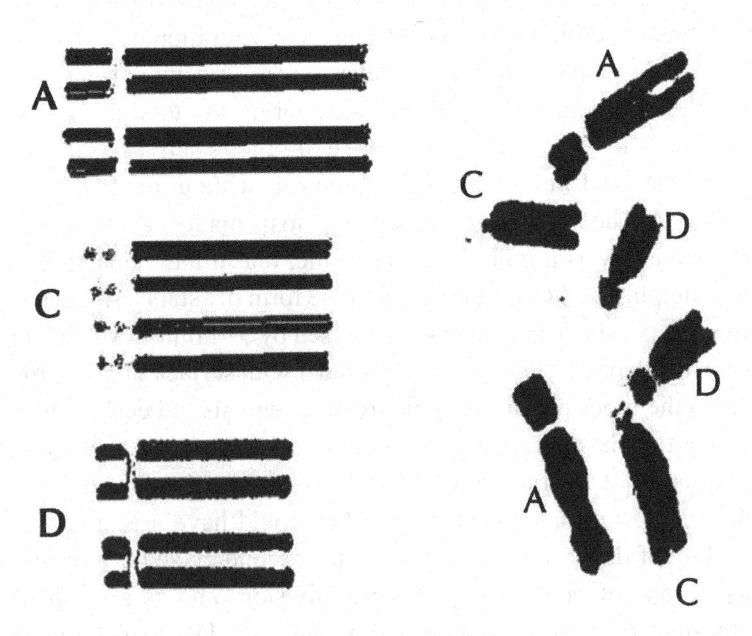

FIGURE 11.2 Karyotype *Crepis capillaris*

It was assumed, the results of experiments with ionizing radiation, the cells of seeds (not seedlings!) *Cr. capillaris* are in phase G_1. The phase of G1 in germinal cells of seed of *Crepis* is heterogeneous, as was shown [18] that first mark cellular nuclei appear in seedlings through 10hrs after treatment of grain of thymidine H^3.

As far as the seeds germinate at first in mitosis comes the cells from the phase of S then from the phase of G1. If the chemical mutagen interacts with the chromosome before the start of germination, that is, before the start of DNA synthesis—in phase G_1, then in metaphases of germinating seedling must appear rearrangements of chromosome type. The data of our experiments presented below, show that after treatment by phosphemid of seeds in the $2n$-metaphases appear rearrangements of chromatid type.

11.1.2 MATERIALS AND METHODS

Air-dried seeds of *Crepis capillaris* of crop of 1967 were analyzed in 1968: April (8 months of storage after harvest), June (10 months of storage), and July (11 months of storage). In the control and in the experiments used distilled water. The chromosome aberrations were analyzed in seedlings (meristem of root tip).

A certain amount of seeds (100 pieces) were treated in an aqueous solution phosphemid of concentrations: $1 \cdot 10^{-2}$M (22.4 mg was dissolved in 10ml water), $2 \cdot 10^{-2}$ M (22.4 mg was dissolved in 50ml) or $2 \cdot 10^{-3}$M (2.24 mg dissolved in 50 ml of water). The treatment was carried out at room temperature (19–21°C) for 3hrs. Then the seeds were washed with running water for 45min. The washed seeds were placed in Petri dishes on filter-paper moistened with a solution of colchicine (0.01%). Seeds were germinating in a thermostat at 25°C, but in July 1968 because of hot weather the temperature in the thermostat could reach 27°C. In parallel control experiments, the seeds were treated with aqua distillate.

After 24, 27, and 31hrs after soaking of seeds we choose "arisings". The term "*arising*" refers to those seedlings that emerge after the beginning of soaking of dry seeds and have a size of less than 1mm. These seedlings were selected in a Petri dish for further germination and subsequent fixation. In Russian the term "arising" is called "proklev".

For the analysis of chromosomes in metaphase plates, root tips (0.5–1cm) were cut off with a razor and placed in the solution: 96% ethanol 3 parts + 1 part of glacial acetic acid. Solution poured out through 3–4 hrs. Seedlings were washed 45 min in 70% alcohol. These seedlings were kept in 70% alcohol. We were preparing temporary pressure preparations: fixed root tips were stained acetous carmine and crushing in a solution of chloral hydrate between the slide and cover slip. We analyzed chromosome aberrations in metaphase plates of seedlings in the first division after treatment of seeds (2*n*-karyotype). In each seedling were counted up all metaphases. Intact seeds of harvest of 1966, 1967, and 1969 years served as control. The seedlings were fixed at different time intervals from 3 to 24 hrs.

Air-dried seeds of *Crepis capillaris* of harvest of 1967 were analyzed in 1968: April (8 months of storage after harvest), June (10 months of storage), and July (11 months of storage). We analyzed rearrangements of chromosomal type (damage of chromosomes before beginning of synthesis of DNA) and chromatid type (damage of chromatids with beginning of synthesis of DNA). A certain amount of seeds (100 pieces) were treated with an aqueous solution phosphemid in next concentrations: $1 \cdot 10^{-2}$M (22.4 mg was dissolved in 10 ml water), $2 \cdot 10^{-3}$ M (22.4 mg was dissolved in 50 ml or 2.24 mg dissolved in 5 ml of water). The treatment was carried out at room temperature (19–21°C) for 3hrs. Then the seeds were washed with running water for 45min. The washed seeds were placed in Petri dishes on filter-paper moistened with a solution of colchicine (0.01%). Seeds were germinating in thermostat at 25°C, but in July 1968 because of hot weather the temperature in the thermostat could reach 27°C.

Mitotic activity in the seedlings was determined on two criteria:

- On the criterion of the number of metaphase plates, depending on the number of nuclei in seedlings in the control and experiment (after seed treatment phosphemid). We used the seedlings after 2, 4, 6, 8, 10, and 20 hrs after the "arising". In each seedling, we counted between 500 and 1,000 nuclei; in an each seedling counted from 500 to 1,000 nuclei;
- On the criterion of the number of seedlings with metaphase to all watched seedlings at all stages of fixation.
- In all the experiments was estimated standard deviation from the mean.

11.1.3 RESULTS AND DISCUSSION

The natural level of mitotic activity—control was estimated by the criterion of the number of metaphases in seedlings: of the 18,000 counted nuclei metaphase plates were 2.83 ± 0.12%.

Table 11.1 shows data within 3 years of the evaluation of natural mitotic activity on the criterion of the number of metaphases in all investigated roots, in different years (1966, 1967, and 1969) ranged 90—99%.

TABLE 11.1 The natural level of chromosome rearrangements in the 2*n*-meristem cells of *Crepis capillaris* after soaking the seeds in water and germinating in 0.01% solution of colchicine. Harvest of 1966, 1967, 1969 years

Year of harvest	Month, year studies	Number of investigating seedlings		Metaphases		Rearrangements of chromosomal type	
		Σ	With metaphases	Σ	With rearrangements, %±	Σ	%±
1966	XII,1966	25	25	1,125	0.44±0.198	0	-
	I, 1967	27	27	2,120	0.24±0.105	2	-
	III, 1967	43	42	1,692	0.24±0.118	0	-
Average:		*95*	94/**99.0%**	*4,937*	***0.31***±0.008	2	0.04±0.029
1967	IV, 1968	*49*	48	1,169	0.77±0.256	0	-
	VI,VII,1968	20	14	350	1.14±0.569	3	0.86±0.492
	III, 1969	20	18	1,316	2.96±0.468	24	1.73±0.369
Average:		89	80/**89.9%**	2,835	**1.62**±0.214	27	**0.81**±0.155
1969	IV,1970	*33*	32	1,384	1.59±0.006	21	1.52±0.329
Total (1996-1969гг):		*217*	*206(**94.9** ±1.49%)*	*9,156*	***1.05***±0.134	*50*	***0.55***±0.080

We do not take into account, single metaphase with numerous damages of the spindle and chromosomes.

On the average, over the years and months, were studied 217 seedlings. Of these, 94.9% contained metaphases. The frequency of metaphases with rearrangements in 1966 was in the range 0.24–0.44%. In 1967, there was a clear tendency to increase the frequency of rearrangements, as they are stored. Mitotic activity in 1967 was about 90%. Number of rearrangements of chromosomal type was various and small. For all the years discovered 50 such changes (see Table 11.1). It should be noted a tendency to increase their number during the aging of seeds (harvest 1967, 19 months of storage). The average level of metaphases with rearrangements in different years was less than 2%, rearrangements of chromosomal type in average of 0.55%; the largest number of them was in 1967.

As the count the nuclei in the seedlings after treatment by phosphemid at concentrations of $1 \cdot 10^{-2}$ to $5 \cdot 10^{-4}$ M of 68,500 counted nuclei were 743 metaphases, an average of 1.08 ± 0.06%, thus phosphemid suppressed mitotic activity more than doubled (see above 2.82% in control).

After exposure to seeds by phosphemid the average for all concentrations and of fixations analyzed 17,513 seedlings with metaphases, among them were discover 2,306 metaphases with rearrangements (13.17%). Alterations of chromosomal type made less than 1–0.113%. This magnitude is similar to the natural frequency. At a concentration of phosphemid of $2 \cdot 10^{-3}$ M mitotic activity was on the average 51.5% (Table 11.2), that is, two times lower than in the controls (94.9%) (cf Table 11.1).

TABLE 11.2 Aberrations of chromosomes in 2*n*-meristem cells of *Crepis capillaris* after 24 and 27 hrs from the start of treatment of seeds in a solution of phosphemid $2 \cdot 10^{-3}$M (2.24 mg, 50 ml of water). Fixation through 3–8 hrs after "arising", Seeds were germinating in 0.01% solution of colchicine. Harvest of 1967 year. The analysis in April 1968

Time, in hours		Number of investigating seedlings			Metaphases		Rearrangements of chromosomal type
From soaking up to "arising"	From soaking up to fixation	Σ	With metaphases		Σ	With rearrangements, %±	
			Σ	%			
24	3	57	24	42.1	741	9.4±1.08	0
	6	56	32	57.1	1,270	17.2±1.06	1
	8	54	36	66.7	1,815	14.0±0.82	0
27	3	47	16	34.0	373	20.6±2.10	1
	5	19	12	63.2	313	15.3±2.04	0
Total		233	120	51.5±3.28	4,512	14.8±0.53	0.043±0.031%

Data of Table 11.2 show that at weak concentration of preparation—$2 \cdot 10^3$ with the increase of term of fixation in every "arising" mitotic activity increases, and frequency of alterations diminishes.

At maximum concentration—$1 \cdot 10^{-2}$M (Table 11.3), we registered the large number of rearrangements—nearly 60% through 27hr after "arising" in term 24hr, of which 16.4% were mitosis with multiple rearrangements (it is marked an asterisk). On average, significant increase the number of strongly damaged metaphases was 13.3±2.9%. Rearrangements of chromosomal type at this dilution also did not find.

Average frequency of mitotic activity of phosphemid in a concentration of $1 \cdot 10^{-2}$M at the same quantity of seeds, as in Table 11.2, was lower almost twice—25% (Table 11.3). Average number of metaphases with rearrangements increased—nearly 22% versus 14.8%.

Thus the average level of metaphases with chromatid aberrations was increasing with increasing of dose of preparation.

TABLE 11.3 Rearmaments of the chromosomes in the 2*n*-meristem cells of *Crepis capillaris* seedlings at 24 and 27hrs from the start of treatment of seeds in a solution of phosphemid $1 \cdot 10^{-2}$M (22.4mg, 10ml water). Fixation through 3–24hr after "arising". Seeds were germinating in 0.01% solution of colchicine. Harvest 1967. The analysis: April 1968

Time, in hours		Number of investigated seedlings		Metaphases		
From soaking up to "arising"	From soaking up to fixation	Σ	With metaphases	Σ	With rearrangements	
					Σ	%±
24	3	21	6	148	25	16.9±3.09
	6	19	3	365	24(2*)	6.6±1.23
27	3	20	2	73	7(2*)	9.59±3.47
	24	12	7	185	110(18*)	59.46±3.62
Total		72	16/**25.0**±5.14%	771	166(22*)	**21.53**±2.85

*Number of greatly damaged metaphases with multiple rearrangements.

At maximum concentration ($1 \cdot 10^{-2}$ M), we registered the large number of rearrangements—59.46% through 27hr after "arising", of which 16.4% were mitosis with multiple rearrangements (it is marked an asterisk). On average, significant increase the number of strongly damaged metaphases was 13.3±2.9%. Rearrangements of chromosomal type do not arise in spite of increasing concentration of preparation.

During storage of untreated seeds and after subsequent treatment of seeds—namely in June and July of 1968 patterns of mitotic activity and frequencies of rearrangements were different, despite the use of the same concentration of phosphemid: $2 \cdot 10^{-2}$M (Figures 11.3a, b and 4a, b). Mitotic activity was above than in April; with increasing time from "arising" to fixation after 12hrs frequency of seedlings with mitoses close to 90% and was slightly below of control.

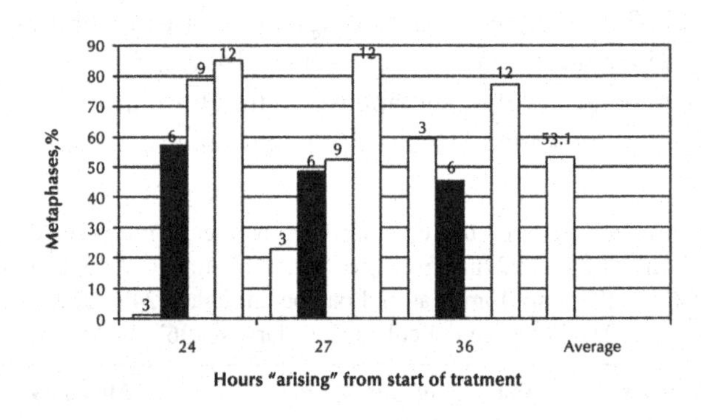

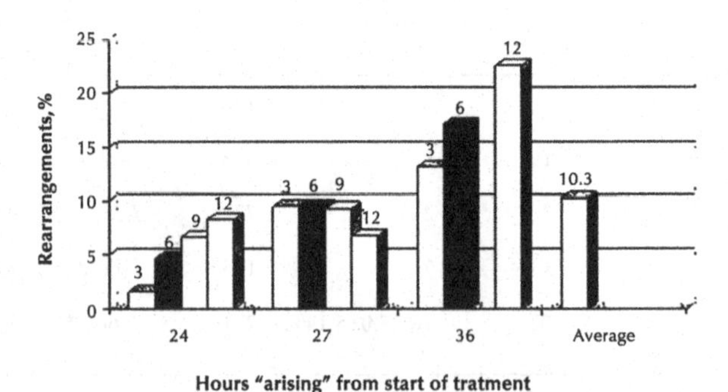

FIGURE 11.3 *Crepis capillaris*: the mitotic indexes in seedlings (*a*) and rearrangements of chromosomes in metaphase (*b*) after 24, 27, and 36 hrs from the start of seed soaking in a solution of phosphemid $2 \cdot 10^{-2}$ M (22.4 mg, 50 ml of water); Fixation: through 3–12 hrs after "arising" Seeds germinate in 0.01% colchicine. Seeds are of harvest 1967. Analysis: June 1968

Through 24 and 27hrs from the start of soaking of seeds the mitotic activity steadily was increasing to depending on term of fixation in each "arising". Later "arising"—36 hrs (see Figure 11.3*a*) or 31 hrs (see Figure 11.4a) mitotic activity was fluctuating at a highest level as compared as at the early stages of "arising" and terms of fixation from "arising" These data on Figures 11.3 and 4 are statistically significant. The frequency of metaphases with rearrangements increased steadily in the fraction of the 24 hr from "arising"; from 3 to 12 hrs (see Figure 11.3*b*) and from 3 to 9 hrs (see Figure 11.4*b*).

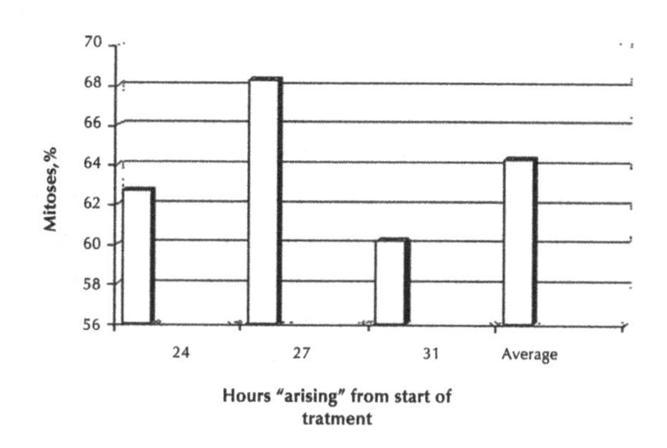

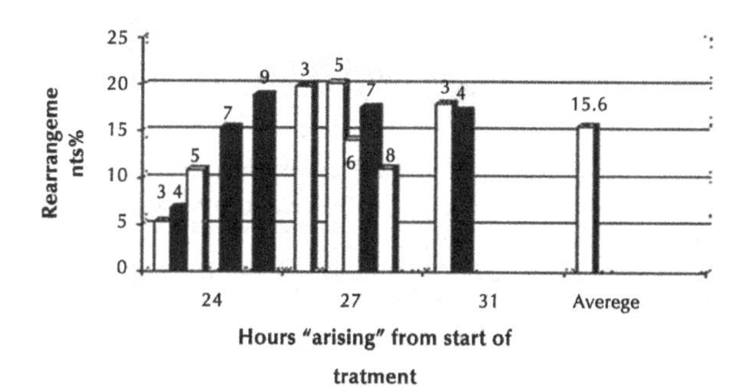

FIGURE 11.4 *Crepis capillaris*: the mitotic indexes in seedlings (*a*) and rearrangement of chromosomes in metaphase 2*n*-meristem cells (*b*) after 24, 27, and 36 hrs from the start of seed soaking in a solution of phosphemid $2{\cdot}10^{-2}$ M (22.4mg, 50ml of water). Fixation: through 3–9 hrs after seedling. Seeds were germinating in 0.01% colchicine. Seeds are of harvest 1967. Analysis: July 1968

Through 24 hr after the soakage of seeds the frequency of metaphases with rearrangements was growing steadily with time of fixing after "arising", average frequency of rearrangements was higher through 27, 31, and 36 hrs (see Figure 11.3*b* and 11.4*b*).

At the 36 hr "arising", frequency of metaphases with rearrangements increases with time from "arising" before fixation (see Figure 11.3*b*). Through 12 hr after "arising", the level was almost 23% in 31 hr "arising" (see Figure 11.4*b*), at 31 hr "arising" (see Figure 11.4*b*), at a fixations through 3 and 4 hr frequency of metaphases with rearrangements was at a high level—17–18%.

At all terms of fixation was observing rearrangements of chromatid type, and only solitary rearrangements of the chromosomal type was found—an average of less than 1%, what is similar to the control level. This fact is important, because shows, that the chemical preparation do not breaks chromosomes before synthesis DNA.

From the data of Figures 11.3 and 11.4 at the first "arising"—24hr under any fixation and concentration the frequency of aberrations increased from 3hr and more. The level of rearrangements in the early fixations was lower than in the later them, apparently due to the fact that the mutagen affected the chromosome the smallest period of time.

Facts reflected in Figures 11.3 and 11.4, it is possible to explain 1) that preparation does not influence on chromosomes during phase G_1, but remains in seeds; 2) chromosomes are damaged by preparation regardless of the stage of mitotic cycle, but this damage shows up only during synthesis of DNA (presence exclusively chromatid rearrangements). The number of the broken areas of chromosomes increases with the increase of terms of fixation; 3) existence of both factors is possible. Works of B. N. Sidorov and N. N. Sokolov will be described below, convincingly explaining an origin exceptionally of chromatid alterations.

Increased mitotic activity of the seedlings with increasing terms of fixation indicates the possibility of washout of the preparation from the cells during the growth of seedlings. Increasing number of rearrangements in "arisings" in conformance with increasing time before fixation and kipping of high level of metaphases with rearrangements in the seedlings obliges to suggest that phosphemid penetrating in the seeds from the beginning of their treatment, is included in the metabolism of cells and stores there in a long time.

Increased number of rearrangements in “arising” in conformance with increasing time before fixation and kipping of high level of metaphases with rearrangements in the seedlings obliges to suggest that phosphemid penetrates in the seeds from the beginning of their treatment, is included in the metabolism of cells and remains there during a long time.

Phosphemid suppressed mitotic activity then stronger than higher is its concentration (see Tables 11.2 and 11.3, Figures 11.3 and 11.4). Phosphemid also interacts with proteins of spindle, since at higher concentrations of the preparation significantly increased the number of heavily damaged mitoses (see Table 11.3). Often such metaphases form a sort of star at the center of the cell. A similar pattern sometimes was observed in the culture of fibroblasts. In addition to these disruptions, in some cells we have seen all the chromosomes were fragmented.

Mitotic activity relatively of the number of metaphases and of number of non-dividing nuclei through 2–10hrs after “arising” in the control amounted to 2.11% at 16,000 nuclei. After treatment of the seeds the number of metaphases in the seedlings over the same period (2–10 hrs) was lower and depended on the concentration of phosphemid. At a high concentration of $1 \cdot 10^{-2}$ M at an average around 16,000 nuclei was 0.73% of metaphases, with preparation concentration $5 \cdot 10^{-3}$ M— 0.82% of metaphases around 26,500 nuclei, at a concentration of $5 \cdot 10^{-4}$ M observed 1.42% of metaphases around 21,000 nuclei. However, by 20 hr after the start of treatment increased the frequency of mitoses both in control—3.40%, and in the experience at concentration of phosphemid $5 \cdot 10^{-3}$ M—2.38%. These data also reflect the decline of mitotic activity with the increase of concentration of phosphemid and her increasing with reduction of concentration of preparation.

Mutagen may be remains in the seeds in conjunction with other cellular proteins, by that braking advancement of phases of mitotic cycle, thus influencing on the chromosomes longer and therefore stronger is damaged the synthesis of larger number of loci of chromosomes.

In the 60–70th year’s outstanding scientists N. N. Sokolov, B. N. Sidorov [19-21] realized a series of studies on effects of ethylene imine on seedlings *Cr. capillaris*. They cultivated seedlings in a solution of colchicine for five cell generations. They found in tetraploid and higher polyploidy cells rearrangements of chromatid type—"not multiplied" under influence of colchicines.

In the 60—70th year's outstanding scientists N. N. Sokolov, B. N. Sidorov [19-21] was conducting workings about effects of ethylene imine on seedlings *Cr. capillaris*. They cultivated seedlings in a solution of colchicine for five cell generations [19]. They found in tetraploid and higher polyploidy cells new rearrangements of chromatid type, which were arising de novo, but not as result duplicating.

The authors explain this phenomenon is the fact that the mutagen is saved in the cells and there are new rearrangements. In a next article [20] seedlings were treating of ethylene imine, these seedlings were washed in running water within 2 hr. From these through 48hr were preparing "thin gruel". Intact seedlings of *Cr. capillaris* were treating by that thin gruel. In these seedlings treated by thin gruel were appearing rearrangements of chromatid type.

The authors suggested that ethylene imine formed active secondary mutagens, connecting with the components of the cell, including with nucleic acids. In the third article of this series [21], the authors *in vitro* added ethylene imine to amino acids: glycine and histidine, to the hexamine (hexamethylenetetramine), to vitamins: thiamin (vitamin B_1), nicotinic acid.

Consequent treatment of seedlings of these preparations did not cause aberrations. The frequency of them was at the level of control. Ethylene imine (concentration 0.05%) caused about 15% of rearrangements, while in mixture with thiamine caused significantly more rearrangements (+22.27%). Treatment by ethylene imine in mixture with nicotinic acid, glycine, and histidine showed even some protective effect. Treatment by ethylene imine together with adenine, guanine (derivative of purine), cytosine (derivative of pyrimidine) showed absence of effect or a small excess of rearrangements above the level of rearrangements of pure ethylene imine.

Received significant excess frequency of rearrangements under the influence of mixtures: uracil + ethylene imine gave an increase of nearly 19%, thymine + ethylene imine caused nearly 61% of rearrangements (+37.25%); guanine and cytosine in a mixture with ethylene imine gave an insignificant action. In the mixture with thio-TEPA (three ethylene imine groups), only thiamine gave significant excess frequency of rearrangements (+25.97%) over control (ethylene imine). After treatment of seedlings by a mixture of ethylene imine with thymine, the excess frequency rearrangements not found. All the rearrangements were chromatid type. The authors explain this phenomenon of the formation of secondary mutagens in cells. Thusly these experiments [19-21] were showing that mutagens do not cause aberrations of chromosomes before or after phases of the mitotic cycle G_2, G_1 and causes only chromatid aberrations effecting on the chromosomes in the course of DNA synthesis. Modern microscopy suggests that mutagen penetrating into the cell is remaining in "space around a chromosome" (see review [22]) in bonding with proteins or DNA, but its effect is revealing when mutagen passes through phase of DNA synthesis and becomes discovered in the form of chromosome rearrangements during mitosis. The same can be evidence of the phenomenon of fragmentation of chromosomes, strong destruction of mitosis and the spindle during treatment by high doses of the alkylating agent (in our case by phosphemid) in culture of fibroblasts of human and mouse or in germinal cells of seeds *Cr. capillaris*. Perhaps the same principle is working in the course of "chain process" of Dubinin. Someday, 4D-microscopy will reveal the mechanism of interaction chromosomes with the chemical mutagens.

Understanding the mechanism of chemical mutagenesis in the present time is fundamentally important in view of the global contamination of the surrounding nature by different chemicals that damage the genetic structure of organisms.

11.1.4 CONCLUSION

1. We have previously shown that in cultivated fibroblasts phosphemid was suppressing mitotic index, induces rearrangements of chromosomes.
2. In seedlings *Crepis capillaris* phosphemid also causes inhibition of the mitotic cycle. The average number of metaphases on the number of nuclei in seedlings after treatment phosphemid decreased twice.

3. Phosphemid after treatment of seeds *Cr. capillaris* causes rearrangements in the cells of seedlings, regardless of age treated seeds, but depending on the concentrations of drug. The greatest number of rearrangements occurs when using the highest concentration of the preparation.
4. After treatment of the seeds *Cr. capillaris* by phosphemid were found rearrangements only chromatid type in seedlings. The number of rearrangements of chromosomal type was at the level of controls or smaller. This means that the preparation works as well as other chemical mutagens, i.e., chromosomes break during DNA synthesis.
5. Phosphemid in treatment seeds *Cr. capillaris* showed heterogeneity of germinal cells in the seeds during the G_1 phase. The frequency of chromosomal rearrangements varies depends on the time between fixation and "arising".
6. At high concentrations ($1 \cdot 1^{-2}$ M) phosphemid caused the destruction of mitotic spindle and multiple fractures of the chromosomes.

KEYWORDS

- **Acetous carmine**
- **Chromosomes**
- ***Crepis capillaris***
- **Metaphases**
- **Phosphemidum**

REFERENCES

1. Ross, W. (1962). *Biological alkylating agents, Fundamental chemistry and design of compounds for selective toxicity* (p. 259). London: Moscow Medicine. Translation edited by A.Ja. Berlin's (1964).
2. Loveless, A. (1966). *Genetic allied effects of alkylating agents* (p. 255). London. Moscow: Nauka. Translation edited by N.P. Dubinin's (1970).

3. Stroeva, O. G. (1912–1990). *Josef Abramowitz Rapoport* (p. 215). Moscow: Nauka.
4. Josef Abramowitz Rapoport. (2003). *Scientist, warrior, citizen. Essays, memoirs, materials* (p. 335). Compiled by O. G. Stroeva. Moscow: Nauka.
5. Kihlman, B. A. (1963). *Aberrations induced by radiomimetic compounds and their relations to radiation induced aberrations. In: Radiation-induced chromosome aberrations* (p. 260). New York.Springer
6. Eiges, N. S. (2008). *Characteristic features of chemical mutagenesis method I.A. Rapoport and its use in breeding of winter wheat* (pp. 14–17). Penza.
7. Eiges, N. S., Vaysfel'd, L. I., Volchenko, G. A., & Volchenko, S. G. (2010). *Some aspects of the securities chemomutant characters a collection of winter wheat and characterization of these characters*. International teleconference number 1: Basic Science and Practice. January 2010. Website: http://tele-conf.ru/nasledstvennyie-morfologicheskie-kletochnyie-fakto/aspektyi-ispolzovaniya-tsennyih-hemomutantnyih-priznakov-kollektsii-ozimoy-pshenitsyi-i-ih-har-ka-chast-1.html.
8. Kalaev, V. N., Butorina, A. K., & Sheluhina, O. Yu. (2006). Assessment of anthropogenic pollution areas Staryj Oskol on cytogenetic parameters of birch seed family. *Ecological Genetics, IV*(2), 9–21.
9. Dubinina, L. G. (1978). *Structural mutations in the experiments with Crepis capillaris* (p. 187). Moscow: Nauka.
10. Dubinin, N. P. (1966). Some key questions of the modern theory of mutations. *Genetika, 7,* 3–20.
11. Dubinin, N. P. (1971). Unresolved issues of modern molecular theory of mutations. *Proc USSR,* (2), 165–178.
12. Dubinin, N. P. (1971). Unresolved issues of modern molecular theory of mutations. *Proc USSR,* (3), 333–344.
13. Chernov, V. A. (1964). Cytotoxic substances in chemotherapy of malignant tumors (p. 320). Moscow: Medicine.
14. Chernov, V. A., Grushina, A. A., & Lytkina, L. G. (1963). Antineoplastic activity of phosphasin. *Pharmacology and Toxicology, 26*(1), 102–108.
15. Weisfeld, L. I. (1965). Cytogenetic phosphasin effect on human cells and mice in tissue culture. *Genetika, 4,* 85–92.
16. Weisfeld, L. I. (1968). Influence phosphasin on the duration of phases of the mitotic cycle of human cells and mice in culture. *Genetika, IV*(7), 119–125.
17. Navashin, M. S. (1985). Problems karyotype and cytogenetic studies in the genus *Crepis* (p. 349). Moscow: Nauka.
18. Protopopova, E. M., Shevchenko, V. V., & Generalova, M. V. (1967). Beginning of DNA synthesis in seeds *Crepis capillaris. Genetika, 6,* 19–23.
19. Sidorov, B. N., Sokolov, N. N., & Andreev, V. A. (1965). Mutagenic effect of ethylene imine in a number of cell generations. *Genetika, 1,* 121–122.
20. Andreev, V. S., Sidorov, B. N., & Sokolov, N. N. (1966). The reasons for long-term mutagenic action of ethylenimine. *Genetika, 4,* 28–36.
21. Sidorov, B. N., Sokolov, N. N., Andreev, V.,A. (1966). Highly active secondary alkylating mutagens. *Genetika, 7,* 124–133.
22. Rubtsov, N. B. (2007). Human chromosome in four dimensions. *Priroda, 8,* 1–8

CHAPTER 12

MATHEMATICAL MODELS ON THE TRANSPORT PROPERTIES OF ELECTROSPUN NANOFIBERS

A. K. HAGHI and G. E. ZAIKOV

CONTENTS

Electrospinning comprises an efficient and versatile technique for fabrication of very thin fibers from polymers or composites. During the recent years more attentions have focused on optimization of this method to solve the problems which make electrospinning uncontrollable. Modeling can help us to achieve this approach. This chapter gives a general outlook of mathematical models for electrospinning and their applications.

Symbols	Definition
ρ_m	Mass density
P	Pressure
t	Time
γ	Surface tension
η	Viscosity
τ	Stress tensor
ζ	Surface deflection
Q	constant volume flow rate
v	Jet velocity
I	The current carried by the jet
K	Electrical conductivity
E	Electric field
σ	Surface charge density
ε	The dielectric constants of the jet
$\tilde{\varepsilon}$	The dielectric constants of the ambient air
r_i	The position of bead i
F_C	The net coulomb force
F_E	The electric field force
F_{ve}	The viscoelastic force
F_B	The surface tension force
F_q	The Lorenz force

12.1 INTRODUCTION

Electrospinning (e.g., vibration-electrospinning, magneto-electrospinning and bubble-electrospinning) is a simple and relatively inexpensive mean of manufacturing high volume production of very thin fibers (more typically 100nm to 1μm) and lengths up to kilometers from a vast variety of materials including polymers, composites and ceramics [1, 2]. Electrospinning technology was first developed and patented by Formhals [3] in the 1930s, and a few years later the actual developments were triggered by Reneker and co-workers [4]. To satisfy the increasing needs for the refined nanosize hybrid fibers based on commercial polymers, various electrospinning techniques have been investigated and developed [5]. Presently, there are two standard electrospinning setups, vertical and horizontal. With the development of this technology, several researchers have developed more intricate systems that can fabricate more complex nanofibrous structures in a more controlled and efficient style [6]. The unique properties of nanofibers are extraordinarily high surface area per unit mass, very high porosity, tunable pore size, tunable surface properties, layer thinness, high permeability, low basic weight, ability to retain electrostatic charges and cost effectiveness [2]. In this method nanofibers produce by solidification of a polymer solution stretched by an electric field [7–9] which can be applied in different areas including wound dressing, drug or gene delivery vehicles, biosensors, fuel cell membranes and electronics, tissue-engineering processes [6, 8, 10]. Electrospinning has proven to be the best nanofiber manufacturing process because of simplicity and material compatibility (Figure 1.1, [11]).

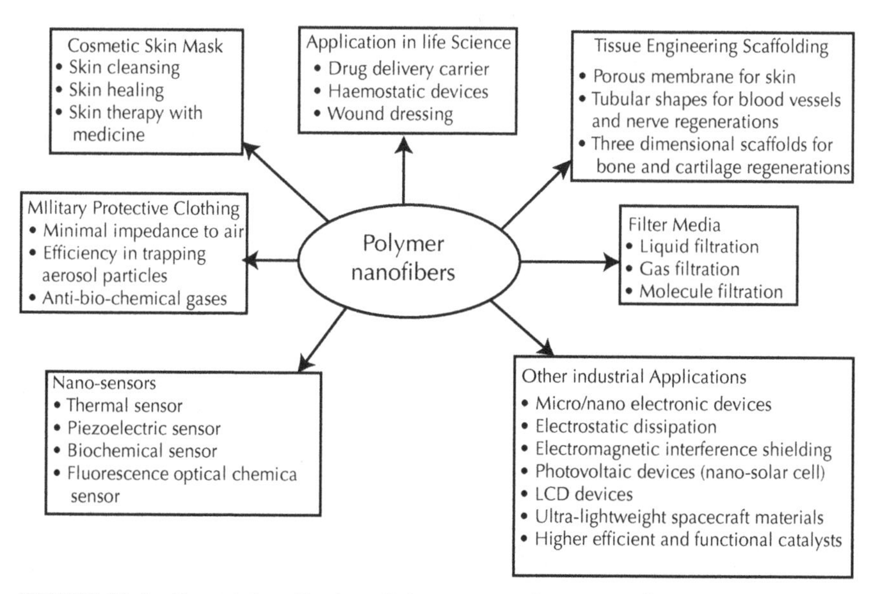

FIGURE 12.1 Potential application of electrospun polymer nanofibers

Generally in this process (Figure 1.2), a polymer solution or melt is supplied through a syringe ~10–20 cm above a grounded substrate. The process is driven by an electrical potential of the order of kilovolts applied between the syringe and the substrate [10]. Electrospinning of polymer solutions involves, to a first approximation, a rapid evaporation of the solvent. The evaporation of the solvent thus will happen on a time scale well below the second-range [12]. The elongation of the jet during electrospinning is initiated by electrostatic force, gravity, inertia, viscosity and surface tension [1]. Unlike, traditional spinning process which principally uses gravity and externally applied tension, electrospinning uses externally applied electric field, as driving force [5, 11].

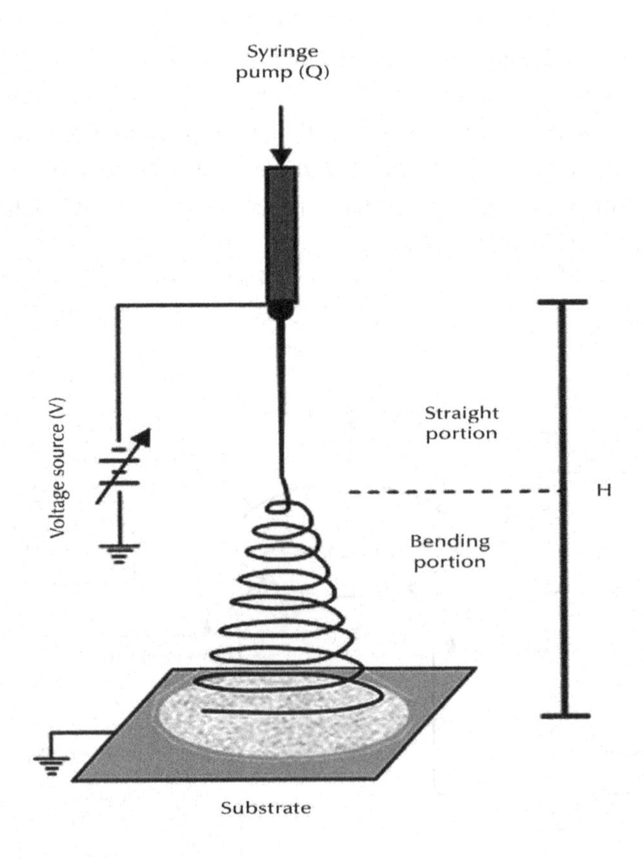

FIGURE 12.2 A scheme of electrospinning set up for nanofiber production

The applied voltage induces a high electric charge on the surface of the solution drop at the needle tip. The drop now experiences two major electrostatic forces: The columbic force which is induced by the electrical field, and electrostatic repulsion between the surface charges. The intensity of these forces causes the hemispherical surface of the drop to elongate and form a conical shape, which is also known as the Taylor cone. By further increasing the strength of the field, the electrostatic forces in the drop will overcome the surface tension of the fluid/air interface and an electrically charged jet will be ejected [13]. After the ejection, the jet elongates and the solvent evaporates, leaving an electrically charged fiber which, during the elongation process, becomes very thin [1, 2, 12, 14]. Study of effects of various electrospinning parameters is important to improve the rate of nanofiber processing. In addition, several applications demand well-oriented nanofibers [11]. Theoretical studies about the stability of an isolated charged liquid droplet predicted that it becomes unstable and fission takes place when the charge becomes sufficiently large compared to the stabilizing effect of the surface tension [15, 16].

The characteristic feature of this process is the onset of a chaotic oscillation of the elongating jet which is due to the electrostatic interactions between the external electric field and the surface charges on the jet as well as the electrostatic repulsion of mutual fiber parts. The fiber can be spun directly onto the grounded (conducting) screen or on an intermediate deposit material. Because of the oscillation, the fiber is deposited randomly on the collector, creating a so called "non-woven" fiber fabric [1, 17]. An important stage of nanofibers formation in electrospinning include fluid instabilities such as whipping instabilities [17]. The applied voltage, V, fluid flow rate, Q, and separation distance, H, are manipulated such that a steady, electrostatically driven jet of fluid is drawn from the capillary tip and is collected upon the grounded substrate [2, 10, 17]. These instabilities depend on fluid parameters and equipment configuration such as location of electrodes and the form of spinneret [17].

In melt or dry/wet solution spinning the shape and diameter of the die, as well as mechanical forces inducing specific draw ratios and drawing speeds highly determine dimensional and structural properties of the final fibers [18]. Properties that are known to significantly affect the electrospinning process are the polymer molecular weight, the molecular-weight distribution, the architecture (branched, linear, etc.) of the polymer, temperature and humidity and air velocity in the chamber and processing parameters (like applied voltage, flow rate, types of collectors, tip to collector distance) as well as the rheological and electrical properties of the solution (viscosity, conductivity, surface tension, etc.) and finally motion of target screen [2, 6, 8, 19].

An important weak point of this method is a convective instability in the elongating jet. The jet will start rapidly whipping as it travels towards the collector. Therefore, during the electrospinning process, the whole substrate is covered with a layer of randomly placed fiber. The created fabric has a chaotic structure and it is difficult to characterize its properties [1, 11]. Electrospun fibers often have beads as "by products" [20]. Some polymer solutions are not readily electrospun when the polymer solution is too dilute, due to limited solubility of the polymer. In these cases, the lack of elasticity of the solution prevents the formation of uniform fibers; instead, droplets or necklace-like structures know as 'beads-on-string' are formed [21]. The electrospun beaded fibers are related to the instability of the jet of polymer solution. The bead diameter and spacing were related to: the fiber diameter, solution viscosity, net charge density carried by the electrospinning jet and surface tension of the solution [20, 21]. Important findings which were obtained during studies are: (i) fibers of different sizes, that is, consisting of different numbers of parent chains, exhibit almost identical hyperbolic density profiles at the surfaces, (ii) the end beads are predominant and the middle beads are depleted at the free surfaces, (iii) there is an anisotropy in the orientation of bonds and chains at the surface, (iv) the centre of mass distribution of the chains exhibits oscillatory behavior across the fibers and (v) the mobility of the chain in nanofiber increases as the diameter of the nanofiber decreases [19].

It is necessary to develop theoretical and numerical models of electrospinning because of demanding a different optimization procedure for each material [8]. Modeling and simulation of electrospinning process will help to understand the following:

a) The cause for whipping instability.
b) The dependence of jet formation and jet instability on the process parameters and fluid properties, for better jet control and higher production rate.

c) The effect of secondary external field on jet instability and fiber orientation [2, 11].

Several techniques such as dry rotary electrospinning [22] and scanned electrospinning nano-fiber deposition system [23] control deposition of oriented nanofibers. In this chapter, some basic and necessary theories for electrospinning modeling are reviewed.

12.2 MATERIALS AND METHODS

12.2.1 THE BASICS OF ELECTROSPINNING MODELING

Modeling of the electrospinning process will be useful for the factors perception that cannot be measured experimentally [24]. Although electrospinning gives continuous fibers, mass production and the ability to control nanofibers properties are not obtained yet. In electrospinning the nanofibers for a random state on the collector plate while in many applications of these fibers such as tissue engineering well-oriented nanofibers are needed. Modeling and simulations will give a better understanding of electrospinning jet mechanics [11]. The development of a relatively simple model of the process has been prevented by the lack of systematic, fully characterized experimental observations suitable to lead and test the theoretical development [17]. The governing parameters on electrospining process which are investigated by modeling are solution volumetric flow rate, polymer weight concentration, molecular weight, the applied voltage and the nozzle to ground distance [1, 7, 25]. The macroscopic nanofiber properties can be determined by multiscale modeling approach. For this purpose, at first the effective properties determined by using modified shear lag model then by using of volumetric homogenization approach, the macro scale properties concluded [26].

Till date two important modeling zones have been introduced. These zones are: a) The zone close to the capillary (jet initiation zone) outlet of the jet and b) The whipping instability zone where the jet spirals and accelerates towards the collector plate [11, 24, 27].

The parameters influence the nature and diameter of the final fiber so obtaining the ability to control them is a major challenge. For selected applications it is desirable to control not only the fiber diameter, but also the internal morphology [28]. An ideal operation would be: the nanofibers diameter to be controllable, the surface of the fibers to be intact and a single fiber would be collectable. The control of the fiber diameter can be affected by the solution concentration, the electric field strength, the feeding rate at the needle tip and the gap between the needle and the collecting screen (Figure 1.3, [1, 12, 29]).

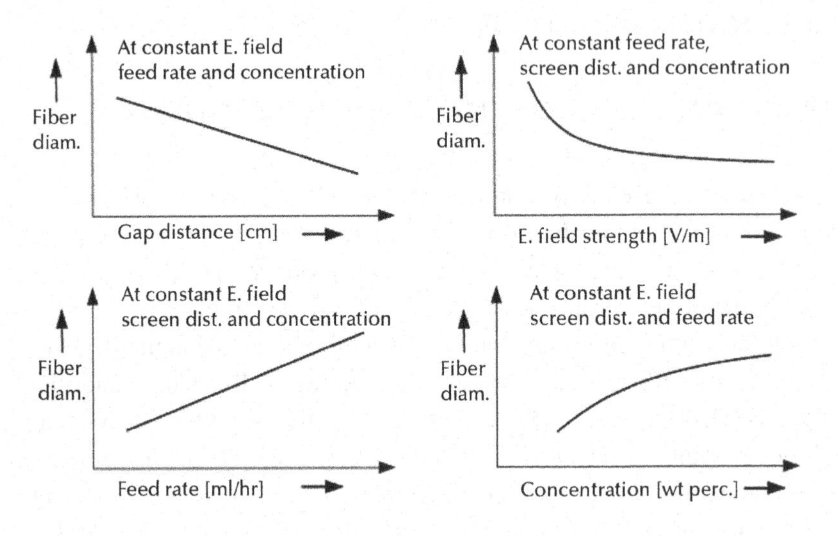

FIGURE 12.3 Effect of process parameters on fiber diameter

Control over the fiber diameter remains a technological bottleneck. However A cubic model for mean fiber diameter was developed for samples by A. Doustgani et al. A suitable theoretical model of the electrospinning process is one that can show a strong-moderate-minor rating effects of these parameters on the fiber diameter. Some disadvantages of this method are low production rate, non oriented nanofiber production, difficulty in diameter prediction and controlling nanofiber morphology, absence of enough information on rheological behavior of polymer solution and difficulty in precise process control that emphasis necessity of modeling [25, 29, 30].

VISCOELASTIC FLOWS ANALYSIS

Recently, significant progress has been made in the development of numerical algorithms for the stable and accurate solution of viscoelastic flow problems, which exits in processes like electrospinning process. A limitation is made to mixed finite element methods to solve viscoelastic flows using constitutive equations of the differential type [31].

GOVERNING SET OF EQUATIONS

The analysis of viscoelastic flows includes the solution of a coupled set of partial differential equations: The equations depicting the conservation of mass, momentum and energy, and constitutive equations for a number of physical quantities present in the conservation equations such as density, internal energy, heat flux, stress, and so on depend on process [31].
Approaches to Viscoelastic Finite Element Computations

There are fundamentally different approaches like: A mixed formulation may be adopted different parameters like velocity, pressure, and so on including the constitutive equation, is multiplied independently with a weighting function and transformed in a weighted residual form [31]. The constitutive equation may be transformed into an ordinary differential equation (ODE). For transient problems this can, for instance, be achieved in a natural manner by adopting a Lagrangian formulation [32].

TIME DEPENDENT FLOWS

By introducing a selective implicit/explicit treatment of various parts of the equations, a certain separating at each time step of the set of equations may be obtained to improve computational efficiency. This suggests the possibility to apply devoted solvers to sub-problems of each fractional time step [31].

BASICS OF HYDRODYNAMICS

Due to the reason that nanofibers are made of polymeric solutions forming jets, it is necessary to have a basic knowledge of hydrodynamics [33]. According to the effort of finding a fundamental description of fluid dynamics, the theory of continuity was implemented. The theory describes fluids as small elementary volumes which still are consisted of many elementary particles.

The equation of continuity:

$$\frac{\partial \rho_m}{\partial t} + div(\rho_m v) = 0 \text{ (For incompressible fluids } div(v) = 0\text{)}$$

The Euler's equation simplified for electrospinning:

$$\frac{\partial v}{\partial t} + \frac{1}{\rho_m} \nabla p = 0$$

The equation of capillary pressure:

$$P_c = \frac{\gamma \partial^2 \zeta}{\partial x^2}$$

The equation of surface tension:

$$\Delta P = \gamma \left(\frac{1}{R_X} + \frac{1}{R_Y} \right)$$ R_x and R_y are radii of curvatures

The equation of viscosity:

$$\tau_{ij} = \eta \left(\frac{\partial v_i}{\partial x_j} + \frac{\partial v_j}{\partial x_i} \right)$$ (For incompressible fluids, $\tau_{i,j}$ = Stress tensor)

$$V = \frac{\eta}{\rho_m}$$ (Kinematic viscosity)

ELECTROHYDRODYNAMIC (EHD) THEORY

In 1966 Taylor discovered that finite conductivity enables electrical charge to accumulate at the drop interface, permitting a tangential electric stress to be generated. The tangential electric stress drags fluid into motion, and thereby generates hydrodynamic stress at the drop interface. The complex interaction between the electric and hydrodynamic stresses causes either oblate or prolate drop deformation, and in some special cases keeps the drop from deforming [34, 35].

Feng has used a general treatment of Taylor–Melcher for stable part of electrospinning jets by one dimensional equations for mass, charge and momentum. In this model a cylindrical fluid element is used to show electrospinning jet kinematic measurements [10].

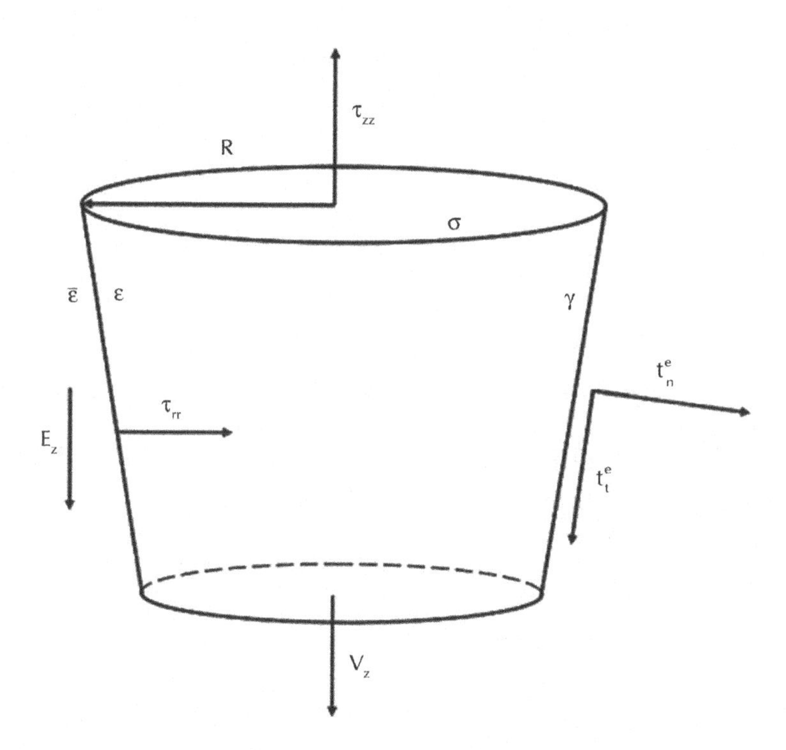

FIGURE 12.4 Scheme of the cylindrical fluid element used in electrohydrodynamic modeling

In Figure 1.4, the essential parameters are: radius, R, velocity, v_z, electric field, E_z, total path length, L, interfacial tension, γ, interfacial charge, σ, tensile stress, τ, volumetric flow rate, Q, conductivity, K, density, ρ, dielectric constant, ε, and zero-shear rate viscosity, η_0. The most important equation that Feng used are [10]:

$$\tilde{R}^2\tilde{v}_z = 1$$

$$\tilde{R}^2\tilde{E}_z + Pe_e\tilde{R}\tilde{v}_z\tilde{\sigma} = 1$$

$$\tilde{\upsilon}_z\tilde{\upsilon}'_z = \frac{1}{Fr} + \frac{\tilde{T}'}{\mathrm{Re}_j\tilde{R}^2} + \frac{1}{We}\frac{\tilde{R}'}{\tilde{R}^2} + \varepsilon(\tilde{\sigma}\tilde{\sigma}' + \beta\tilde{E}_z\tilde{E}'_z + \frac{2\tilde{\sigma}\tilde{E}_z}{\tilde{R}})$$

$$\tilde{E}_z = \tilde{E}_0 - \ln\chi\left[(\tilde{\sigma}\tilde{R})' - \frac{\beta}{2}(\tilde{E}\tilde{R}^2)''\right]$$

$$E_0 = {}^{I}{}_0V_0/R_0 \quad \beta = (\varepsilon/\tilde{\varepsilon}) - 1 \quad \tau = R^2(\tilde{\tau}_{zz} - \tilde{\tau}_{rr}).$$

Feng solved above equation under different fluid properties, particularly for non-Newtonian fluids with extensional thinning, thickening, and strain hardening but Helgeson et al. developed a simplified understanding of electrospinning jets based on the evolution of the tensile stress due to elongation [10].

ELECTRIC FORCES IN FLUIDS

The initialization of instability on the surfaces of liquids should be caused using the application of the external electric field that induces electric forces on surfaces of liquids. A localized approximation was developed to calculate the bending electric force acting on an electrified polymer jet, which is an important element of the electrospinning process for manufacturing of nanofibers. Using this force, a far reaching analogy between the electrically driven bending instability and the aerodynamically driven instability was established. The description of the wave's instabilities is expressed by equations called dispersion laws. The dependence of wavelength on the surface tension γ is almost linear and the wavelengths between jets are a little bit smaller for lower depths. Dependency of wavelength on electric field strength is exponential. The dispersion law is identified for four groups of dielectrics liquids with using of Clausius-Mossotti and Onsager's relation (nonpolar liquids with finite and infinite depth and weakly polar liquids with finite and infinite depth). According to these relations relative permittivity is a function of parameters like temperature, square of angular frequency, wave length number and reflective index [2, 33, 36, 37].

DIMENSIONLESS NON-NEWTONIAN FLUID MECHANICS

The best way for analyzing fluid mechanics problems is converting parameters to dimensionless form. By using this method the numbers of governing parameters for given geometry and initial condition reduce. The non-dimensionalization of a fluid mechanics problem generally starts with the selection of a characteristic velocity then because the flow of non-Newtonian fluids the stress depends non-linearly on the flow kinematics, the deformation rate is a main quantity in the analysis of these flows. Next step after determining different parameters is to evaluate characteristic values of the parameters. Then the non-dimensionalization procedure is to employ the selected characteristic quantities to obtain a dimensionless version of the conservation equations and get to certain numbers like Reynolds number and the Galilei number. The excessive number of governing dimensionless groups poses certain difficulties in the fluid mechanics analysis. Finally by using these results in equation and applying boundary conditions it can be achieved to study different properties [38].

DETECTION OF X-RAY GENERATED BY ELECTROSPINNING

Electrospinning jets producing nanofibres from a polymer solution by electrical forces are fine cylindrical electrodes that create extremely high electric-field strength in their vicinity at atmospheric conditions. However, this quality of electrospinning is only scarcely investigated, and the interactions of the electric fields generated by them with ambient gases are nearly unknown. Pokorny et al. reported on the discovery that electrospinning jets generate X-ray beams up to energies of 20 keV at atmospheric conditions [39]. Mikes et al. investigated on the discovery that electrically charged polymeric jets and highly charged gold-coated nanofibrous layers in contact with ambient atmospheric air generate X-ray beams up to energies of hard X-rays [40, 41].

The first detection of X-ray produced by nanofiber deposition was observed using radiographic films. The main goal of using these films understands of Taylor cone creation. The X-ray generation is probably dependent on diameters of the nanofibers that are affected by the flow rate and viscosity. So it is important to find the ideal concentration (viscosity) of polymeric solutions and flow rate to spin nanofibers as thin as possible. The X-ray radiation can produce black traces on the radiographic film. These black traces had been made by outer radiation generated by nanofibers and the radiation has to be in the X-ray region of electromagnetic spectra, because the radiation of lower energy is absorbed by the shield film. Radiographic method of X-ray detection is efficient and sensitive. It is obvious that this method did not tell us anything about its spectrum, but it can clearly show its space distribution. The humidity, temperature and rheological parameters of polymer can affect on the X-ray intensity generated by nanofibers [33]. The necessity of modeling in electrospinning process and a quick outlook of some important models will be discussed as follows.

12.2.2 MODELING ELECTROSPINNING OF NANOFIBERS

Using theoretical prediction of the parameter effects on jet radius and morphology can significantly reduce experimental time by identifying the most likely values that will yield specific qualities prior to production [25]. All models start with some assumptions and have short comings that need to be addressed [14]. The basic principles for dealing with electrified fluids that Taylor discovered, is impossible to account for most electrical phenomena involving moving fluids under the seemingly reasonable assumptions that the fluid is either a perfect dielectric or a perfect conductor. The reason is that any perfect dielectric still contains a nonzero free charge density. The presence of both an axisymmetric instability and an oscillatory "whipping" instability of the centerline of the jet; however, the quantitative characteristics of these instabilities disagree strongly with experiments [19, 36]. During steady jetting, a straight part of the jet occurs next to the Taylor cone, where only axisymmetric motion of the jet is observed. This region of the jet remains stable in time. However, further along its path the jet can be unstable by non-axisymmetric instabilities such as bending and branching, where lateral motion of the jet is observed in the part near the collector [10].

Branching as the instability of the electrospinning jet can happen quite regularly along the jet if the electrospinning conditions are selected appropriately. Branching is a direct consequence of the presence of surface charges on the jet surface, as well as of the externally applied electric field. The bending instability leads to drastic stretching and thinning of polymer jets towards nano-scale in cross-section. Electrospun jets also caused to shape perturbations similar to undulations, which can be the source of various secondary instabilities leading to nonlinear morphologies developing on the jets [18]. The bending instabilities that occur during electrospinning have been studied and mathematically modeled by Reneker et al by viscoelastic dumbbells connected together [42]. Both electrostatic and fluid dynamic instabilities can contribute to the basic operation of the process [19].

Different stages of electrospun jets have been investigated by different mathematical models during last decade by one or three dimensional techniques [8, 43].

Physical models which study the jet profile, the stability of the jet and the cone-like surface of the jet have been develop due to significant effects of jet shape on fiber qualities [7]. Droplet deformation, jet initiation and, in particular, the bending instability which control to a major extent fiber sizes and properties are controlled apparently predominantly by charges located within the flight jet [18]. An accurate, predictive tool using a verifiable model that accounts for multiple factors would provide a means to run many different scenarios quickly without the cost and time of experimental trial-and error [25].

AN OUTLOOK TO SIGNIFICANT MODELS

The models typically treat the jet mechanics using the localized-induction approximation by analogy to aerodynamically driven jets. They include the viscoelasticity of the spinning fluid and have also been augmented to account for solvent evaporation in the jet. These models can describe the bending instability and fiber morphology, Because of difficulty in measure model variables they cannot accurately design and control the electrospinning process [10]. Here are some current models:

LEAKY DIELECTRIC MODEL

The principles for dealing with electrified fluids were summarized by Melcher and Taylor [36]. Their research showed that it is impossible to explain the most of the electrical phenomena involving moving fluids given the hypothesis that the fluid is either a perfect dielectric or a perfect conductor, since both the permittivity and the conductivity affect the flow. An electrified liquid always includes free charge. Although the charge density may be small enough to ignore bulk conduction effects, the charge will accumulate at the interfaces between fluids. The presence of this interfacial charge will result in an additional interfacial stress, especially a tangential stress, which in turn will modify the fluid dynamics [36, 44].

The electrohydrodynamic theory proposed by Taylor as the leaky dielectric model is capable of predicting the drop deformation in qualitative agreement with the experimental observations [34, 35].

Although Taylor's leaky dielectric theory provides a good qualitative description for the deformation of a Newtonian drop in an electric field, the validity of its analytical results is strictly limited to the drop experiencing small deformation in an infinitely extended domain. Extensive experiments showed a serious difference in this theoretical prediction [34].

Some investigations have been done to solve this defect. For example, to examine electrokinetic effects, the leaky dielectric model was modified by consideration the charge transport [44, 45]. When the conductivity is finite, the leaky dielectric model can be used [45]. By use of this mean,

Saville indicated that the solution is weakly conductive so the jet carries electric charges only on its surface [44, 45].

A MODEL FOR SHAPE EVALUATION OF ELECTROSPINNING DROPLETS

Comprehension of the drops behavior in an electric field is playing a critical role in practical applications. The electric field-driven flow is of practical importance in the processes in which improvement of the rate of mass or heat transfer between the drops and their surrounding fluid [34]. Numerically investigations about the shape evolution of small droplets attached to a conducting surface depended on strong electric fields (weak, strong and super electrical) have done and indicated that three different scenarios of droplet shape evolution are distinguished, based on numerical solution of the Stokes equations for perfectly conducting droplets by investigation of Maxwell stresses and surface tension [13, 46]. The advantages of this model are that the non-Newtonian effect on the drop dynamics is successfully identified on the basis of electrohydrostatics at least qualitatively. In addition, the model showed that the deformation and breakup modes of the non-Newtonian drops are distinctively different from the Newtonian cases [34].

NONLINEAR MODEL

A simple two-dimensional model can be used for description of formation of barb electrospun polymer nanowires with a perturbed swollen cross-section and the electric charges "frozen" into the jet surface. This model was integrated numerically using the Kutta–Merson method with the adoptable time step. The result of this modeling is explained theoretically as a result of relatively slow charge relaxation compared to the development of the secondary electrically driven instabilities which deform jet surface locally. When the disparity of the slow charge relaxation compared to the rate of growth of the secondary electrically driven instabilities becomes even more pronounced, the barbs transform in full scale long branches. The competition between charge relaxation and rate of growth of capillary and electrically driven secondary localized perturbations of the jet surface is affected not only by the electric conductivity of polymer solutions but also by their viscoelasticity. Moreover, a nonlinear theoretical model was able to resemble the main morphological trends recorded in the experiments [18].

A MATHEMATICAL MODEL FOR ELECTROSPINNING PROCESS UNDER COUPLED FIELD FORCES

There is not a theoretical model which can describe the electrospinning process under the multi-field forces so a simple model might be very useful to indicate the contributing factors. Modeling this process can be done in two ways: a) the deterministic approach which uses classical mechanics like Euler approach and Lagrange approach. b) The probabilistic approach uses E-infinite theory and quantum like properties. Many basic properties are harmonious by adjusting electrospinning parameters such as voltage, flow rate and others, and it can offer in-depth inside into physical understanding of many complex phenomena which cannot be fully explain [9].

SLENDER-BODY MODEL

One-dimensional models for inviscid, incompressible, axisymmetric, annular liquid jets falling under gravity have been obtained by means of methods of regular perturbations for slender or long jets, integral formulations, Taylor's series expansions, weighted residuals, and variational principles [27, 47].

Using Feng's theory some familiar assumptions in modeling jets and drops are applied: The jet radius R decreases slowly along z direction while the velocity v is uniform in the cross section of the jet so it is lead to the nonuniform elongation of jet. According to the parameters can be arranged into three categories: process parameters (Q, I and E_∞), geometric parameters (R_0 and L) and material parameters (ρ, η_0 (the zero-shear-rate viscosity), ε, $\tilde{\varepsilon}$, K, and γ). The jet can be represented by four steady-state equations: the conservation of mass and electric charges, the linear momentum balance and Coulomb's law for the E field.

Mass conservation can be stated by:

$$\pi R^2 \upsilon = Q \,.$$

R: Jet radius

The second equation in this modeling is charge conservation that can be stated by:

$$\pi R^2 KE - 2\pi R v c = I \,..$$

The linear momentum balance is:

$$\rho \upsilon \upsilon' = \rho g + \frac{3}{R^2}\frac{d}{dz}(\eta R^2 v) + \frac{\gamma R'}{R^2} + \frac{\sigma c'}{\bar{\varepsilon}} + (\varepsilon - \bar{\varepsilon})EE' + \frac{2cE}{R} \,.$$

The Coulomb's law for electric field:

$$E(z) = E_{\infty}(z) - \ln \chi\left(\frac{1}{\bar{\varepsilon}}\frac{d(\sigma R)}{dz} - \frac{\beta}{2}\frac{d^2(ER^2)}{dz^2}\right).$$

L: The length of the gap between the nozzle and deposition point
R_0: The initial jet radius

$$\beta = \varepsilon / \bar{\varepsilon} - 1$$

$$\chi = L / R_0$$

By these four equations the four unknown functions R, v, E and σ are identified.

At first step the characteristic scales such as R_0 and v_0 are denoted to format dimensionless groups. Inserting these dimensionless groups in four equations discussed above the final dimensionless equations are obtained.

The boundary conditions of four equations which became dimensionless can be expressed (see Figure 1.5) as:

$$\ln z = 0 \quad R(0) = 1 \quad E'(0) = 0$$

$$\ln z = \chi \quad E(\chi) = E_{\infty} .$$

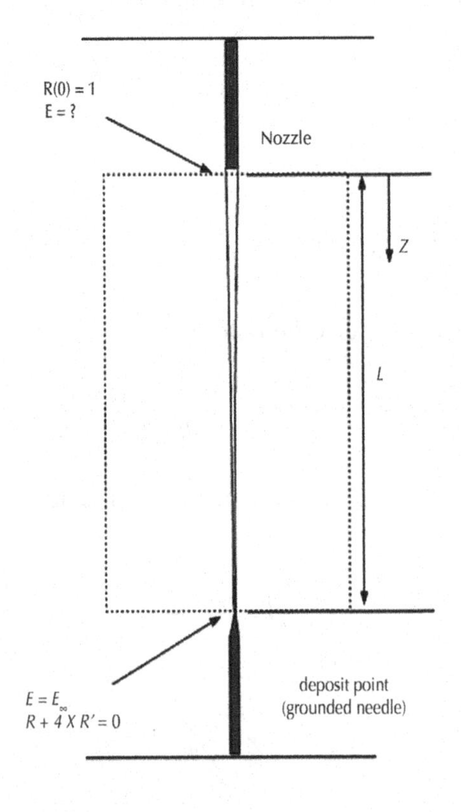

FIGURE 12.5 Boundary conditions

The first step is to write the ODE's as a system of first order ODE's by a numerical relaxation method as for example, *solved* from numerical recipes. The basic idea is to introduce new variables, one for each variable in the original problem and one for each of its derivatives up to one less than the highest derivative appearing. For solving this ODE the Fortran program is used, in the first step an initial guess uses for χ and the other parameters would change according to χ [1, 48]. The limitation of slender-body theory is: avoiding treating physics near the nozzle directly [27].

A MODEL FOR ELECTROSPINNING VISCOELASTIC FLUIDS

When the jet thins, the surface charge density varies, which in turn affects the electric field and the pulling force. Roozemond combined the "leaky dielectric model" and the "slender-body approximation" for modeling electrospinning viscoelastic jet [41]. The jet could be represented by four steady-state equations: the conservation of mass and electric charges, the linear momentum balance and Coulomb's law for the electric field, with all quantities depending only on the axial position z. The equations can be converted to dimensionless form using some characteristic scales and dimensionless groups like most of the models. These equations could be solved by converting to ODE's forms and using suitable boundary conditions [27, 48].

LATTICE BOLTZMANN METHOD (LBM)

Developing lattice Boltzmann method instead of traditional numerical techniques like finite volume, finite difference and finite element for solving large-scale computations and problems involving complex fluids, colloidal suspensions and moving boundaries is so useful [11].

MATHEMATICAL MODEL FOR AC-ELECTROSPINNING

Much of the nanofiber research reported so far was on nanofibers made from DC potential [7]. In DC-electrospinning, the fiber instability or 'whipping' has made it difficult to control the fiber location and the resulting microstructure of electrospun materials. To overcome these limitations, some new technologies were applied in the electrospinning process.

The investigations proved that the AC potential resulted in a significant reduction in the amount of fiber 'whipping' and the resulting mats exhibited a higher degree of fiber alignment but were observed to contain more residual solvent. In AC-electrospinning, the jet is inherently unsteady due to the AC potential unlike DC ones, so all thermal, electrical, and hydrodynamics parameters was considered to be effective in the process [49, 50].

The governing equations for an unsteady flow of an infinite viscous jet pulled from a capillary orifice and accelerated by an AC potential can be expressed as follows:

1. The conservation of mass equation
2. Conservation of charge
3. The Navier-Stokes equation

Using these governing equations, final model of AC electrospinning is able to find the relationship between the radius of the jet and the axial distance from nozzle, and a scaling relation between fiber radius and the AC frequency [50].

MULTIPLE JET MODELING

It was experimentally and numerically exhibited that the jets from multiple nozzles expose higher repulsion by another jets from the neighborhood by Columbic forces than jets spun by a single nozzle process [5]. Yarin and Zussman achieved upward electrospinning of fibers from multiple jets without the use of nozzles; instead using the spiking effect of a magnetic liquid [51]. For large-scale nanofibre production and the increase in production rate, multi-jet electrospinning systems can be designed to increase both productivity and covering area [52, 53]. The linear Maxwell model and nonlinear Upper-Convected Maxwell (UCM) model were used to calculate the viscoelasticity. By using these models the primary and secondary bending instabilities can be calculated. Maxwell model and the non-linear UCM model lead to rather close results in the flow dominated by the electric forces. In a multiple-nozzle set up, not only the external applied electric field and self-induced Columbic interactions influence the jet path, but also mutual-Columbic interactions between different jets contribute [53].

A MATHEMATICAL MODEL OF THE MAGNETIC ELECTROSPINNING PROCESS

For controlling the instability, magnetic electrospinning is proposed. For describing the magnetic electrospun jet, it can be used Reneker's model [42]. This model does not consider the coupling effects of the thermal field, electric field and magnetic field. Therefore, the momentum equation for the motion of the beads is [54]

$$m\frac{d^2 r_i}{dt^2} = F_C + F_E + F_{ve} + F_B + F_q .$$

12.3 CONCLUSION

Electrospinning is a very simple and versatile method of creating polymer-based high-functional and high-performance nanofibers that can revolutionize the world of structural materials. The process is versatile in that there is a wide range of polymer and biopolymer solutions or melts that can spin. The electrospinning process is a fluid dynamics related problem. In order to control the property, geometry, and mass production of the nanofibers, it is necessary to understand quantitatively how the electrospinning process transforms the fluid solution through a millimeter diameter capillary tube into solid fibers which are four to five orders smaller in diameter. When the applied electrostatic forces overcome the fluid surface tension, the electrified fluid forms a jet out of the capillary tip towards a grounded collecting screen. Although electrospinning gives continuous nanofibers, mass production and the ability to control nanofibers properties are not obtained yet. Combination of both theoretical and experimental approaches seems to be promising step for better description of electrospinning process. Applying simple models of the process can be useful in atoning the lack of systematic, fully characterized experimental observations and the theoretical aspects in predicting and controlling effective parameters. The analysis and comparison of model with experiments identify the critical role of the spinning fluid's parameters. The theoretical and quantitative tools developed in different models provide semi- empirical methods for predicting ideal electrospinning process or electrospun fiber properties. In each model, researcher tried to improve the existing models or changed the tools in electrospinning by using another view. Therefore, it is attempted to have a whole view on important models after investigation about basic objects. A real mathematical model, or, more accurately, a real physical model, might initiate a revolution in understanding of dynamic and quantum-like phenomena in the electrospinning process. A new theory is much needed which bridges the gap between Newton's world and the quantum world.

KEYWORDS

- **Electrospinning**
- **Hydrodynamics**
- **Jet Instability**
- **Modeling**
- **Nanofibers**

REFERENCES

1. Solberg, R. H. M. (2007). *Position-controlled deposition for electrospinning*, in Department Mechanical Engineering (p. 67). Eindhoven: University of Technology Eindhoven.
2. Chronakis, I. S. *Processing, Properties and Applications, in Micro-/Nano-Fibers by Electrospinning Technology* (pp. 264–286).
3. Formhals, A. (1934). *Process and apparatus for preparing artificial threads*, U. Patent, Editor: Germany.
4. Reneker, D. H. & Chun, I. (1996). Nanometre diameter fibres of polymer produced by electrospinning. *Nanotechnology, 7,* 216–223.
5. Kim, G., Cho, Y. S., & Kim, W. D. (2006). Macromolecular nanotechnology, stability analysis for multi-jets electrospinning process modified with a cylindrical electrode. *European Polymer Journal, 42,* 2031–2038.
6. Bhardwaj, N. & Kundu, S. C. (2010). Research review paper, electrospinning: A fascinating fiber fabrication technique. *Biotechnology Advances, 28,* 325–347.

7. Theron, S. A., Zussman, E., & Yarin, A. L. (2004). Experimental investigation of the governing parameters in the electrospining of polymer solutions. *Polymer, 45,* 2017–2030.
8. Kowalewski, T. A., Barral, S., & Kowalczyk, T. (2009). Modeling Electrospinning of Nanofibers. *IUTAM Symposium on Modelling Nanomaterials and Nanosystems, 13,* 279–293.
9. Xu, L. (2009). A mathematical model for electrospinning process under coupled field forces. *Chaos, Solitons and Fractals, 42,* 1463–1465.
10. Helgeson, E. M., et al. (2008). Theory and kinematic measurements of the mechanics of stable electrospun polymer jets. *Polymer, 49,* 2924–2936.
11. Karra, S. (2007). *Modeling electrospinning process and a numerical scheme using lattice Boltzmann method to simulate viscoelastic fluild flows, in Mechanical Engineering* (p. 60). Madras: Indian Institute of Technology.
12. Bognitzki, M., et al. (2001). Nanostructured fibers via electrospinning. *Advanced Materials, 13,* 70–73.
13. Basaran, O. A. & Suryo, R. (2007). Fluid dynamics: The invisible jet. *Nature Physics, 3,* 679–680.
14. Titchenal, N. & Schrepple, W. Modeling of electro-spinning. *Materials Sciense and Engineering.*
15. Yarin, A. L., Koombhongse, S., & Reneker, D. H. (2001). Taylor cone and jetting from liquid droplets in electrospinning of nanofibers. *Journal of Applied Physics, 90,* 4836–4846.
16. Papageorgous, D. T. & Petropoulos, P. G. (2004). Generation of interfacial instabilities in charged electrified viscous liquid films. *Journal of Engineering Mathematics, 50,* 223–240.
17. Shin, Y. M., et al. (2001). Experimental characterization of electrospinning: The electrically forced jet and instabilities. *Polymer, 42,* 9955–9967.
18. Holzmeister, A., Yarin, A. L., & Wendorff, J. H. (2010). Barb formation in electrospinning: Experimental and theoretical investigations. *Polymer, 51,* 2769–2778.
19. Frenot, A. & Chronakis, I. S. (2003). Polymer nanofibers assembled by electrospinning. *Current Opinion in Colloid and Interface Science, 8,* 64–75.
20. Fong, H., Chun, I., & Reneker, D. H. (1999). Beaded nanofibers formed during electrospinning. *Polymer, 40,* 4585–4592.
21. Yu, H. J., Fridrikh, S. V., & Rutledge, G. C. (2006). The role of elasticity in the formation of electrospun fibers. *Polymer, 47,* 4789–4797.
22. El-Auf, A. K. (2004). *Nanofibers and nanocomposites poly (3,4-ethylene dioxythiophene)/poly(styrene sulfonate) by electrospinning* in *Department of Materials Science and Engineering* (p. 261). Philadelphia: Drexel University.
23. Czaplewski, D., Kameoka, J., & Craighead, H. G. (2003). Nonlithographic approach to nanostructure fabrication using a scanned electrospinning source. *Journal of Vacuum Science & Technology B (Microelectronics and Nanometer Structures), 21,* 2994–2997.
24. Patanaik, A., Jacobs, V., & Anandjiwala, R. D. (2008). *Experimental study and modeling of the electrospinning process, 86th Textile Institute World Conference* (pp. 1160–1168). Hong Kong.

25. Thompson, C. J., et al. (2007). Effects of parameters on nanofiber diameter determined from electrospinning model. *Polymer, 48,* 6913–6922.
26. Agic', A. (2012). Multiscale Modeling electrospun nanofiber structure. *Materials Science Forum, 714,* 33–40.
27. Feng, J. J. (2002). The stretching of an electrified non-Newtonian jet: A model for electrospinning. *Physics of Fluid, 14,* 3912–3926.
28. Helgeson, M. E. & Wagner, N. J. (2007). A correlation for the diameter of electrospun polymer nanofibers. *American Institute of Chemical Engineers,* 53, 51–55.
29. Doustgani, A., et al. (2012). Optimizing the mechanical properties of electrospun polycaprolactone and nanohydroxyapatite composite nanofibers. *Composites: Part B. 43,* 1830–1836.
30. Fridrikh, S. V., et al. (2003). Controlling the fiber diameter during electrospinning. *Physical Review Letters, 90,* 144502
31. Baaijens, P. T. F., (2001). *Mixed finite element methods for viscoelastic flow analysis: A review,* in *Faculty of Mechanical Engineering* (p. 37). Eindhoven: Eindhoven University of Technology Center for Polymers and Composites.
32. Rasmussen, H. K. & Hassager, O. (1993). Simulation of transient viscoelastic flow. *Journal of Non-Newtonian Fluid Mechanic, 46,* 298–305.
33. Mikeš, I. P. (2011). *Physical principles of electrostatic spinning.* In *Physical engineering* (p. 122). Liberec: Technical University in Liberec.
34. Ha, J. W. & Yang, S. M. (2000). Deformation and breakup of Newtonian and non-Newtonian conducting drops in an electric field. *Journal of Fluid Mechanics, 405,* 131–156.
35. Taylor, G. I. (1996). Studies in electrohydrodynamics. I. The circulation produced in a drop by an electric field. Proceedings of the Royal Society of London. *Series A, Mathematical and Physical Sciences, 291,* 159–166.
36. Melcher, J. R. & Taylor, G. I. (1969). Electrohydrodynamics: A review of the role of interfacial shear stresses. *Annual Review of Fluid Mechanics, 1,* 111–146.
37. Yarin, A. L., Koombhongse, S., & Reneker, D. H. (2001). Bending instability in electrospinning of nanofibers. *Journal of Applied Physics, 89,* 3018–3026.
38. De Souza, R. & Mendes, P. (2007). Dimensionless non-Newtonian fluid mechanics. *Journal Non-Newtonian Fluid Mechanics, 147,* 109–116.
39. Pokorn'y, P., Mikes, P., & Luk'aˇs, D. (2010). Electrospinning jets as X-ray sources at atmospheric conditions. *A Letter Journal Exploring the Frontiers of Physics, 92,* 47002–47007.
40. MIikes, P., et al. (2010). *High energy radiation emitted from nanofibers,* in *7th International Conference—TEXSCI.* Liberec, Czech Republic.
41. Kornev, K. G. (2011). Electrospinning: Distribution of charges in liquid jets. *Journal of Applied Physics, 110,* 124910–124915.
42. Reneker, D. H., et al. (2000). Bending instability of electrically charged liquid jets of polymer solutions in electrospinning. *Journal of Applied Physics, 87,* 4531–4547.
43. He, J. H., et al. (2007). Review, mathematical models for continuous electrospun nanofibers and electrospun nanoporous microspheres. *Polymer International, 56,* 1323–1329.
44. Saville, D. A. (1997). Electrohydrodynamics: The Taylor-Melcher leaky dielectric model. *Annual Review of Fluid Mechanics, 29,* 27–64.

45. Parageorgiou, T. D. & Broeck, J. M. V. (2007). Numerical and analytical studies of non-linear gravity-capillary waves in fluid layers under normal electric fields. *IMA Journal of Applied Mathematics, 72,* 832–853.
46. Reznik, S. N., et al. (2004). Transient and steady shapes of droplets attached to a surface in a strong electric field. *Journal of Fluid Mechanics, 516,* 349–377.
47. Ramos, J. I. (1996). One-dimensional models of steady, inviscid, annular liquid jets. *Applied Mathematical Modelling, 20,* 593–607.
48. Roozemond, P. C. (2007). *A model for electrospinning viscoelastic fluids in department of mechanical engineering* (p. 26). Eindhoven: Eindhoven University of Technology.
49. Shin, Y. M., et al. (2001). Electrospinning: A whipping fluid jet generates submicron polymer fibers. *Applied Physics Letters, 78,* 3–7.
50. Ji-Huan, H., Yue, W., & Ning, P. (2005). A mathematical model for preparation by AC-electrospinning process. *International Journal of Nonlinear Sciences and Numerical Simulation, 6,* 243–248.
51. Yarin, A. L. & Zussman, E. (2004). Upward electrospinning of multiple nanofibers without needles/nozzles. *Polymer, 45,* 2977–2980.
52. Varesano, A., Carletto, R. A., & Mazzuchetti, G. (2009). Experimental investigations on the multi-jet electrospinning process. *Journal of Materials Processing Technology, 209,* 5178–5185.
53. Theron, S. A., et al. (2005). Multiple jets in electrospinning: Experiment and modeling. *Polymer, 46,* 2889–2899.
54. Xu, L., Wu, Y., & Nawaz, Y. (2011). Numerical study of magnetic electrospinning processes. *Computers and Mathematics with Applications, 61,* 2116–2119.

CHAPTER 13

A PROMISING NEW CLASS OF POLYMER MATERIAL WITH PARTICULAR APPLICATION IN NANOTECHNOLOGY

A. K. HAGHI and G. E. ZAIKOV

CONTENTS

13.1 INTRODUCTION

In recent years there has been considerable research to improve poly(ethylene terephthalate) (PET) fabrics dyeability. It is known that PET fabrics are usually dyed using disperse dyes in the presence of a carrier or at elevated temperatures. Many attempts have been made during the last two decades to replace environmentally unfriendly carriers with non toxic chemicals [35]. Modification of PET fabrics by attaching additives or functional groups to the polymer molecules via grafting and copolymerization are the common methods to enhance the dyeability of PET fabrics [1–15].

Literature review showed that there has not been a previous report regarding the treatment of amine terminated hyperbranched polymers on PET fabric and study of its dyeability with acid dyes [16–28]. In the most recent investigation in this field, fiber grade PET was compounded with polyesteramide hyperbranched polymer and dyeability of resulted samples with disperse dyes was studied [29–35]. The results showed that the dyeability of dyed modified samples comprised of fiber grade PET films and a hyperbranched polymer (Hybrane H1500) were better than the neat PET and this was increased by increasing amount of hyperbranched polymer in presence or absence of a carrier. The dyeability of the samples was attributed to decrease in glass transition temperature for blended PET or hyperbranched polymer in comparison with neat PET [29].

In the first part of this chapter, novel hyperbranched polymer (HBP) with amine terminal group was synthesized from methyl acrylate and diethylene triamine by melt polycondensation. The obtained hyperbranched polymer was characterized using Fourier transform infrared spectroscopy (FTIR) and nuclear magnetic resonance (NMR). In this contribution for the first time, the synthesized hyperbranched polymer applied to PET fabric and the dyeability of HBP treated PET fabrics and untreated PET fabric with C.I. Acid Red 114 was investigated. Dyeing of PET fabric with acid dye is very helpful and attractive object.

In the second part, a different case study was presented to show how the treatment conditions can be affected the dyeability of PET fabrics. Response surface methodology (RSM) was used to obtain a quantitative relationship between selected treatment conditions and PET fabrics dyeability.

13.2 MATERIALS AND METHODS

13.2.1 CASE 1: SYNTHESIS, CHARACTERIZATION, AND APPLICATION OF HBP TO PET FABRICS

MATERIALS

Methyl acrylate, diethylene triamine of pure grade were obtained from Merck, Germany. All reagents used were analytically pure. Poly(ethylene terephthalate) (PET) fabric (28×19 count/cm^2) used throughout this work and before use it was treated with a solution containing 5g/l Na_2CO_3 and 1g/l of a non ionic detergent at 60°C for 30min to remove undesired materials. Distilled water was used for the treatments and washings. Acid dye (C.I. Acid Red 114) was provided by the Ciba Ltd. (Tehran, Iran) and used to evaluate the dye absorption behavior (dyeability) of samples. The chemical structure of AR 114 is shown in Figure 13.1.

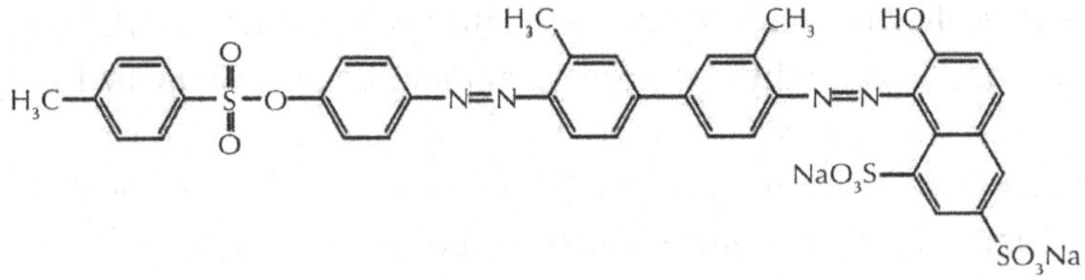

FIGURE 13.1 The chemical structure of C. I. Acid Red 114

13.2.2 MEASUREMENTS

The IR spectra of samples were recorded by Nicolet 670 FTIR spectrophotometer in the wave number range 500–4,000cm^{-1}. Nominal resolution for all spectra was 4cm^{-1}.

The ^1H and ^{13}C NMR spectra were recorded by Bruker AVANCE (500MHz for ^1H and 125MHz for ^{13}C) using DMSO-d_6 as a solvent at room temperature. All chemical shifts (δ) are expressed in ppm, the following abbreviations are used to explain the multiplicities: s = singlet, d = doublet, t = triplet, q = quartet, m = multiplet.

X-ray diffraction (XRD) patterns of hyperbranched-treated and un-treated polyester fabric were carried out with Equinox 3000 X-ray diffractometer with Cu Kα radiation (λ=0.154 nm) and operated at 40kV voltage and 30mA electric current.

The ultraviolet–visible (UV–vis) absorption spectrum of hyperbranched polymer was taken with a Carry 100 UV–vis spectrometer.

The zeta potential of the untreated polyester fabric and the HBP treated polyester fabrics were measured by a Molvern Zetasizer 3600 device. Fabric suspension prepared by pulling out the fibers from the fabric and cutting the fibers into lengths of approximately 0.5mm. Then the small piece of fibers soaked in deionized water for 24hr.

The color parameters of dyed samples were obtained under illuminant D_{65} at 10° standard observer in the visible range using Color-Eye 7000A spectrophotometer. The Kubelka–Munk single constant theory was used to calculate K/S values at the wavelength of maximum absorption (λ_{max}) for each fabric as following equation

$$\frac{K}{S}=\frac{(1-R)^2}{2R}-\frac{(1-R_0)^2}{2R_0} \tag{1}$$

where, R is the reflectance value at the wavelengths of maximum absorbance (λ_{max}), K is the absorption coefficient and S is the scattering coefficient.

Washing fastness test was performed according to ISO/R 105/IV, Part 10. Rubbing fastness test was performed according to ISO/R 105/IV, Part 18.

Contact angle measurement of treated samples was carried out by the following methods.

13.2.3 SYNTHESIS

SYNTHESIS OF AB_2 TYPE MONOMER

Diethylene triamine (0.5 mol, 52 ml) was added in a three-necked flask equipped with a constant-voltage dropping funnel, condenser, and a nitrogen inlet. The flask was placed in an ice bath and solution of methyl acrylate (0.5 mol, 43 ml) in methanol (100 ml) was added dropwise into the flask. The reaction mixture was stirred with a magnetic stirrer. Then the mixture was removed from the ice bath and left to react with a flow of nitrogen at room temperature (Scheme 1). After stirring for 4hrs, the nitrogen flow was removed and AB_2 type monomer was obtained.

IR ν_{max} (cm^{-1}, KBr), 3300–3500 (N–H), 2956, 2832, 1735 (C=O).

SYNTHESIS OF HYPERBRANCHED POLYMER

The obtained light yellow and viscous mixture was transferred to an eggplant-shaped flask for an automatic rotary vacuum evaporator to remove the methanol under low pressure. Then the temperature was raised to 150°C using an oil bath and condensation reaction was carried out for 6hrs. A pale yellow viscous hyperbranched polymer was obtained. The ^{1}H and ^{13}C-NMR spectrum was in accordance with the proposed structure.

IR ν_{max} (cm^{-1}, KBr), 3,300–3,500 (N–H), 1,631 (C=O). ^{1}H NMR (ppm, DMSO-d_6), 4.95 (s), 3.45 (m), 2.64 (m). ^{13}C NMR (ppm, DMSO-d_6), 171.98, 59.86, 57.71, 40.68–39.68 (group).

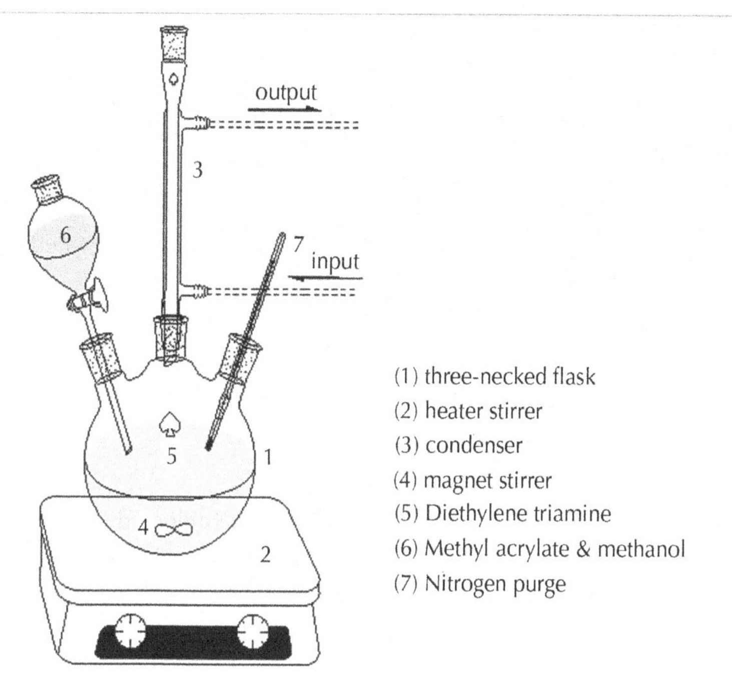

SCHEME 13.1 Schematic of reactor for monomer synthesis

13.2.4 PET FABRIC TREATMENT WITH HYPERBRANCHED POLYMER

PET fabrics were immersed in sodium hydroxide solution (10% w/v) at 94°C for 1hr, using a liquor ratio of 40:1. After the alkaline treatment of fabrics, the samples were thoroughly rinsed with distilled water and neutralized with acetic acid, then rinsed and dried at room temperature. Application of hyperbranched polymer to original PET fabric and alkali-treated PET fabric (HPET) was carried out using exhaust method. PET samples were immersed in a hyperbranched polymer solution (20g/dl) and the temperature was raised at a rate of 2.5°C/min. After 60min, the samples were thoroughly rinsed with distilled water to remove unfixed hyperbranched polymer and dried at room temperature. Figure 2.2 shows this application profile.

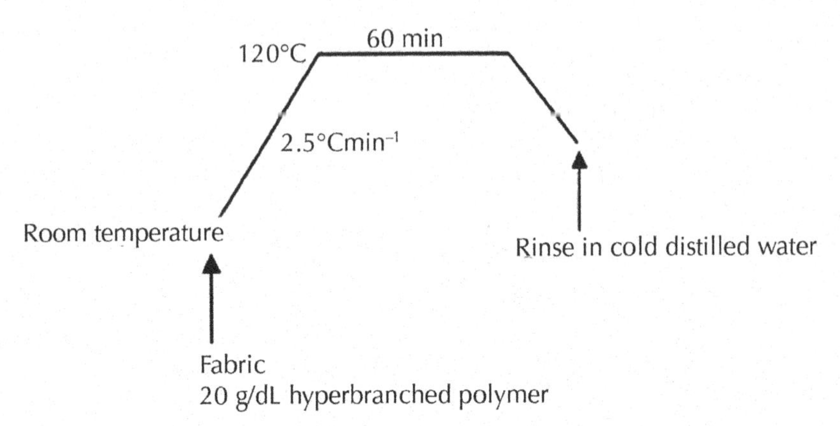

FIGURE 13.2 Hyperbranched polymer application profile

The HBP treated and untreated PET fabrics were subjected to five repeated washing fastness test for 30 min using sodium carbonate (2g/l) and detergent (5g/l) at 60°C, as shown in Figure 2.3.

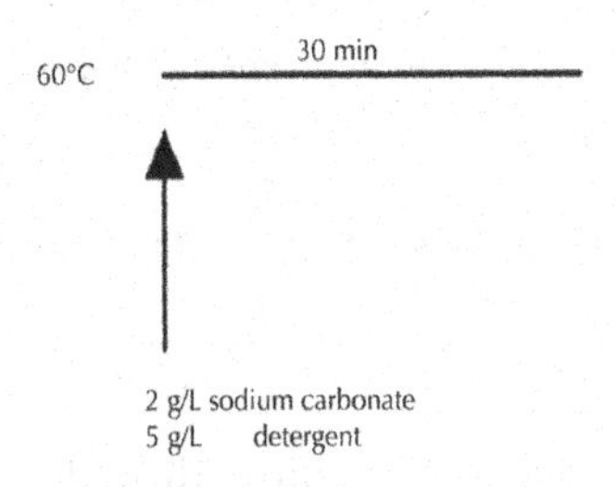

FIGURE 13.3 Washing fastness test profile

13.2.5 DYEING PROCEDURE WITH ACID DYE

The dyeability of the hyperbranched treated PET fabric and untreated polyester fabric were examined using C.I. Acid Red 114 (AR 114). All samples were dyed competitively in the same dye bath. The treated PET fabrics were introduced into the dye bath at 40°C, then the temperature was gradually raised up to boiling point within 20 min, and then dyeing was continued for 60 min with occasional stirring. Dyeing process was carried out using a liquor ratio of 60:1. At the end of dyeing the dyed samples were rinsed with cold water, then with hot water at about 50°C and finally rinsed with tap water. Figure 2.4 show the dyeing profile.

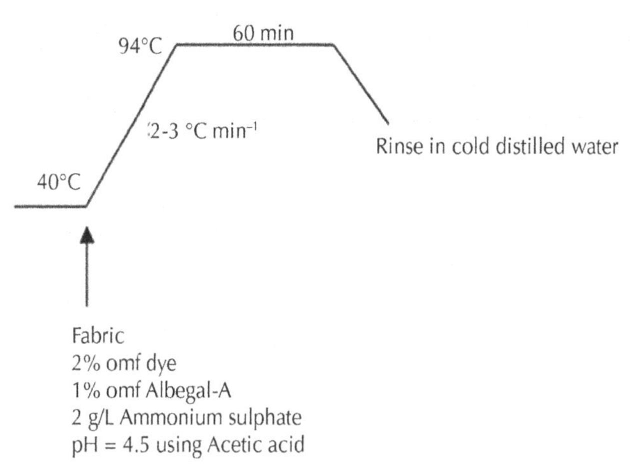

FIGURE 13.4 Dyeing method

13.2.6 CASE 2: EXPERIMENTAL DESIGN AND OPTIMIZATION OF TREATMENT CONDITIONS

Response surface methodology (RSM) is a combination of mathematical and statistical techniques used to evaluate the relationship between a set of controllable experimental factors and observed results. This optimization process involves three major steps: (i) performing statistically designed experiments, (ii) estimating the coefficients in a mathematical model, and (iii) predicting the response and checking the adequacy of the model [36]. RSM is used in situations where several input variables influence some output variables (responses) of the system. The main goal of RSM is to optimize the response, which is influenced by several independent variables, with minimum number of experiments. RSM, which is a powerful and useful experimental design tool, is capable of fitting a second-order (quadratic) prediction equation for the response. Central Composite Design (CCD), first introduced by Box and Wilson is the most common type of second-order designs that used in RSM and is appropriate for fitting a quadratic surface [37].

In the present study CCD was employed for the optimization of PET fabrics treated with hyperbranched polymer as shown in Table 2.1. The experiment was performed for at least three levels of each factor to fit a quadratic model. Based on preliminary experiments, hyperbranched polymer solution concentration (X_1), treatment temperature (X_2), and time (X_3) were determined as critical factors with significance effect on treated PET fabrics. These factors were three independent variables and chosen equally spaced, while the fabrics dyeability (K/S value) was dependent variable (response). A schematic of experimental design is shown in Figure 2.5. -1, 0, and 1 are coded variables corresponding to low, intermediate and high levels of each factor, respectively. The actual design experiment for three independent variables listed in Table 2.2.

TABLE 13.1 Design of experiment (factors and levels)

Factor	Variable	Unit	Factor level		
			-1	0	1
X_1	Hyperbranched polymer concentration	(wt.%)	2	6	10
X_2	Temperature	(°C)	90	110	130
X_3	Time	(min)	45	60	75

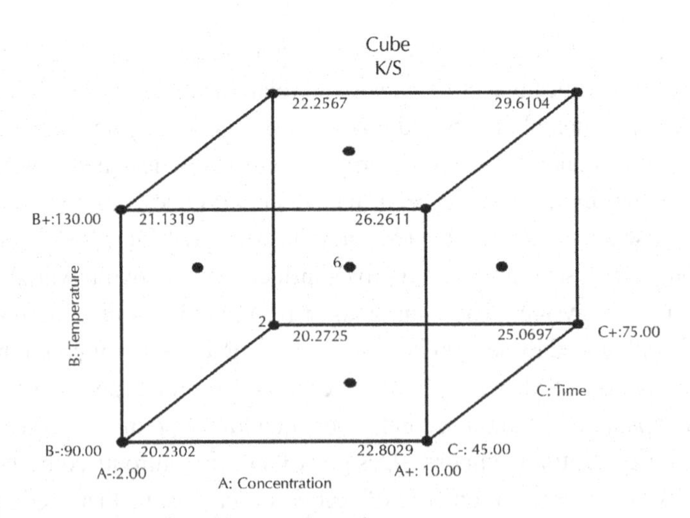

FIGURE 13.5 Design of experiment schematic

The following quadratic model, which also includes the linear model, was fitted to the data presented in Table 2.2:

$$Y = \beta_0 + \sum_{i=1}^{k} \beta_i . x_i + \sum_{i=1}^{k} \beta_{ii} . x_i^2 + \sum_{i<j=2} \sum^{k} \beta_{ij} . x_i x_j + \varepsilon \quad (2)$$

where, Y is the predicted response, x_i and x_j are coded variables, β_0 is constant coefficient, β_i is the linear coefficients, β_{ii} is the quadratic coefficients, β_{ij} is the second-order interaction coefficients, k is the number of factors, and ε is the approximation error.

The experimental data in Table 2.2 were analyzed using Design-Expertò software including analysis of variance (ANOVA). The values of coefficients for parameters (β_0, β_i, β_{ii}, β_{ij}) in equation 2 and p-values are calculated by regression analysis. The coefficient of determination (R^2) has been explained by the regression model.

TABLE 13.2 Design of experiment (coded values)

Expt. no.	Coded value		
	X_1	X_2	X_3
1	-1	-1	1
2	-1	1	1
3	-1	0	0
4	-1	1	-1
5	-1	-1	-1
6	0	0	0
7	0	0	0
8	0	0	0
9	0	0	0
10	0	0	0
11	0	0	0
12	0	0	-1
13	0	0	1
14	0	-1	0
15	0	1	0
16	1	-1	-1
17	1	-1	1
18	1	1	1
19	1	1	-1
20	1	0	0

13.3 RESULTS AND DISCUSSION

13.3.1 SYNTHESIS OF HYPERBRANCHED POLYMER

As mentioned earlier, preparation of hyperbranched polymer requires two-step reaction comprise preparation of AB2 type monomers (1 and 2) by Michael addition reaction of methyl acrylate and diethylene triamine and synthesis of hyperbranched polymer by polycondensation reaction respectively (Scheme 2). It is well known that in hyperbranched polymers, the structural units include terminal (T), dendritic (D), and linear (L) units that are shown in Scheme 2.

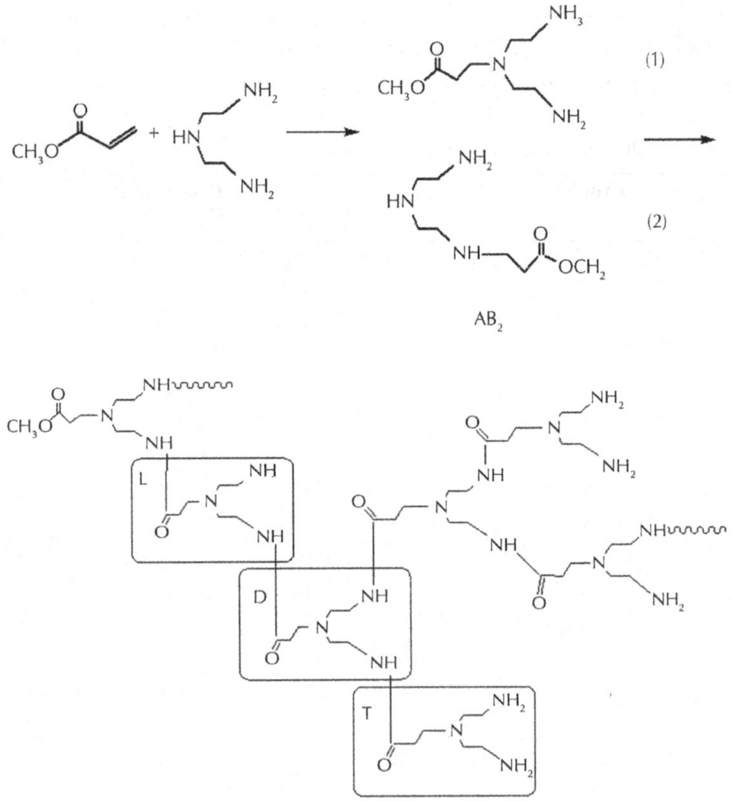

SCHEME 13.2 Chemical structure of amine terminated hyperbranched polymer

13.3.2 FTIR SPECTROSCOPY CHARACTERIZATION

HYPERBRANCHED POLYMER ANALYSIS

To confirm the polycondensation reaction progress, AB_2 type monomer and hyperbranched polymer were examined by FTIR spectroscopy. Spectral differences between AB_2 type monomer and hyperbranched polymer are observed in the fingerprint region between 1800 and 900 cm^{-1}. The absorption bands in FTIR spectra are assigned according to literature [38]. FTIR spectra of the AB_2 type monomer and the hyperbranched polymer are given in Figure 2.6. The FTIR spectra of both monomer and polymer show the peak position at above 3,300cm^{-1} due to N–H stretching vibration. AB_2 type monomer had absorption band at around 1,463cm^{-1} (bending vibration of CH_2) and bands between 1,000 and 1,400cm^{-1}. The peaks can be observed at 1,735cm^{-1} corresponding to C=O stretching vibration of esters.

FTIR spectrum of hyperbranched polymer is characterized by absorption at around 1,631 cm^{-1} (C=O stretching vibration of amides) and 1,463 cm^{-1} (bending vibration of CH_2). The absorption at 1,735 cm^{-1} that is attributed to the C=O stretching vibration of esters is generally weak in hyperbranched polymer. Melt-polycondensation reaction and synthesis can be responsible for this change in the band intensity.

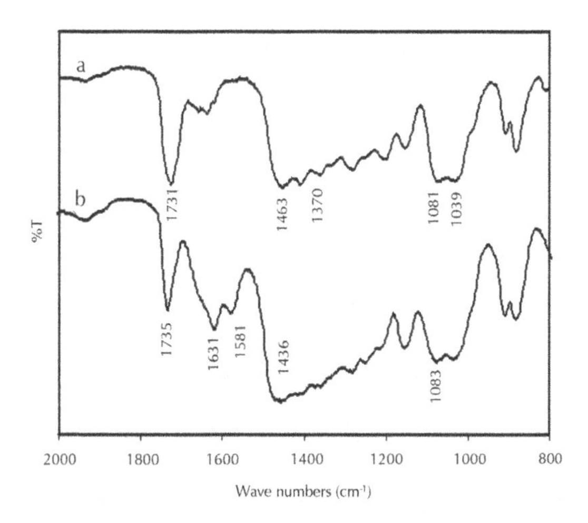

FIGURE 13.6 FTIR spectra of (a) AB_2 type monomer and (b) hyperbranched polymer

HYPERBRANCHED TREATED PET FABRICS ANALYSIS

In order to confirm the preparation process, we studied the FTIR spectroscopic data. FTIR spectra of untreated PET fabric and hyperbranched polymer treated PET fabric are shown in Figure 2.7. The FTIR of both samples shows that the peak positions are at 2,929; 2,867; 1,714; 1,238; and 1,093 cm^{-1}. The band at 1,238 and 1,093 cm^{-1} is due to C-O stretching. While the band at 1,714 cm^{-1} reflects the carbonyl group stretching. The characteristics absorption band of hyperbranched polymer at around 1,600 cm^{-1}, which reflects the N–H bending of hyperbranched polymer, appeared after the treatment of PET fabric.

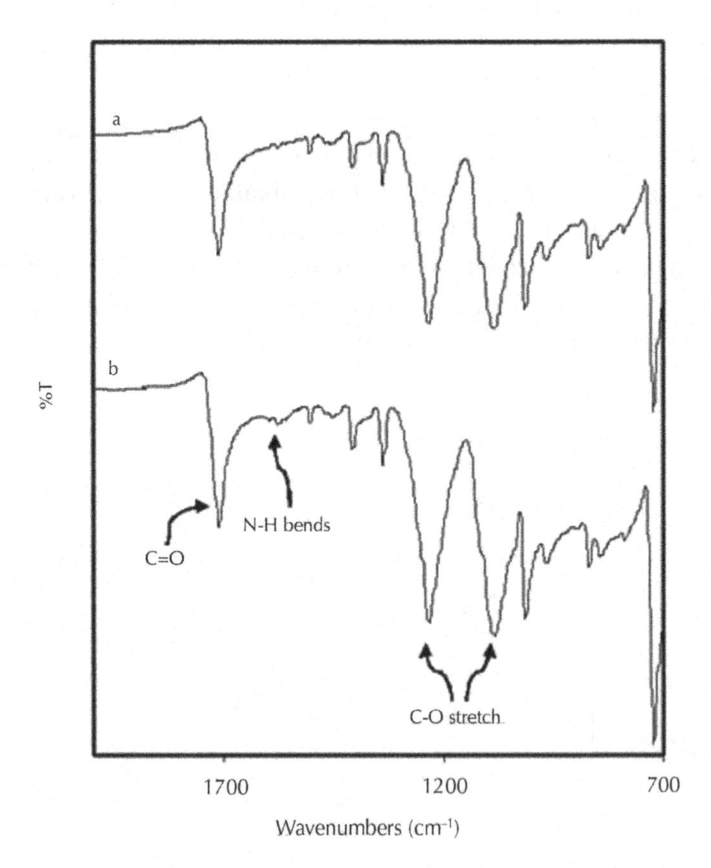

FIGURE 13.7 FTIR spectra of (a) untreated PET fabric and (b) HBP treated PET fabric

13.4 NMR SPECTROSCOPY CHARACTERIZATION

13.4.1 ^{1}H NMR ANALYSIS

The possible structure of the novel hyperbranched polymer was supported with NMR spectroscopy. Figure 2.8 shows the ^{1}H NMR spectrum and its assignment of amine terminated hyperbranched polymer. A good correlation is found between the ^{1}H NMR spectrum of hyperbranched polymer and its expected structure.

The chemical shifts ranged from δ = 2.80–3.70 ppm were assigned to protons of methylene. The methylene protons adjacent to the carbonyl unit give rise to a signal at δ = 2.90 ppm. The presence of the amine unit is confirmed by the resonance at δ = 2.74ppm for the NH_2 of the amine group. The chemical shift at δ = 2.50ppm are related to the DMSO-d_6 as a solvent [39, 40]. The proton absorption of amid group (-CONH-) was found between δ =5.15 and 5.40 ppm with its peak at δ =5.28 ppm [41].

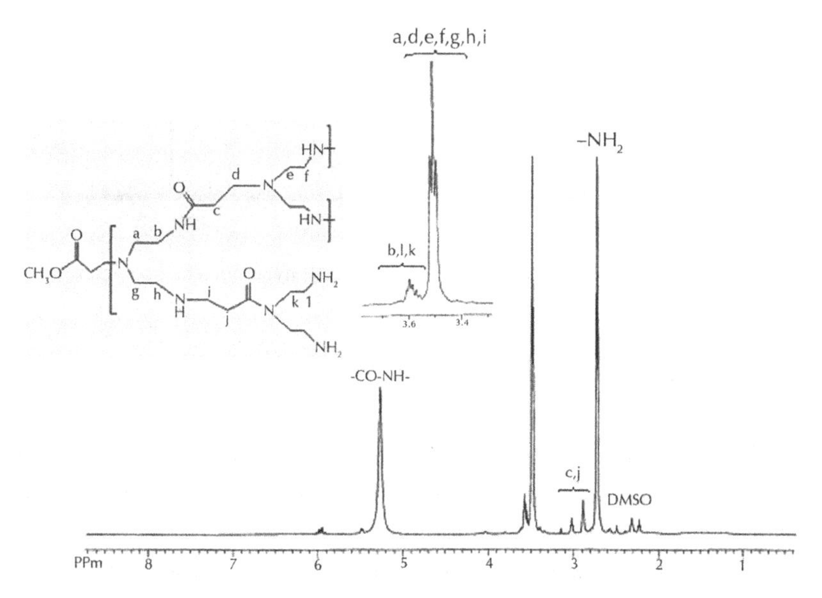

FIGURE 13.8 ^{1}H NMR spectrum of amine terminated hyperbranched polymer

13.4.2 ^{13}C NMR ANALYSIS

Both ^{1}H NMR and ^{13}C NMR have been utilized to confirm the structure of hyperbranched polymer. The ^{13}C NMR spectrum of hyperbranched polymer in DMSO-d_6 is shown in Figure 2.9. This spectrum is in conformity with the expected branched structure. The methylene carbon appears at δ = 50–60 ppm. The chemical shift at δ = 40 ppm are due to the solvent (DMSO) and the carbonyl carbon (-C=O) resonance is downfield at δ = 171.97 ppm.

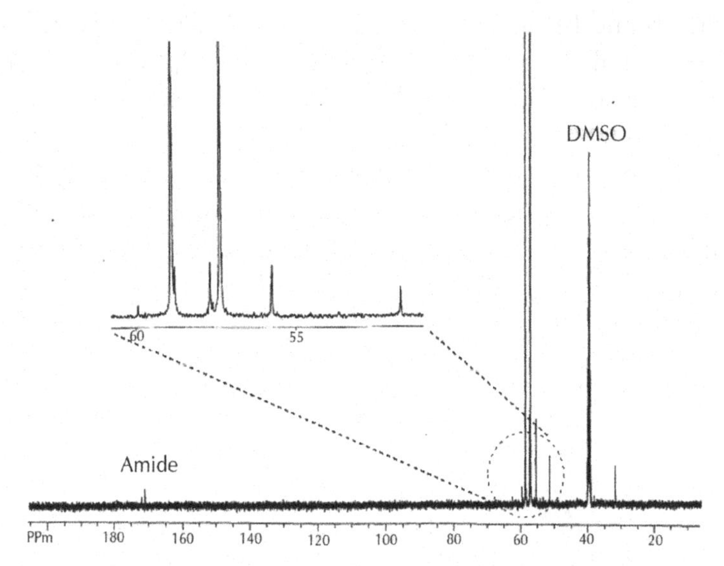

FIGURE 13.9 ^{13}C NMR spectrum of amine terminated hyperbranched polymer

13.5 UV-VIS SPECTROSCOPY

Absorption characteristic of hyperbranched polymer was investigated by UV–vis spectrophotometry. The UV-vis absorption spectrum of hyperbranched polymer in water solution (concentration: 20g/dl) is shown in Figure 2.10. It is obvious that the synthesized hyperbranched polymer gives strong absorption in the range of 200–350nm. Subsequently the hyperbranched polymer treatment can improve the ultraviolet protection property of PET fabric.

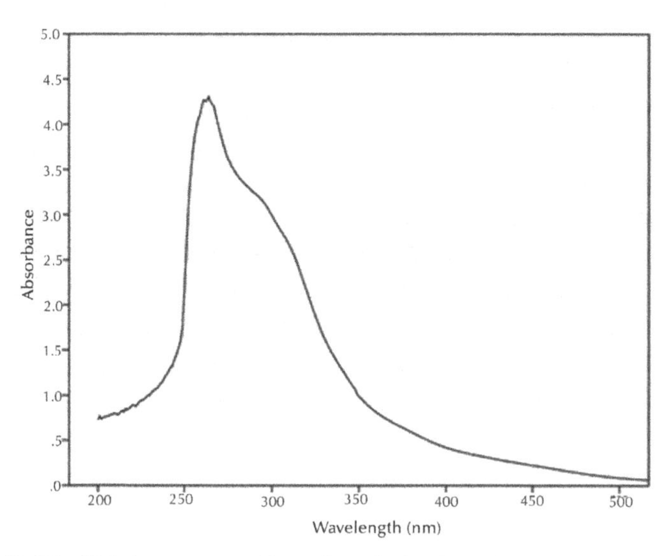

FIGURE 13.10 UV-vis spectrum of hyperbranched polymer in water

13. 6 SOLUBILITY PROPERTIES

Solubility properties of hyperbranched polymer are determined in different organic solvents included polar and nonpolar solvents. Hyperbranched polymer is found to be soluble in H_2O, DMSO, and DMF and insoluble in THF, xylene and toluene. Because of its highly branched structure and the nature of end functional groups, this polymer was highly soluble in polar solvents such as H_2O, DMSO, and DMF, as shown in Table 2.3.

TABLE 13.3 Solubility* properties of hyperbranched polymer

Polymer	Solvent						
	H_2O	DMSO	DMF	THF	Toluene	Xylene	Ethanol
Hyper-branched polymer	+	+	+	–	–	–	±

*The amount of polymer sample and solvents are 20 mg and 2 ml, +: soluble at room temperature (25°C), ±: partially soluble, -: insoluble at room temperature

13.6.1 CONTACT ANGLE MEASUREMENT

HBP CONTACT ANGLE

Contact angle measurements were carried out by experimental equipment consisting of a camera, computer, and monitor. The contact angle of hyperbranched polymer aqueous solution droplet placed on a polyester plate is shown in Figure 2.11. Comparison of the droplet by contact angle measurements shows that the contact angle decreased dramatically with increasing concentration of hyperbranched polymer solution from 2.5 to 20 g/dl.

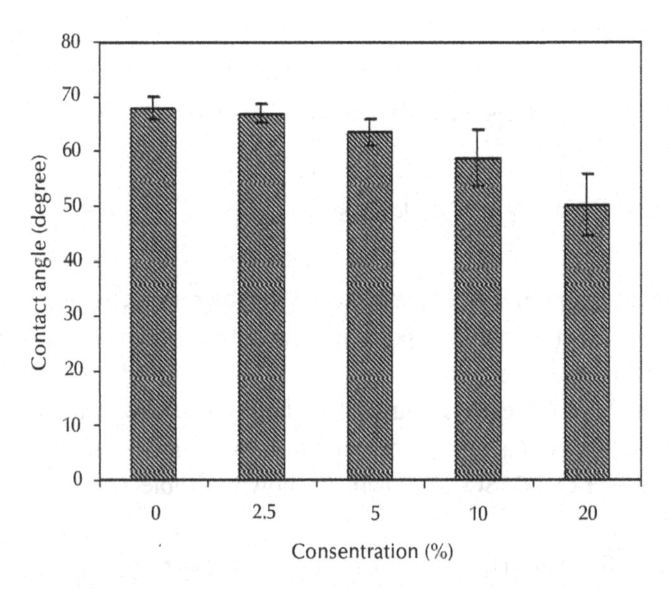

FIGURE 13.11 Contact angle of hyperbranched polymer solutions on PET plate

13.6.2 HBP TREATED PET FABRICS CONTACT ANGLE

Table 2.4 shows the corresponding values of the contact angle untreated and HBP treated PET fabrics. It is obvious that the alkali treatment considerably influenced the contact angle of PET fabrics and when the HPET fabric was treated with HBP, the contact angle was not change significantly. But when the PET fabric treated with HBP (PET-HBP), the contact angle was considerably decrease and HBP treated fabrics become hydrophilic. This confirms the presence of the HBP on the PET fabrics surface, which remains more hydrophilic than the untreated PET fabric.

Complete wetting of fiber surfaces is critical in all wet processes for textiles, particularly for dyeing. Uneven fiber wetting in dyeing invariably leads to uneven dye absorption. Therefore, HBP treatment cause PET fabric surface become hydrophilic and dyeing properties of PET fabrics with acid dyes become modified.

TABLE 13.4 Contact angle of untreated and HBP treated PET fabrics

No.	Sample	Contact angle (°)	
		Advancing	Receding
1	PET	88 ±6	85 ±4
2	HPET	56 ±6	50 ±8
3	PET-HBP	58 ±6	55 ±4
4	HPET-HBP	61 ±6	59 ±6

13.7 X-RAY DIFFRACTION (XRD)

Figure 2.12 illustrate wide-angle X-ray patterns of polyester samples including untreated PET, HPET and HBP treated fabrics. All the samples have characteristic peaks around $2\theta=17$, 22 and 25° corresponding to plans with Miller indices (010), (110) and (100), which are in good agreement with other papers [42]. As can be seen from the Figure 2.12, the HBP treated PET fabrics exhibit a similar XRD pattern to untreated PET fabric, but the intensity of the 2θ =15–30° peak was decreased for PET-HBP sample, which suggested a loss of crystalline order with applying hyperbranched polymer.

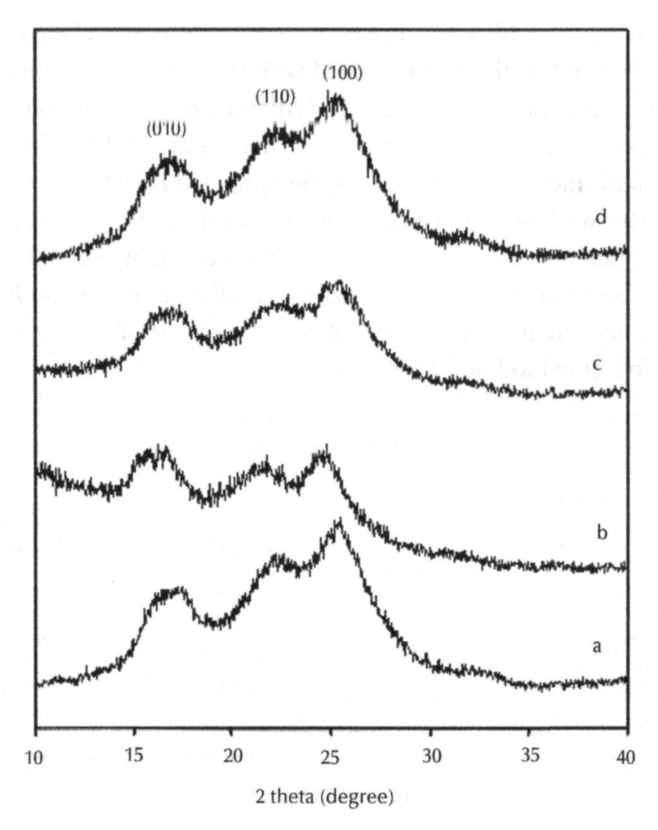

FIGURE 13.12 XRD patterns of (a) PET, (b) HPET, (c) PET-HBP, and (d) HPET-HBP fabrics

The d-sapacing of crystalline part of the polymer was calculated using the Bragg equation:

$$d = \frac{\lambda}{2\sin\theta} \tag{3}$$

where, d is Bragg spacing, θ is the diffraction angle and λ is the wavelength (1.54Å) of X-ray radiation used.

A change in the crystallite size was observed after HBP treatment. The Scherrer formula [42] was used to calculate the crystallite size as given by

$$l_{(hkl)} = \frac{k\lambda}{\Delta_{(hkl)} \cos\theta} \tag{4}$$

where, $l_{(hkl)}$ is the dimension of crystallite size, Δ is the full width at half maximum intensity (FWHM) of the diffraction peak and k is a constant assigned as a 1. Table 2.5 shows the XRD parameters for untreated and HBP treated PET fabrics.

TABLE 13.5 XRD parameters for untreated and HBP treated PET fabrics

No.	Sample	Reflection index	*hkl*		
			010	110	100
1	PET	Crystallite size (Å)	30.9	33.6	25.8
		2θ (deg.)	17.22	22.58	25.48
		Bragg spacing (Å)	5.14	3.93	3.49
2	HPET	Crystallite size (Å)	25.6	33.6	30.5
		2θ (deg.)	16.25	21.57	24.81
		Bragg spacing (Å)	5.45	4.11	3.59
3	PET-HBP	Crystallite size (Å)	25.3	29.8	25.3
		2θ (deg.)	16.83	22.41	25.48
		Bragg spacing (Å)	5.26	3.96	3.49
4	HPET-HBP	Crystallite size (Å)	26.0	33.9	25.4
		2θ (deg.)	16.86	22.28	25.39
		Bragg spacing (Å)	5.25	3.99	3.50

Table 2.5 shows that the d-spacing of crystalline part of HBP treated samples are higher than untreated PET fabrics.

The broadening in the peak and change in the crystallite size due to treatment HBP to PET fabrics may be due to formation of disordered system in the HBP treated fabric.

13.8 ZETA POTENTIAL

The surface charge of samples has been investigated by zeta potential measurement to understand how HBP act on the PET fabrics. The zeta potential values of the untreated (HPET) and the HBP treated PET fabrics (HPET-HBP) as a function of the pH of the liquid phase are shown in Figure 2.13. From the results, it can be seen that the zeta potentials for the untreated PET fabrics over the entire pH range are negative. While the HBP treated PET fabrics exhibited a positive charge on the surface at a low pH. This is probably attributed to the presence of numerous terminal primary amino groups on HBP treated PET fabrics, that will protonate in the liquid phase and give rise to positive charge at lower pH values. It implies that HBP enhances the adsorption of acid dye on PET fabrics.

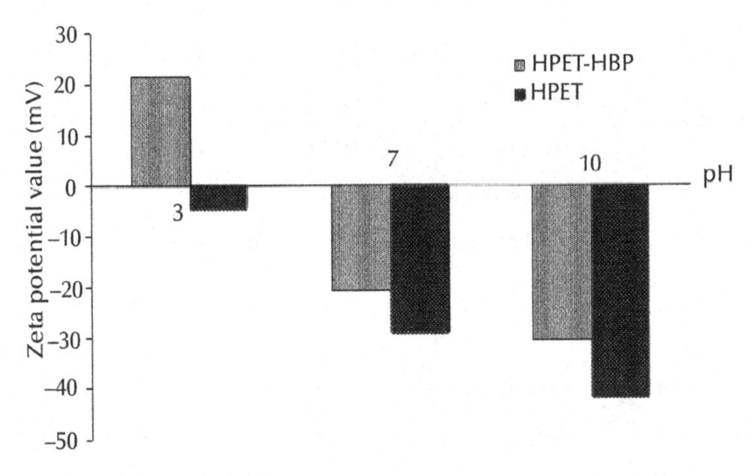

FIGURE 13.13 Zeta potential values of the untreated (HPET) and HBP treated PET fabrics (HPET-HBP)

13.9 DYEING PROPERTIES OF HBP TREATED POLYESTER FABRIC

The effect of HBP treatment on dye absorbance behavior (dyeability) of PET fabrics is evaluated by measuring its optical properties. The reflectance spectra of HBP treated and untreated PET fabrics are shown in

Figure 2.14 for dyed samples. As shown in Figure 2.14 the reflectance spectrum of HBP treated PET fabric is less than other sample. Subsequently, the treatment of polyester fabrics by HBP significantly changes the reflectance of samples.

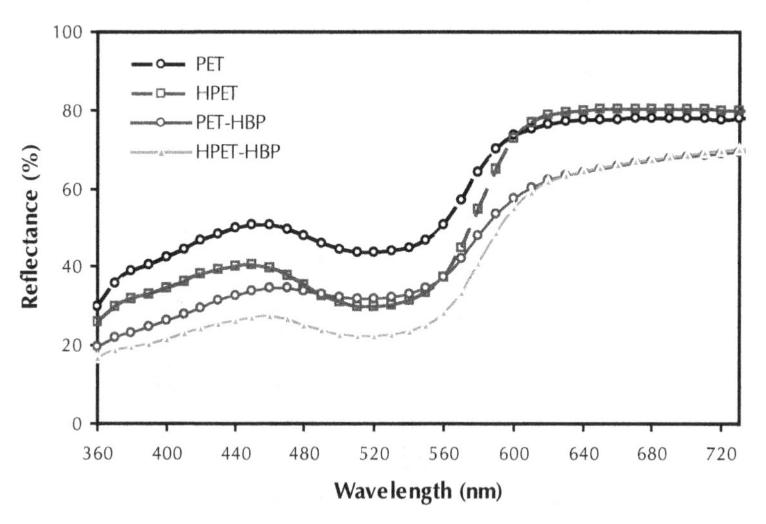

FIGURE 13.14 Reflectance spectra of dyed HBP treated and untreated PET fabrics

The CIELAB color parameter (L*, a*, b*, C*, and h°) of HBP treated and untreated PET fabrics are shown in Table 2.6. As shown in this table, the lightness (L*) of untreated polyester fabric is more than other sample. Also the hue (h°) of HBP treated PET fabric is more than untreated sample. Subsequently, the HBP treatment has significant change on color parameters of PET fabrics.

TABLE 13.6 Color parameters of dyed samples at 2% omf acid dye

No.	Samples	L*	a*	b*	C*	h°
1	PET	78.951	18.079	5.954	19.034	0.318
2	HPET	72.780	30.075	7.577	31.015	0.247
3	PET-HBP	70.291	18.432	10.525	21.225	0.519
4	HPET-HBP	64.861	27.547	11.558	29.873	0.397

The color strength (K/S) results of the HBP treated PET fabric and untreated PET fabric dyed with 2% omf Acid Red 114 using a competitive dyeing method, are shown in Figure 2.15. Higher K/S values indicate greater dye uptake and higher color strengths. It is clear that the color strength of the HBP treated PET fabric was very higher than that of the corresponding untreated PET fabric. The increase in the color strength may be because of the positively charged amino groups in the HBP treated PET fabric.

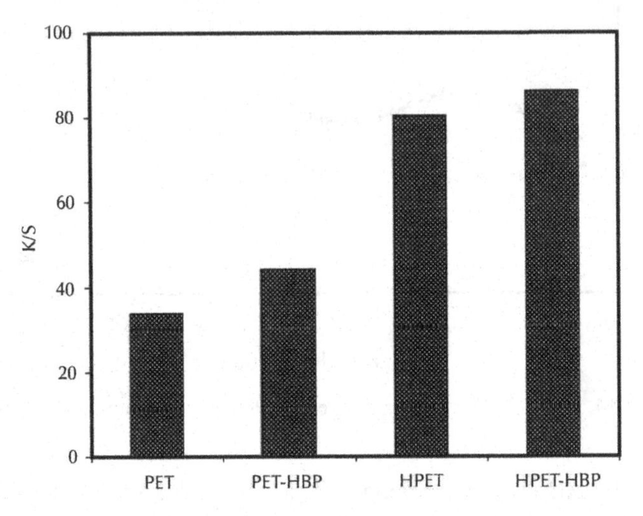

FIGURE 13.15 Effect of fabric treatment on color strength achieved using 2% omf C.I. Acid Red 114

13.10 FASTNESS PROPERTIES

13.10.1 HBP TREATMENT FASTNESS

The treated samples were subjected to five repeated washing fastness test at 60°C. Figure 2.16 shows the washing fastness result of the undyed HBP treated and untreated PET fabric and its effect on dyeability of PET fabrics. It is obvious that the extent of color strength change was relatively small for treated and untreated PET fabrics.

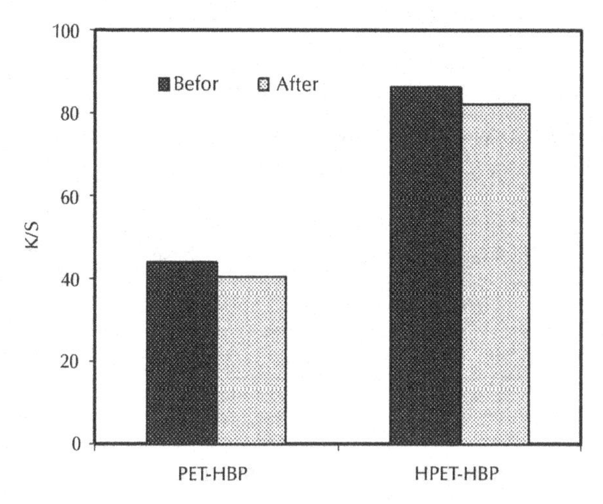

FIGURE 13.16 Effect of washing fastness test on dyeability of treated and untreated PET fabrics

13.10.2 DYEING FASTNESS

Table 2.7 shows the washing fastness and rubbing fastness of untreated PET fabrics (PET), hydrolyzed PET fabrics (HPET), and HBP treated PET fabrics (PET-HBP and HPET-HBP) dyed with acid dye. Fastness results show that the rubbing fastness of treated samples were good; the washing fastness of all sample were, however, poor and unaffected by treatment with HBP.

TABLE 13.7 Fastness properties of HBP treated and untreated PET fabrics

No.	Samples	Washing fastness*	Rubbing fastness**	
			Dry	Wet
1	PET	1	4	3-4
2	HPET	1	5	4-5
3	PET-HBP	1	4-5	4
4	HPET-HBP	1	4-5	4

*With gray scale

**With stain scale

13.11 THE ANALYSIS OF VARIANCE (ANOVA) AND OPTIMIZATION

All 20 experimental runs of central composite design (CCD) were performed in accordance with Table 2.8.

TABLE 13.8 The actual design of experiments and response

Un-coded value			Response
HBP concentration, (wt.%)	Temperature, (°C)	Time, (min)	K/S value
2	90	75	20.38
2	130	75	21.57
2	110	60	19.86
2	130	45	21.09
2	90	45	19.78
6	110	60	22.68
6	110	60	22.18
6	110	60	22.67
6	110	60	22.97
6	110	60	22.64
6	110	60	22.26
6	110	45	22.59
6	110	75	24.47
6	90	60	22.58
6	130	60	26.07
10	90	45	23.35
10	90	75	24.97
10	130	75	29.92
10	130	45	26.01
10	110	60	23.23

A significance level of 5% was selected; that is, statistical conclusions may be assessed with 95% confidence. In this significance level, the factor has significant impact on fabrics dyeability if the p-value is less than 0.05. And when p-value is greater than 0.05, it is concluded the factor has no significant impact on fabrics dyeability. The results of analysis of variance (ANOVA) are shown in Table 2.9. The regression equation was obtained from the analysis of variance.

TABLE 13.9 Analysis of variance for response surface

Source	Sum of squares	F-value	p-value
Model	104.71	25.98	<0.0001
X_1 -concentration	61.58	137.49	<0.0001
X_2 -temperature	18.51	41.33	<0.0001
X_3 -time	7.19	16.05	0.0025
X_1X_2	3.27	7.30	0.0223
X_1X_3	2.48	5.52	0.0406
X_2X_3	0.59	1.31	0.2794
X_1^2	4.59	10.24	0.0095
X_2^2	6.10	13.62	0.0042
X_3^2	1.31	2.93	0.1177
Lack of Fit	3.41	31.17	Not significant
Std Dev: 0.67	R^2 : 0.959	Adj-R^2 0.922	

$$K/S = +22.57 + 2.48X_1 + 1.36X_2 + 0.85X_3 + 0.64X_1X_2 + 0.56X_1X_3 + 0.27X_2X_3 - 1.29X_1^2 + 1.49X_2^2 + 0.69X_3^2 \quad (5)$$

From the p-values presented in Table 2.3, it can be concluded that p-value of term X_3 is greater than the p-values for terms X_1 and X_2. Also p-value of term X_3^2 and X_2X_3 is much greater than the significance level of 0.05. The interaction between temperature and time is not significant. But p-values for terms related to X_1 and X_2 are less than 0.05. Therefore, HBP concentration and treatment temperature have a significant impact on the HBP treated fabrics dyeability.

As terms related to treatment time have no significant impact on samples K/S values, we removed term X_3^2 and X_2X_3 and fitted the equation by regression analysis again. The fitted equation in coded unit is given by

$$K/S = +22.65 + 2.48X_1 + 1.36X_2 + 0.85X_3 + 0.64X_1X_2 + 0.56X_1X_3 - 1.03X_1^2 + 1.75X_2^2 \quad (6)$$

Now, all the p-values are less than the significance level of 0.05.

The effects of three variables on K/S value of dyed samples are shown in Figure 2.17. Figure 2.17 (a) shows K/S value of dyed samples at different HBP concentration and temperature for constant treatment time. The surface plot shows that at any given temperature the K/S value of samples increase with increasing the HBP concentration. Maximum K/S value was observed at high HBP concentration and high temperature. Figure 2.17 (b) shows the response surface plot of interactions between HBP concentration and treatment time. It can be seen the increase in K/S value with increase in concentration at any given time. Moreover, it should be noted that at higher temperature, the increase in concentration and time result the higher K/S value.

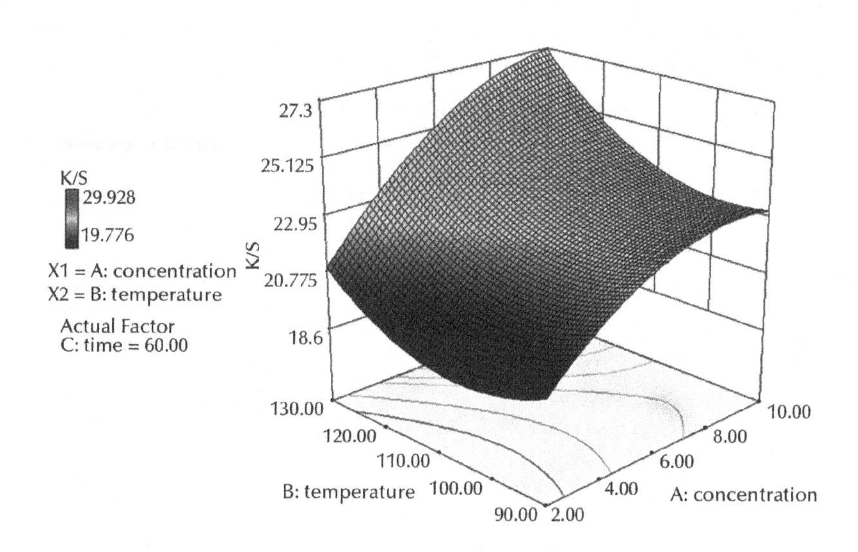

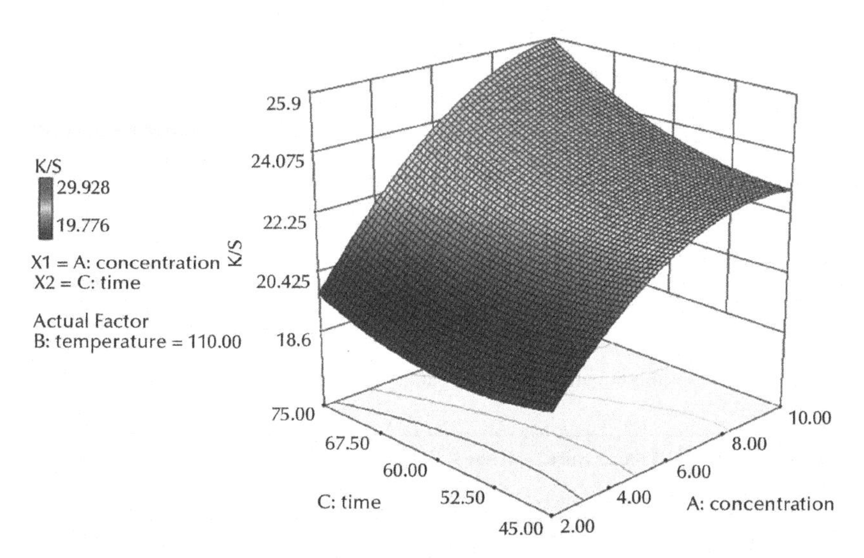

FIGURE: 13.17 Response surface for K/S value of dyed samples in term of: (a) HBP concentration and treatment temperature, (b) HBP concentration and treatment time

The actual and the predicted K/S value of samples plot with correlation coefficient of 0.91 are shown in Figure 2.18. Actual values are the measured response data for a particular run, and the predicted values evaluated from the model. It can be observed that experimental values are in good agreement with the predicted values. The optimal conditions to obtain the maximum dyeability (K/S value) efficiency are shown in Table 2.10.

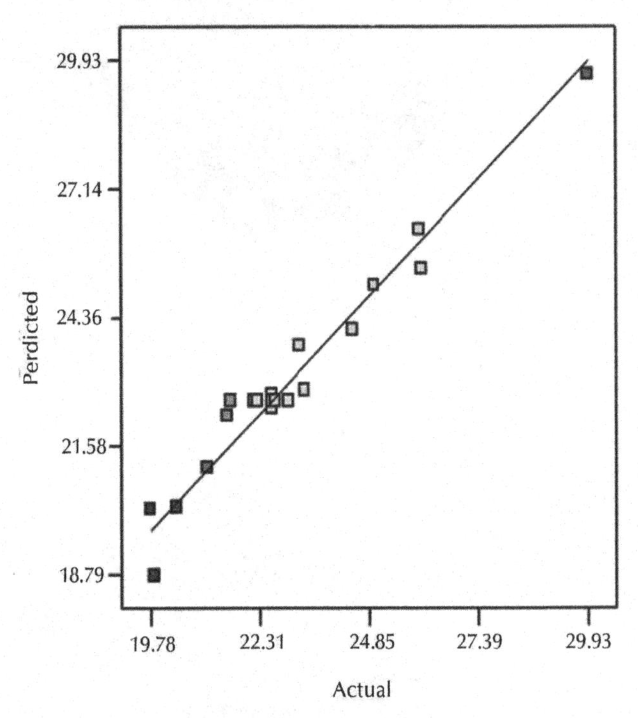

FIGURE 13.18 The actual and predicted plot for K/S value of dyed samples

TABLE 2.10 Optimum values of the treatment conditions for maximum dyeability efficiency

HBP concentration (wt.%)	Temperature (°C)	Time (min)	Predicted K/S	Experimental K/S	Desirability
10	130	75	29.25	29.93	0.93

13.12 CONCLUSION

In recent years, dendritic polymers have attracted increasing attention due to their unique chemical and physical properties. These polymers consist of three subsets namely hyperbranched polymers, dendrigraft polymers, and dendrimers. Hyperbranched polymers (HBPs) are highly branched, polydisperse and three-dimensional macromolecules synthesized from a multifunctional monomer to produce a molecule with dendritic structure [1–10].

Dendrimers are well-defined and needed a stepwise route to construct the perfectly symmetrical structure. Hence, synthesis of dendrimers is time-consuming and expensive procedure. Although hyperbranched polymers are irregularly shaped and not perfectly symmetrical like dendrimers, hyperbranched polymers rapidly prepared and generally synthesized by one-step process via polyaddition, polycondensation, radical polymerization, and so on of AB_x type monomers and no purification steps are needed for their preparation. Therefore, HBPs are attractive materials for industrial applications due to their simple production process [1-4, 11-19]. In general, according to molecular structures and properties, hyperbranched polymers represent a transition between linear polymers and perfect dendrimers. Comparison of hyperbranched polymers with their linear analogues indicated that HBPs have remarkable properties, such as low melt and solution viscosity, low chain entanglement, and high solubility, as a result of the large amount of functional end groups and globular structure. [10–16].

In recent years, many functional hyperbranched polymers with various terminal groups, such as hydroxyl, amine, carboxyl, acetoxy, and vinyl have been suggested as excellent candidate for use in drug delivery [20, 21], gene therapy vectors [21, 22], coatings [23,24], additives [25], catalysis [26], gas separation [27], nanotechnology, and supramolecular science [28], and many more.

Some examples of modification of fibers with hyperbranched polymers were successful. For instance, the dyeability of modified polypropylene (PP) fibers by hyperbranched polymer was investigated [29]. The results showed that the incorporation of hyperbranched polymer prior to fiber spinning considerably improved the color strength of polypropylene fiber with C.I. Disperse Blue 56 and has no significant effect on physical properties of the PP fibers.

The synthesized amino-terminated hyperbranched polymer was found to be used as salt-free dyeing auxiliary for treated cotton fabrics [30].

The washing fastness, rubbing fastness, and leveling properties of hyperbranched polymer treated cotton fabrics were better than untreated cotton fabrics. In the study on applying amino-terminated hyperbranched polymer to cotton fabric, it was demonstrated that hyperbranched polymer treatment on cotton fabrics has no undesirable effect on mechanical properties of fabrics. Dyeing of treated cotton fabrics with direct and reactive dyes in the absence of electrolyte showed that the color strength of treated samples were better than untreated cotton fabrics. Furthermore, application of hyperbranched polymer to cotton fabrics reduced UV transmission and showed good antibacterial activities [31-34].

In this chapter, a novel hyperbranched, functional, water-soluble, and amine terminated polymer was synthesized by Melt-polycondensation reaction of methyl acrylate and diethylene triamine. The hyperbranched polymer sample prepared was characterized by FTIR and NMR spectroscopy. The PET fabrics and hydrolyzed PET fabrics with sodium hydroxide were treated with the synthesized hyperbranched polymer. Color strength of hyperbranched polymer treated PET fabrics exhibited improvements when dyed with acid dye (C.I. Acid Red 114) in conventional dyeing system. The color characteristics of the dyed fabrics were evaluated using CIELAB method. The results of the measurement on the fabrics after competitive dyeing with acid dyes indicated that the hyperbranched polymer, as indicated by zeta potential measurement, offer positively charged amino groups for differential-dyeing. The treated samples exhibited good wash fastness. The resulting hyperbranched polymer is readily soluble in polar solvent, such as water, DMSO, and DMF.

To optimize and predict the dyeability of PET fabrics which treated with hyperbranched polymer, CCD experimental design was employed. A more systematic understanding of process parameters was obtained and a quantitative relationship between treatment conditions and fabrics dyeability was established by response surface methodology (RSM).

In this chapter, a hyperbranched, functional, and water-soluble polymer was successfully synthesized by Melt-polycondensation reaction of methyl acrylate and diethylene triamine. The FTIR, ^{1}H and ^{13}C NMR spectroscopy measurements of hyperbranched polymer indicated that this polymer comprised terminal amine group and branching was occurred. The solubility properties of resulting hyperbranched polymer indicated that this polymer is readily soluble in polar solvent, such as water, DMSO, and DMF. The hyperbranched polymer was applied to alkali-treated PET fabric using exhaust method. Using Bragg's equation it was found that d-spacing increase after applying HBP to PET fabrics. The study of dyeability of treated samples with C.I. Acid Red 114 indicated that the color strength of HBP treated PET fabrics is more than untreated PET fabric due to the presence of terminal primary amino groups in the molecular structure of the HBP, that will protonate in the liquid phase and give rise to positive charge at lower pH values, as shown by zeta potential measurement. Samples contact angles tends to decrease with the HBP treatment, and the hydrophobic untreated PET fabrics surface becomes hydrophilic. This confirms the presence of the HBP on the PET fabrics surface and its effect on hydrophilic behavior of PET fabrics. It is noteworthy that in our attempt PET fabric dyed with acid dye effectively using HBP. Compared with previously reported article, our attempt has an innovation in PET dyeing with acid dyes.

The concentration, temperature, and time are the most important factors for HBP treatment to PET fabrics and dyeability efficiency. Response surface methodology was successfully applied to find out the optimum level of the above factors using central composite design (CCD). The optimum concentration of HBP, temperature, and time were found to be 10 wt.%, 130°C and 75min, respectively, for maximum K/S value of HBP treated samples.

KEYWORDS

- **Fourier transform infrared spectroscopy**
- **Melt-polycondensation reaction**
- **Nuclear magnetic resonance**
- **Poly(ethylene terephthalate)**
- **Zeta potential**

REFERENCES

1. Jikei, M., & Kakimoto, M. (2001). Hyperbranched polymers: A promising new class of materials. *Progress in Polymer Science, 26,* 1233–1285.
2. Gao, C., & Yan, D. (2004). Hyperbranched polymers: From synthesis to applications. *Progress in Polymer Science,* 29, 183–275.
3. Voit, B. I., & Lederer, A. (2009). Hyperbranched and highly branched polymer architectures synthetic strategies and major characterization aspects. *Chemical Reviews, 109,* 5924–5973.
4. Kumar, A., & Meijer, E. W. (1998). Novel hyperbranched polymer based on urea linkages. *Chemical Communications,* 1629–1630.
5. Grabchev, I., Petkov, C., & Bojinov, V. (2004). Infrared spectral characterization of poly(amidoamine) dendrimers peripherally modified with 1,8-naphthalimides. *Dyes and Pigments, 62,* 229–234.
6. Qing-Hua, C., Rong-Guo, C., Li-Ren, X., Qing-Rong, Q., & Wen-Gong, Z. (2008). Hyperbranched poly (amide-ester) mildly synthesized and its characterization. *Chinese Journal of Structural Chemistry, 27,* 877–883.
7. Kou, Y., Wan, A., Tong, S., Wang., L., & Tang J. (2007). Preparation, characterization and modification of hyperbranched polyester-amide with core molecules. *Reactive and Functional Polymers, 67,* 955–965.
8. Schmaljohann D., Pötschke, P., Hässler, R., Voit, B. I., Froehling, P. E., Mostert, B., & Loontjens, J. A. (1999). Blends of amphiphilic, hyperbranched polyesters and different polyolefins. *Macromolecules, 32,* 6333–6339.
9. Kim, Y.H. (1998). Hyperbranched polymers 10 years after. *Journal of Polymer Science Part A: Polymer Chemistry, 36,* 1685–1698.
10. Liu, G., & Zhao, M. (2009). Non-isothermal crystallization kinetics of AB_3 hyperbranched polymer/polypropylene blends. *Iranian Polymer Journal, 18,* 329–338.
11. Seiler, M. (2006). Hyperbranched polymers: Phase behavior and new applications in the field of chemical engineering. *Fluid Phase Equilibria, 241,* 155–174.
12. Yates, C. R., & Hayes, W. (2004). Synthesis and applications of hyperbranched polymers. *European Polymer Journal, 40,* 1257–1281.
13. Voit, B. (2000). New developments in hyperbranched polymers. *Journal of Polymer Science: Part A: Polymer Chemistry, 38,* 2505–2525.
14. Nasar, A.S., Jikei, M., & Kakimoto, M. (2003). Synthesis and properties of polyurethane elastomers crosslinked with amine-terminated AB2-type hyperbranched polyamides. *European Polymer Journal, 39,* 1201–1208.
15. Froehling P. E. (2001). Dendrimers and dyes-a review. *Dyes and Pigments, 48,* 187–195.
16. Jikei, M., Fujii, K. & Kakimoto, M. (2003). Synthesis and characterization of hyperbranched aromatic polyamide copolymers prepared from AB_2 and AB monomers. *Macromolecular Symposia, 199,* 223–232.
17. Radke, W., Litvinenko, G., & Müller A. H. E. (1998). Effect of core-forming molecules on molecular weight distribution and degree of branching in the synthesis of hyperbranched polymers. *Macromolecules, 31,* 239–248.

18. Maier, G., Zech. C., Voit, B. & Komber. H. (1998). An approach to hyperbranched polymers with a degree of branching of 100%. *Macromolecular Chemistry and Physics, 199,* 2655–2664.
19. Voit, B., Beyerlein, D., Eichhorn, K., Grundke, K., Schmaljohann, D. & Loontjens, T. (2002). Functional hyper-branched polyesters for application in blends, cations, and thin films. *Chemical Engineering and Technology, 25,* 704–707.
20. Gao, C., Xu, Y., Yan, D., & Chen, W. (2003). Water-soluble degradable hyperbranched polyesters: novel candidates for drug delivery? *Biomacromolecules, 4,* 704–712.
21. Paleos, C. M., Tsiourvas, D., & Sideratou, Z. (2007). Molecular engineering of dendritic polymers and their application as drug and gene delivery systems. *Molecular Pharmaceutics, 4,* 169–188.
22. Wu, D., Liu, Y., Jiang, X., He, C., Goh, S. H., & Leong, K. W. (2006). Hyperbranched poly (amino ester) s with different terminal amine groups for DNA delivery. *Biomacromolecules, 7,* 1879–1883.
23. Rolf A.T.M. van Benthem. (2000). Novel hyperbranched resins for coating applications. *Progress in Organic Coatings, 40,* 203–214.
24. Tang, W., Huang, Y., Meng, W., & Qing, F. (2010). Synthesis of fluorinated hyperbranched polymers capable as highly hydrophobic and oleophobic coating materials. *European Polymer Journal, 46,* 506–518.
25. Zhang, W., Zhang, Y., & Chen, Y. (2008). Modified brittle poly(lactic acid) by biodegradable hyperbranched poly (ester amide). *Iranian Polymer Journal, 17,* 891–898.
26. Astruc, D., & Chardac, F. (2001). Dendritic catalysts and dendrimers in catalysis. *Chemical Reviews, 101,* 2991–3023.
27. Fang, J., Kita, H., & Okamoto, K. (2000). Hyperbranched polyimides for gas separation applications 1. Synthesis and characterization. *Macromolecules, 33,* 4639–4646.
28. Liu, C., Gao, C., & Yan, D. (2006). Synergistic supramolecular encapsulation of amphiphilic hyperbranched polymer to dyes. *Macromolecules, 39,* 8102–8111.
29. Burkinshaw, S. M., Froehling, P. E., & Mignanelli, M. (2002). The effect of hyperbranched polymers on the dyeing of polypropylene fibres. *Dyes and Pigments, 53,* 229–235.
30. Zhang, F., Chen, Y., Lin, H., & Lu, Y. (2007). Synthesis of an amino-terminated hyperbranched polymer and its application in reactive dyeing on cotton as a salt-free dyeing auxiliary. *Coloration Technology, 123,* 351–357.
31. Zhang, F., Chen, Y., Lin, H., Wang, H., & Zhao, B. (2008). HBP-NH_2 grafted cotton fiber: Preparation and salt-free dyeing properties. *Carbohydrate Polymers, 74,* 250–256.
32. Zhang, F., Chen, Y. Y., Lin, H., Zhang, D. S. (2008). Performance of cotton fabric treated with an amino-terminated hyperbranched polymer. *Fibers and Polymers, 9,* 515–520.
33. Zhang, F., Chen, Y., Ling, H., & Zhang, D. (2009). Synthesis of HBP-HTC and its application to cotton fabric as an antimicrobial auxiliary. *Fibers and Polymers, 10,* 141–147.
34. Zhang, F., Zhang, D., Chen, Y., & Lin, H. (2009). The antimicrobial activity of the cotton fabric grafted with an amino-terminated hyperbranched polymer. *Cellulose, 16,* 281–288.

35. Khatibzadeh, M., Mohseni, M., & Moradian, S. (2010). Compounding fibre grade polyethylene terephthalate with a hyperbranched additive and studying its dyeability with a disperse dye. *Coloration Technology, 126,* 269–274.
36. Murugesan, K., Dhamija, A., Nam, I., Kim, Y., & Chang, Y. (2007). Decolourization of reactive black 5 by laccase: Optimization by response surface methodology. *Dyes and Pigments, 75,* 176–184.
37. Myers, R. H. & Montgomery, D. C., & Anderson-cook, C. M. (2009). *Response surface methodology: process and product optimization using designed experiments* (3rd edn.). USA: John Wiley and Sons.
38. Pavia, D. L., Lampman, G. M., Kriz, G. S., & Vyvyan, J. R. (2009). *Introduction to Spectroscopy* (4th edn.). US: Brooks & Cole.
39. Tang, W., Huang, Y., Meng, W., & Qing, F. (2010). Synthesis of fluorinated hyperbranched polymers capable as highly hydrophobic and oleophobic coating materials. *European Polymer Journal, 46,* 506–518.
40. Reul, R., Nguyen, J., & Kissel, T. (2009). Amine-modified hyperbranched polyesters as non-toxic, biodegradable gene delivery systems. *Biomaterials, 30,* 5815–5824.
41. Pan, B., Cui, D., Gao, F., & He, R. (2006). Growth of multi-amine terminated poly (amidoamine) dendrimers on the surface of carbon nanotubes. *Nanotechnology, 17,* 2483–2489.
42. Youssefi, M., Morshed, M., & Kish, M. H. (2007). Crystalline structure of poly (ethylene terephthalate) filaments. *Journal of Applied Polymer Science, 106,* 2703–2709.

CHAPTER 14

RESEARCH OF INFLUENCE OF NANOFILLER – SCHUNGITE ON RELEASE FURACILINUM FROM POLYHYDROXYBUTYRATE FILMS

A. A. OLKHOV, A. L. IORDANSKII, R. YU. KOSENKO, YU. S. SIMONOVA, and G. E. ZAIKOV

CONTENTS

14.1 INTRODUCTION

In last decade has occurred significant number of the works devoted to the quantitative description of processes of liberation of low molecular weight medicinal substances from polymeric matrixes. One of the major problems of modern medicine is creation of new methods of the treatment based on purposeful local introduction of medical products in a certain place with set speed. The basic requirement shown to the carrier of active material, its destruction and a gradual conclusion from an organism [1, 2].

Medicinal forms (m.f.) with controllable liberation (medicinal forms with operated liberation, medicinal forms with programmed liberation)—group of medicinal forms with the modified liberation, characterized by elongation of time of receipt m.f. in a biophase and its liberation, corresponding to real requirement of an organism.

The decision of a problem of controllable liberation m.f. will allow designing a polymeric matrix of various degree of complexity, setting thereby programmed speed of allocation of a medicine in surrounding biological environment. Regulation of transport processes in polymers taking into account their morphological features is one of actual problems in polymers. Research of interaction of polymeric materials with water important for many reasons, but there are two main things: this interaction which plays the important role in the processes providing ability to live of the person. It influences operational properties of polymeric materials.

The mechanism of address delivery m.f. includes diffusive process of its liberation of a polymeric system, therefore research progress and structures of a composite with transport characteristics m.f. is a necessary condition of creation new materials [3]. It is known, that essential impact makes on diffusive properties of polymeric films morphology and crystallinity of the components forming a composite. Besides, at composite formation the nature of the filler changing not only structure of the most polymeric matrix is very important.

The purpose of the present work was studying of influence finely divided schungite on structure, mechanical characteristics and kinetics of liberation of a medicinal drug—furacilinum from films polyhydroxybutyrate.

14.2 MATERIALS AND METHODS

As a polymeric matrix in work used polyhydroxybutyrate (PHB)—biodecomposed polymer. Thanking these properties PHB it is applied as packing, to the biomedical appointment, self-resolving fibers and films, and so on.

Basic properties PHB: melting point of 173–180 wasps, temperature of the beginning of thermal degradation of 150 wasps, degree of crystallinity 65–80%, molecular weight 10^4–10^6 g/mole, ultimate tensile strength 40MPa, an elastic modulus 3.5 GPa, tensile elongation 6–8%.

The basic lack of products from PHB is low specific elongation at the expense of high crystallinity and formation of large sphemlitic aggregates. Introduction in PHB, for example, finely divided mineral particles of filler can reduce fragility at the expense of formation of fine-crystalline polymer structure. As a medical product in work used antibacterial means—furacilinum (m.f.). Filler—fine-grained schungite technical specifications 2169-001-5773937-natural formation, on 30% consisting of carbon and 70% of silicates, the Zagozhinsky deposit [4].

It is necessary to note high probability of presence in schungit carbon of appreciable quantities of, their chemical derivative and molecular complexes which including can play a role of structural plastifiers.

In work investigated a film drug of matrix type which can be used at direct superposition on a wound or internally, for example, at intravitreal in a fabric [5].

Film is made as follows:

Powder PHB filled in with chloroform (500–600 mg PHB on 15–20 ml $CHCl_3$) and agitated on a magnetic stirrer before formation of homogeneous weight (~10 mines). Then a solution lead up to boiling and at a working agitator brought furacilinum, and behind it - schungite, agitated even 10 min). A hot solution filtrated through two beds of kapron and poured out in Petri dish D = 9 sm which seated in a furnace at temperature 25°C, densely covered with the second Petri dish and dried up to constant weight.

For all samples defined the maintenance furacilinum, liberation furacilinum in water, density of samples, physical-mechanical characteristics on corresponding state standard specification, technical specifications and to laboratory techniques [6].

14.3 RESULTS AND DISCUSSION

In Figure 3.1 typical dependence of size of liberation (M_∞/M_t) furacilinum from time is shown at the various maintenance schungite. From it follows, that all values of desorption monotonously increase in due course releases.

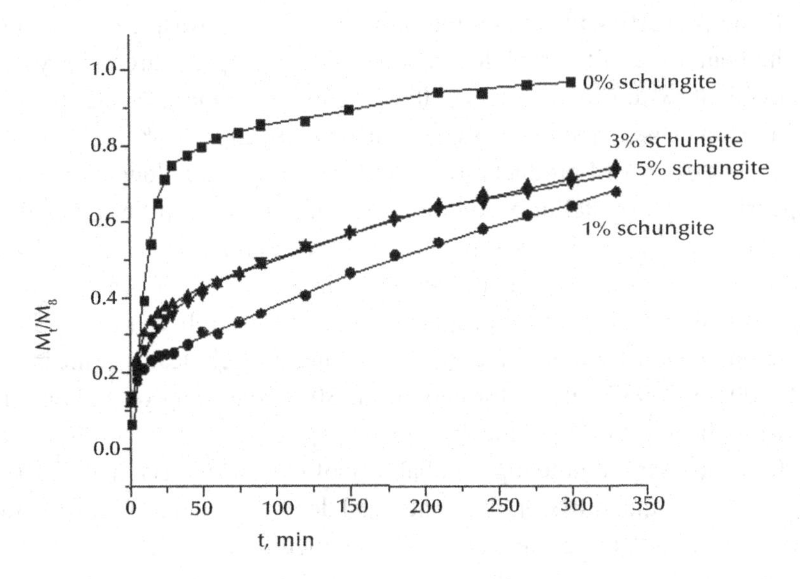

FIGURE 14.1 Kinetic curve releases of a medical product from films PHB at concentration furacilinum 3%

Apparently from the figure, schungite and, hence, appreciable impact on diffusion rate f.m. makes on its speed of liberation. To maintenance growth of schungite there is a significant falling of diffusivities, that is, reduction of speed of allocation of medicinal substance. This effect can be connected with immobilization (interaction) of molecule m.f. with a surface schungite (with oxygen groups $Si\text{-}O_2$). It is follow-up possible to expect, that impenetrable particles schungite represent a barrier to diffusion m.f., and, hence, speed of transport in a composite drops.

In Figure 3.2a, dependence of density of films PHB on the maintenance schungite in absence furacilinum for single-stage and two-phased films is represented.

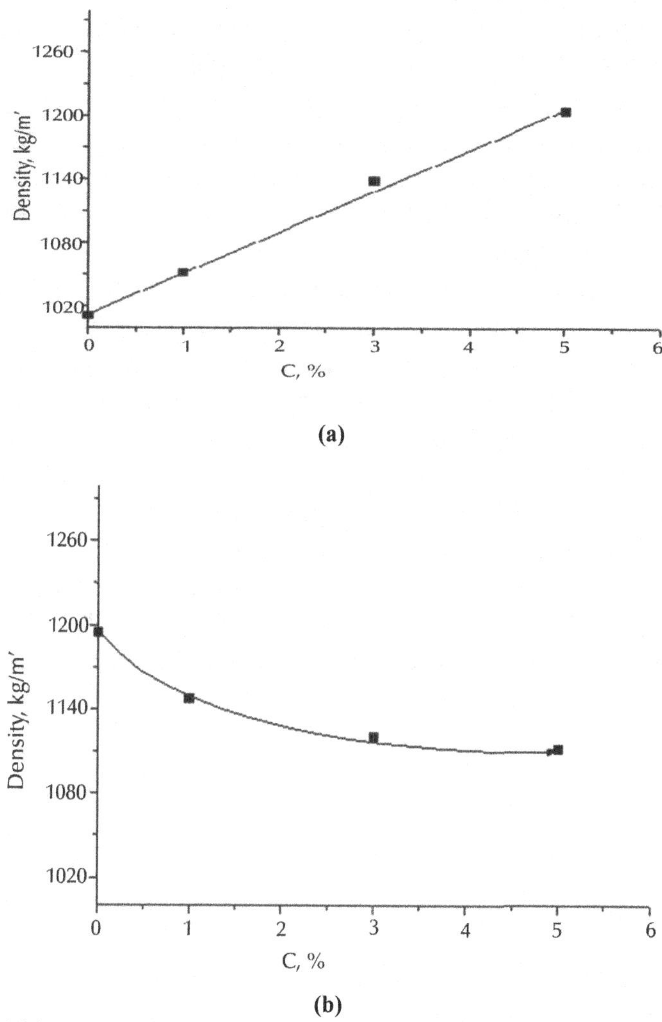

FIGURE 14.2 (a) Dependence of density of films PHB on the maintenance schungite without furacilinum. (b) Dependence of density of films PHB on the maintenance schungite at 5% furacilinuma

From the figure it is visible, that to maintenance growth of schungite there is directly proportional increase in density of films. Increase of density of composite films is caused by increase in their concentration of more dense filler.

In Figure 3.2b, dependence of density of films PHB on the maintenance of schungite is presented at 5% furacilinum. From the figure it is visible, that with maintenance increase of schungite, the density decreases. It is possible to explain density decrease by a synergism resulting interference Schungite and furacilinum. However it is possible to explain formation of large associates' furatsilin-shungit. Associates can arise at the expense of adsorption of molecules furacilinum on particles schungite, having an active surface, in solvent at formation of composite films. Associates can interfere with crystallization PHB and by that to increase quantity of the loosened amorphous phase of polymer.

Under condition of a share constancy schungite in composites with maintenance increase furacilinum the density varies like the curve in Figure 3.2b.

So at introduction 0.5% furacilinum are narrower, there is a reduction of density of a composition by 1%, and then at the further increase the maintenance furacilinum to 5% the density of films drops on 3–4%. Density reduction is caused, possibly, the structures PGB and formation of associates furatsilin-shungit which, increasing quantity of an amorphous phase, create additional free volume in polymer—a matrix.

The establishment of influence of the maintenance schungite and furacilinum on mechanical characteristics of composite films was one of the purposes supplied in work. Dependence of ultimate tensile strength of films PHB on maintenance m.f. is presented in Figure 3.3a.

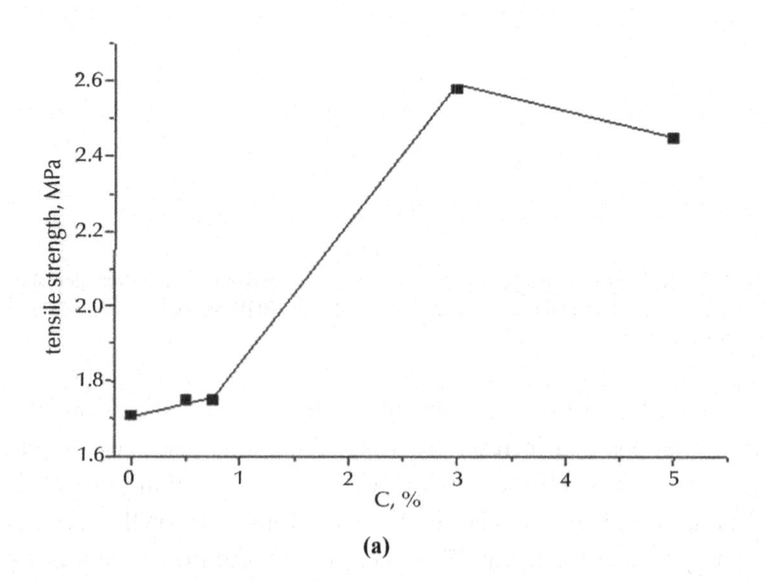

(a)

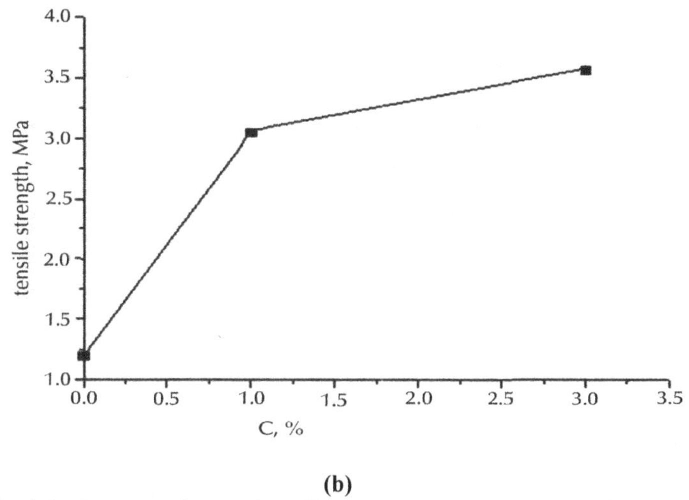

(b)

FIGURE 14.3 (a) Dependence of tensile strength of films PHB on the maintenance furacilinum without schungite. **b** Dependence of hardness of films PHB on the maintenance schungite at 5% furacilinum

As shown in Figure 3.3a, hardness increases from the minimum value 1.71 MPa in absence schungite to 2.5MPa, that is, increases in 1.5 times. The hardness increase, possibly, is caused by formation of hydrogen bridges between furan groups LV and It—groups PHB.

In Figure 3.3b, dependence of hardness of films PHB on the maintenance schungite is presented at 5% furacilinum in films. In Figure 3.3b, the increase in hardness with maintenance increase schungite more, than in 2 times is visible, that also speaks formation of fine-crystalline structure of films a preparation stage. As it was already marked above, additional hardening also follows the account of formation of associates PHB-furacilinum-schungite.

Presence schungite in a composition result ins to hardening of films PHB (Figure 3.4). In Figure 3.4, dependence of hardness of films PHB on the maintenance schungite is presented.

The increase in hardness of films schungite occurs to maintenance increase in 2.5 times. Apparently, at a stage of preparation of films schungite forms more perfect heterogeneous fine-crystalline structure. Schungite, being the nucleation centre, creates set of nuclei, thereby, doing structure of films more ordered, that promotes hardness increase [7].

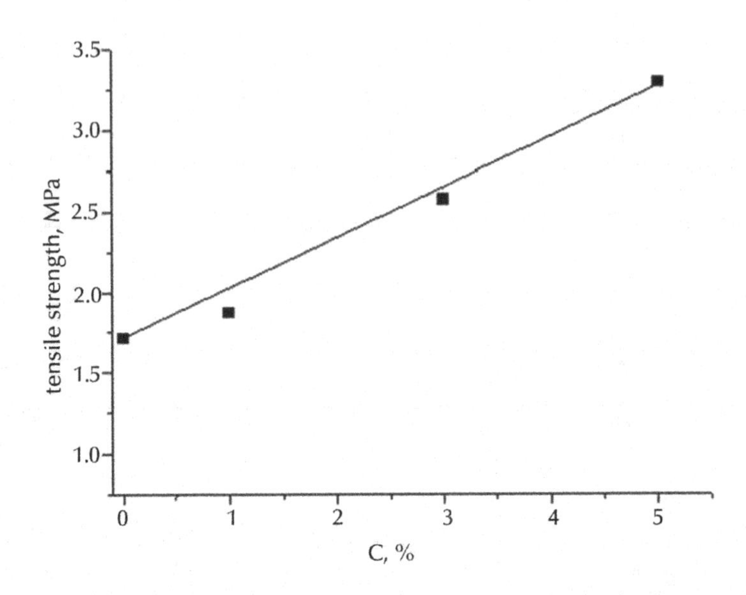

FIGURE 14.4 Dependence of hardness of films PHB on the maintenance schungite without furacilinum

Considering collaterally Figures 3.3a, b and Figure 3.4, it is possible to assume structural changes of a composite at influence on PHB schungite and m.f.

Apparently, in composites PHB-schungite-furacilinum difficult associates, that is, formation of hydrogen bridges between oxygen-containing groups' furacilinum and similar groups PHB are formed. As a result formed communications act in a role of the crosslinking localized in amorphous regions of a polymeric matrix. Presence of such crosslinking should result in to increase in hardness of composite films. Under our assumptions, at first molecules furacilinum form hydrogen bridges with trailer. It groups, and then is absorbed by other end on a surface schungite (see Figure 3.5).

Growth of number of a crosslinking does not occur in direct ratio since there is a restriction by quantity of endgroups PHB. The yielded assumption is confirmed by means of IR-spectroscopy methods, DSC and EPR.

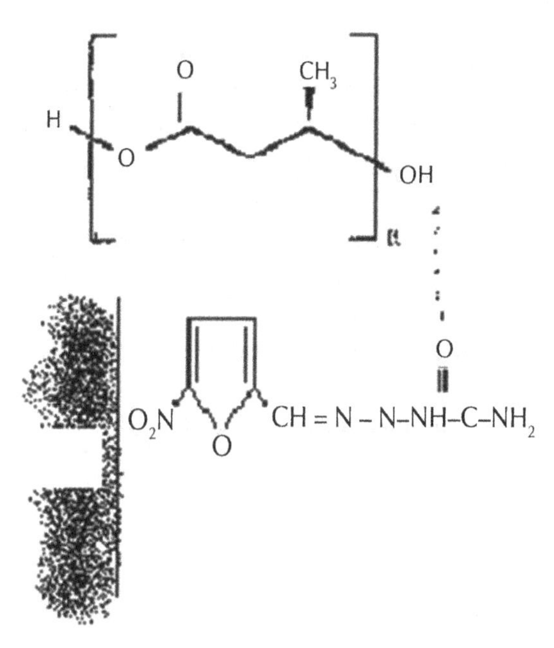

FIGURE 14.5 Schema of formation of associate PHB- furacilinum – schungite

Thus, as a result of the made work the following has been positioned:

1. For the purpose of creation of biodecomposed polymeric materials for delivery of medicinal substances threefold system PHB-schungite-furacilinum has appeared rather perspective;
2. Schungite 0–5% occurs to maintenance growth reduction of values of speed of liberation m.f. from composites in 10 times. That is schungite the prolonged an effect has on liberation kinetics furacilinum.
3. In ternary films maintenance increase furacilinum in absence schungite raises hardness in 1.5 times; the maintenance increase schungite in absence furacilinum raises hardness in 2.5 times;
4. Formation of associates PHB-schungite-furacilinum assumed by us result ins to substantial growth of hardness of composite films (in 2–3 times) in comparison with initial and two - componential films.

14.4 CONCLUSION

For the purpose of creation of biodecomposed polymeric materials for delivery of medicinal substances threefold system PHB-schungite-furacilinum has appeared rather perspective. Schungite 0–5% occurs to maintenance growth reduction of values of speeds liberation furacilinum from composites in 10 times. In ternary films maintenance increase furacilinum in absence schungite raises hardness in 1.5 times; the maintenance increase schungite in absence furacilinum raises hardness in 2.5 times. Formation of associates PHB-schungite-furacilinum assumed by us result ins to substantial growth of hardness of composite films (in 2–3 times) in comparison with initial and two-componental films.

KEYWORDS

- **Furacilinum**
- **Polyhydroxybutyrate**
- **Schungite**
- **Structure**
- **Two-componental films**

REFERENCES

1. Plate, N. A. & Vasiliev, A. E. (1986). *Physiologically active polymers*. Moscow: Khimiya.
2. Feldshtein, M. M & Plate, N. N. (1999). A structure-property relationship and quantitative approach to the development of universal transdermal drug delivery system. In Sohn, T. & Voicu, V. A. (Eds.) *Nuclear, biological, and chemical risks* (Vol. 25, pp. 441–458). Dordrecht: Kluwer Academic.
3. Zaikov, G. E., Iordanskii, A. L., & Markin, V. S. (1988). *Diffusion of electrolytes in polymers, Ser New concepts in polymer science* (p. 321). Utrecht: VSP Science Press.
4. A site "Chemist" (http://www.xumuk.ru/farmacevt/1420.html) – Furacilinum.
5. Koros, W. J. (Ed.) (1990). Barrier Polymers and Structures. In *ACS Symposium* (series 423). Washington: American Chemical Society.

6. Simonova, Ju. S. (2008). (Reg. No. 60, p. 82). Moscow: Moscow State Academy of Fine Chemical Technology.
7. Kuleznev, V. N & Shershenev, V. A. (1988). *Chemistry and physics of polymers: Studies for chemical-tenol, Higher school.* (p. 312). Moscow: Vysshaya Shkola Publishers.
8. Rogers, K. (1968). Rastvorimost and diffusion. In сб *Problems of physics and chemistry of solidity of organic matters* (pp. 229–328). Moacow: The World (The Lane with English).

CHAPTER 15

SOME ASPECTS OF ELECTROSPINNING PARAMETERS

B. HADAVI MOGHADAM, M. HASANZADEH, and A. K. HAGHI

CONTENTS

15.1 INTRODUCTION

The wettability of solid surfaces is a very important property of surface chemistry, which is controlled by both the chemical composition and the geometrical microstructure of a rough surface [1-3]. When a liquid droplet contacts a rough surface, it will spread or remain as droplet with the formation of angle between the liquid and solid phases. Contact angle (CA) measurements are widely used to characterize the wettability of rough surface [3-5]. There are various methods to make a rough surface, such as electrospinning, electrochemical deposition, evaporation, chemical vapor deposition (CVD), plasma, and so on.

Electrospinning as a simple and effective method for preparation of nanofibrous materials have attracted increasing attention during the last two decade [6]. Electrospinning process, unlike the conventional fiber spinning systems (melt spinning, wet spinning, etc.), uses electric field force instead of mechanical force to draw and stretch a polymer jet [7]. This process involves three main components including syringe filled with a polymer solution, a high voltage supplier to provide the required electric force for stretching the liquid jet, and a grounded collection plate to hold the nanofiber mat. The charged polymer solution forms a liquid jet that is drawn toward a grounded collection plate. During the jet movement to the collector, the solvent evaporates and dry fibers deposited as randomly oriented structure on the surface of a collector [8-13]. The electrospun nanofiber mat possesses high specific surface area, high porosity, and small pore size. Therefore, they have been suggested as excellent candidate for many applications including filtration [14], multifunctional membranes [15], biomedical agents [16], and tissue engineering scaffolds [17-18], wound dressings [19], full cell [20] and protective clothing [21].

The morphology and the CA of the electrospun nanofibers can be affected by many electrospinning parameters including solution properties (the concentration, liquid viscosity, surface tension, and dielectric properties of the polymer solution), processing conditions (applied voltage, volume flow rate, tip to collector distance, and the strength of the applied electric field), and ambient conditions (temperature, atmospheric pressure, and humidity) [9-12].

In this study, the influence of four electrospinning parameters, comprising solution concentration, applied voltage, tip to collector distance, and volume flow rate, on the CA of the electrospun PAN nanofiber mat was carried out using response surface methodology (RSM) and artificial neural network (ANN). First, a central composite design (CCD) was used to evaluate main and combined effects of above parameters. Then, these independent parameters were fed as inputs to an ANN while the output of the network was the CA of electrospun fiber mat. Finally, the importance of each electrospinning parameters on the variation of CA of electrospun fiber mat was determined and comparison of predicted CA value using RSM and ANN are discussed.

MAIN OBJECTIVES OF THIS CHAPTER:

In this chapter, effects of four electrospinning parameters, including solution concentration (wt.%), applied voltage (kV), tip to collector distance (cm), and volume flow rate (ml/h), on contact angle (CA) of nanofiber mat are studied. To optimize and predict the CA of electrospun fiber mat, RSM and ANN are employed and a quantitative relationship between processing variables and CA of electrospun fibers is established. It is found that the solution concentration is the most important factor impacting the CA of electrospun fiber mat. The obtained results demonstrated that both the proposed models are highly effective in estimating CA of electrospun fiber mat. However, more accurate results are obtained by ANN model as compared to the RSM model. In ANN model the determination coefficient (R^2) and absolute percentage error between actual and predicted response are obtained as 0.965 and 1.97, respectively.

15.2 MATERIALS AND METHODS

15.2.1 MATERIALS

PAN powder was purchased from Polyacryle Co. (Iran). The average molecular weight (M_w) of PAN was approximately 100,000g/mol. *N-N*, dimethylformamide (DMF) was obtained from Merck Co. (Germany) and was used as a solvent. These chemicals were used as received.

15.2.2 ELECTROSPINNING

The PAN powder was dissolved in DMF and gently stirred for 24 hr at 50°C. Therefore, homogenous PAN/DMF solution was prepared in different concentration ranged from 10 to 14 wt. %. Electrospinning was set up in a horizontal configuration as shown in Figure 4.1. The electrospinning apparatus consisted of 5ml plastic syringe connected to a syringe pump and a rectangular grounded collector (aluminum sheet). A high voltage power supply (capable to produce 0–40 kV) was used to apply a proper potential to the metal needle. It should be noted that all electrospinnings were carried out at room temperature.

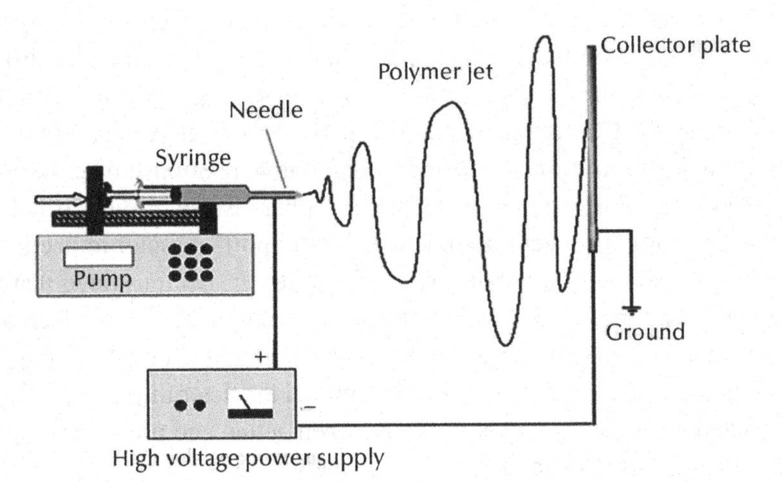

FIGURE 15.1 Schematic of electrospinning set up

15.2.3 MEASUREMENT AND CHARACTERIZATION

The morphology of the gold-sputtered electrospun fibers were observed by scanning electron microscope (SEM, Philips XL-30). The average fiber diameter and distribution was determined from selected SEM image by measuring at least 50 random fibers. The wettability of electrospun fiber mat was determined by CA measurement. The CA measurements were carried out using specially arranged microscope equipped with camera and PCTV vision software as shown in Figure 4.2. The droplet used was distilled water and was 1 μl in volume. The CA experiments were carried out at room temperature and were repeated five times. All contact angles measured within 20 s of placement of the water droplet on the electrospun fiber mat. A typical SEM image of electrospun fiber mat, and its corresponding diameter distribution and CA image are shown in Figure 4.3.

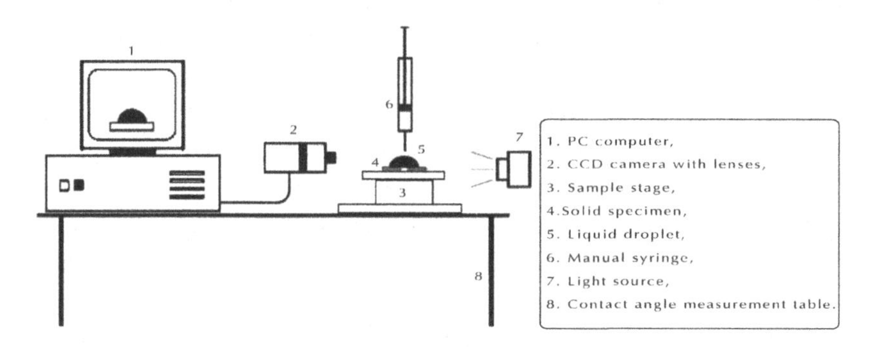

FIGURE 4.2 Schematic of CA measurement set up

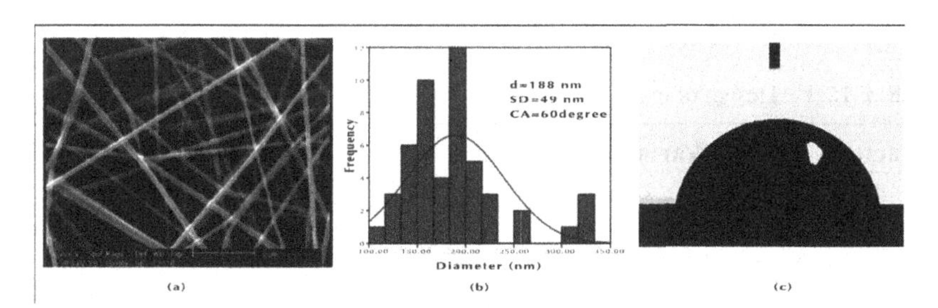

FIGURE 15.3 A typical (a) SEM image, (b) fiber diameter distribution, and (c) CA of electrospun fiber mat

15.3 EXPERIMENTAL DESIGN

15.3.1 RESPONSE SURFACE METHODOLOGY(RSM)

The RSM is a combination of mathematical and statistical techniques used to evaluate the relationship between a set of controllable experimental factors and observed results. This optimization process is used in situations where several input variables influence some output variables (responses) of the system [22–23].

In the present study, central CCD was employed to establish relationships between four electrospinning parameters and the CA of electrospun fiber mat. The experiment was performed for at least three levels of each factor to fit a quadratic model. Based on preliminary experiments, polymer solution concentration (wt. %), applied voltage (kV), tip to collector distance (cm), and volume flow rate (ml/hr) were determined as critical factors with significance effect on CA of electrospun fiber mat. These factors were four independent variables and chosen equally spaced, while the CA of electrospun fiber mat was dependent variable. The values of –1, 0, and 1 are coded variables corresponding to low, intermediate and high levels of each factor respectively. The experimental parameters and their levels for four independent variables are shown in Table 4.1. The regression analysis of the experimental data was carried out to obtain an empirical model between processing variables. The contour surface plots were obtained using Design-Expert software.

TABLE 15.1 Design of experiment (factors and levels)

Factor	Variable	Unit	Factor level		
			-1	0	1
X_1	Solution concentration	(wt.%)	10	12	14
X_2	Applied voltage	(kV)	14	18	22
X_3	Tip to collector distance	(cm)	10	15	20
X_4	Volume flow rate	(ml/h)	2	2.5	3

The quadratic model, Equation (1) including the linear terms, was fitted to the data.

$$Y = \beta_0 + \sum_{i=1}^{4} \beta_i x_i + \sum_{i=1}^{4} \beta_{ii} x_i^2 + \sum_{i=1}^{3} \sum_{j=2}^{4} \beta_{ij} x_i x_j \quad (1)$$

where, Y is the predicted response, x_i and x_j are the independent variables, β_0 is a constant, β_i is the linear coefficient, β_{ii} is the squared coefficient, and β_{ij} is the second-order interaction coefficients [22, 23].

The quality of the fitted polynomial model was expressed by the determination coefficient (R^2) and its statistical significance was performed with the Fisher's statistical test for analysis of variance (ANOVA).

15.3.2 ARTIFICIAL NEURAL NETWORK (ANN)

The ANN is an information processing technique, which is inspired by biological nervous system, composed of simple unit (neurons) operating in parallel. A typical ANN consists of three or more layers, comprising an input layer, one or more hidden layers, and an output layer. Every neuron has connections with every neuron in both the previous and the following layer. The connections between neurons consist of weights and biases. The weights between the neurons play an important role during the training process. Each neuron in hidden layer and output layer has a transfer function to produce an estimate as target. The interconnection weights are adjusted, based on a comparison of the network output (predicted data) and the actual output (target), to minimize the error between the network output and the target [6, 24, 25].

In this study, feed forward ANN with one hidden layer composed of four neurons was selected. The ANN was trained using back-propagation algorithm. The same experimental data used for each RSM designs were also used as the input variables of the ANN. There are four neurons in the input layer corresponding to four electrospinning parameters and one neuron in the output layer corresponding to CA of electrospun fiber mat. Figure 4.4 illustrates the topology of ANN used in this investigation.

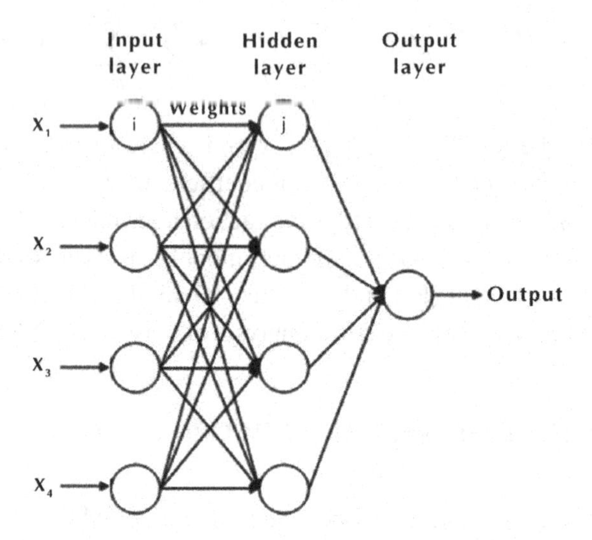

FIGURE 15.4 The topology of artificial neural network used in this study

15.4 RESULTS AND DISCUSSION

This section discusses in details the wettability behavior of electrospun fiber mat concluded from CA measurements. The results of the proposed RSM and ANN models are also presented followed by a comparison between those models.

15.4.1 THE ANALYSIS OF VARIANCE (ANOVA)

All 30 experimental runs of CCD were performed according to Table 4.2. A significance level of 5% was selected; that is, statistical conclusions may be assessed with 95% confidence. In this significance level, the factor has significant effect on CA if the p-value is less than 0.05. On the other hand, when p-value is greater than 0.05, it is concluded the factor has no significant effect on CA.

The results of analysis of variance (ANOVA) for the CA of electrospun fibers are shown in Table 4.3. Equation (2) is the calculated regression equation.

TABLE 15.2 The actual design of experiments and response

No.	Electrospinning parameters				Response
	X_1 Concentration	X_2 Voltage	X_3 Distance	X_4 Flow rate	CA (°)
1	10	14	10	2	44±6
2	10	22	10	2	54±7
3	10	14	20	2	61±6
4	10	22	20	2	65±4
5	10	14	10	3	38±5
6	10	22	10	3	49±4
7	10	14	20	3	51±5
8	10	22	20	3	56±5
9	10	18	15	2.5	48±3
10	12	14	15	2.5	30±3
11	12	22	15	2.5	35±5
12	12	18	10	2.5	22±3
13	12	18	20	2.5	30±4
14	12	18	15	2	33±4
15	12	18	15	3	25±3
16	12	18	15	2.5	26±4
17	12	18	15	2.5	29±3
18	12	18	15	2.5	28±5
19	12	18	15	2.5	25±4
20	12	18	15	2.5	24±3
21	12	18	15	2.5	21±3
22	14	14	10	2	31±4
23	14	22	10	2	35±5
24	14	14	20	2	33±6
25	14	22	20	2	37±4
26	14	14	10	3	19±3
27	14	22	10	3	28±3
28	14	14	20	3	39±5
29	14	22	20	3	36±4
30	14	18	15	2.5	20±3

$$\begin{aligned}CA = 25.80 - 9.89X_1 - 2.17X_2 - 4.33X_3 - 2.33X_4 \\ -1.63X_1X_2 - 1.63X_1X_3 + 1.63X_1X_4 - 0.88X_2X_3 - 0.63X_2X_4 + 0.37X_3X_4 \\ +7.90X_1^2 + 6.40X_2^2 - 0.096X_3^{\ 2} + 2.90X_4^{\ 2}\end{aligned} \quad (2)$$

TABLE 15.3 Analysis of variance for the CA of electrospun fiber mat

Source	SS	DF	MS	F-value	Probe > F	Remarks
Model	4175.07	14	298.22	32.70	<0.0001	Significant
X_1-Concentration	1760.22	1	1760.22	193.01	<0.0001	Significant
X_2-Voltage	84.50	1	84.50	9.27	0.0082	Significant
X_3-Distance	338.00	1	338.00	37.06	<0.0001	Significant
X_4-Flow rate	98.00	1	98.00	10.75	0.0051	Significant
X_1X_2	42.25	1	42.25	4.63	0.0481	Significant
X_1X_3	42.25	1	42.25	4.63	0.0481	Significant
X_1X_4	42.25	1	42.25	4.63	0.0481	Significant
X_2X_3	12.25	1	12.25	1.34	0.2646	
X_2X_4	6.25	1	6.25	0.69	0.4207	Significant
X_3X_4	2.25	1	2.25	0.25	0.6266	
X_1^2	161.84	1	161.84	17.75	0.0008	Significant
X_2^2	106.24	1	106.24	11.65	0.0039	Significant
X_3^2	0.024	1	0.024	0.0026	0.9597	
X_4^2	21.84	1	21.84	2.40	0.1426	
Residual	136.80	15	9.12			
Lack of Fit	95.30	10	9.53	1.15	0.4668	

From the p-values presented in Table 4.3, it can be concluded that the p-values of terms X_3^2, X_4^2, X_2X_3, X_2X_4, and X_3X_4 is greater than the significance level of 0.05, therefore, they have no significant effect on the

CA of electrospun fiber mat. Since the above terms had no significant effect on CA of electrospun fiber mat, these terms were removed. The fitted equations in coded unit are given in Equation (3).

$$\begin{aligned} CA = 26.07 - 9.89X_1 - 2.17X_2 - 4.33X_3 - 2.33X_4 \\ -1.63X_1X_2 - 1.63X_1X_3 + 1.63X_1X_4 \\ +9.08X_1^2 + 7.58X_2^2 \end{aligned} \quad (3)$$

Now, all the p-values are less than the significance level of 0.05. Analysis of variance showed that the RSM model was significant ($p<0.0001$), which indicated that the model has a good agreement with experimental data. The determination coefficient (R^2) obtained from regression equation was 0.958.

15.4.2 ARTIFICIAL NEURAL NETWORK

In this study, the best prediction, based on minimum error, was obtained by ANN with one hidden layer. The suitable number of neurons in the hidden layer was determined by changing the number of neurons. The good prediction and minimum error value were obtained with four neurons in the hidden layer. The weights and bias of ANN for CA of electrospun fiber mat are given in Table 4. The R^2 and mean absolute percentage error were 0.965 and 5.94%, respectively, which indicates that the model showed good fitting with experimental data.

TABLE 15.4 Weights and bias obtained in training ANN

Hidden layer	**Weights**	IW_{11}	IW_{12}	IW_{13}	IW_{14}
		1.0610	1.1064	21.4500	3.0700
		IW_{21}	IW_{22}	IW_{23}	IW_{24}
		-0.3346	2.0508	0.2210	-0.2224
		IW_{31}	IW_{32}	IW_{33}	IW_{34}
		-0.6369	-1.1086	-41.5559	0.0030
		IW_{41}	IW_{42}	IW_{43}	IW_{44}
		-0.5038	-0.0354	0.0521	0.9560
	Bias	b_{11}	b_{21}	b_{31}	b_{41}
		-2.5521	-2.0885	-0.0949	1.5478

Output layer	**Weights**	LW_{11}
		0.5658
		LW_{21}
		0.2580
		LW_{31}
		-0.2759
		LW_{41}
		-0.6657
	Bias	b
		0.7104

15.4.3 EFFECTS OF SIGNIFICANT PARAMETERS ON RESPONSE

The morphology and structure of electrospun fiber mat, such as the nanoscale fibers and interfibrillar distance increases the surface roughness as well as the fraction of contact area of droplet with the air trapped between fibers. It is proved that the CA decrease with increasing the fiber diameter [26], therefore the thinner fibers, due to their high surface roughness, have higher CA than the thicker fibers. Hence, we used this fact for comparing CA of electrospun fiber mat. The interaction contour plot for CA of electrospun PAN fiber mat are shown in Figure 4.5.

As mentioned in the literature, a minimum solution concentration is required to obtain uniform fibers from electrospinning. Below this concentration, polymer chain entanglements are insufficient and a mixture of beads and fibers is obtained. On the other hand, the higher solution concentration would have more polymer chain entanglements and less chain mobility. This causes the hard jet extension and disruption during electrospinning process and producing thicker fibers [27]. Figure 5 (a) show the effect of solution concentration and applied voltage at middle level of distance (15 cm) and flow rate (2.5 ml/hr) on CA of electrospun fiber mat. It is obvious that at any given voltage, the CA of electrospun fiber mat decrease with increasing the solution concentration.

Figure 5 (b) shows the response contour plot of interaction between solution concentration and spinning distance at fixed voltage (18 kV) and flow rate (2.5 ml/hr). Increasing the spinning distance causes the CA of electrospun fiber mat to increase. Because of the longer spinning distance could give more time for the solvent to evaporate, increasing the spinning distance will decrease the nanofiber diameter and increase the CA of electrospun fiber mat [28, 29]. As demonstrated in Figure 5 (b), low solution concentration cause the increase in CA of electrospun fiber mat at large spinning distance.

The response contour plot in Figure 5 (c) represented the CA of electrospun fiber mat at different solution concentration and volume flow rate. Ideally, the volume flow rate must be compatible with the amount of solution removed from the tip of the needle. At low volume flow rates, solvent would have sufficient time to evaporate and thinner fibers were produced, but at high volume flow rate, excess amount of solution fed to the tip of needle and thicker fibers were resulted [28–30]. Therefore, the CA of electrospun fiber mat will be decreased.

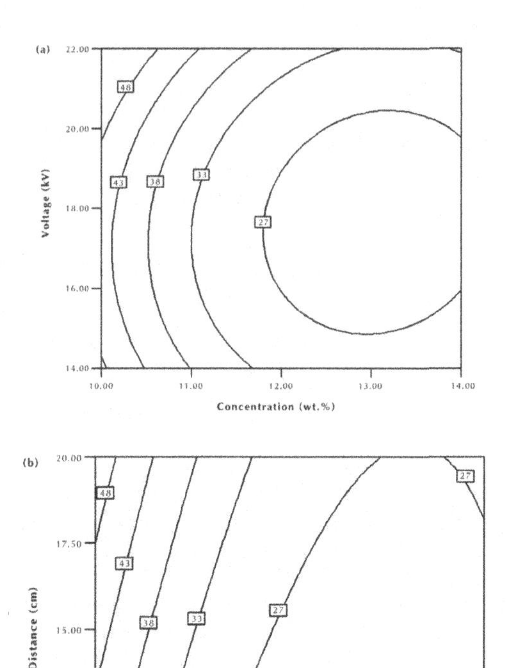

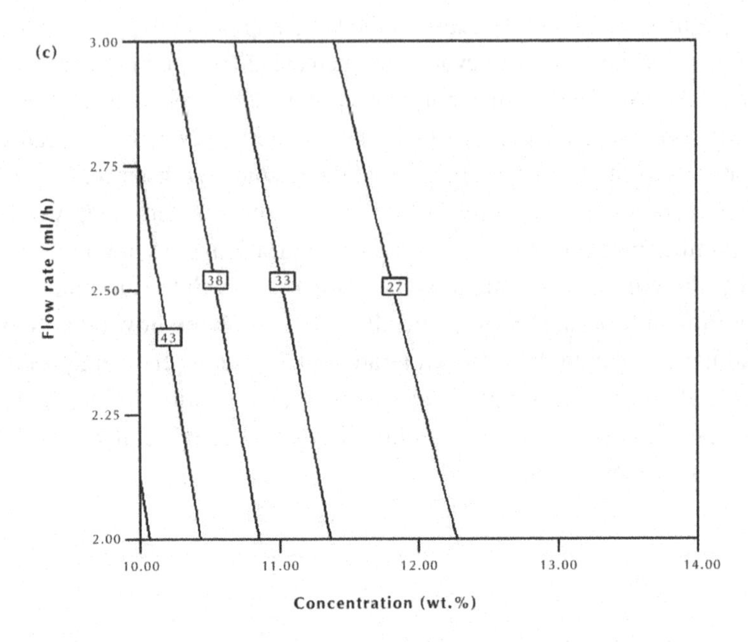

FIGURE 15.5 Contour plots for contact angle of electrospun fiber mat showing the effect of: (a) solution concentration and applied voltage, (b) solution concentration and spinning distance, (c) solution concentration and volume flow rate

As shown by Equation (4), the relative importance (RI) of the various input variables on the output variable can be determined using ANN weight matrix [31].

$$RI_j = \frac{\sum_{m=1}^{N_h} ((|IW_{jm}| / \sum_{k=1}^{N_i} |IW_{km}|) \times |LW_{mn}|)}{\sum_{k=1}^{N_i} \left\{ \sum_{m=1}^{N_h} ((|IW_{km}| / \sum_{k=1}^{N_i} |IW_{km}|) \times |LW_{mn}|) \right\}} \times 100 \tag{4}$$

where, RI_j is the relative importance of the jth input variable on the output variable, N_i and N_h are the number of input variables and neurons in hidden layer, respectively ($N_i = 4$, $N_h = 4$ in this study), IW and LW are the connection weights, and subscript "n" refer to output response (n = 1) [31].

The relative importance of electrospinning parameters on the value of CA calculated by Equation (4) and is shown in Figure 4.6. It can be seen that, all of the input variables have considerable effects on the CA of electrospun fiber mat. Nevertheless, the solution concentration with relative importance of 49.69% is found to be most important factor affecting the

CA of electrospun nanofibers. These results are in close agreement with those obtained with RSM.

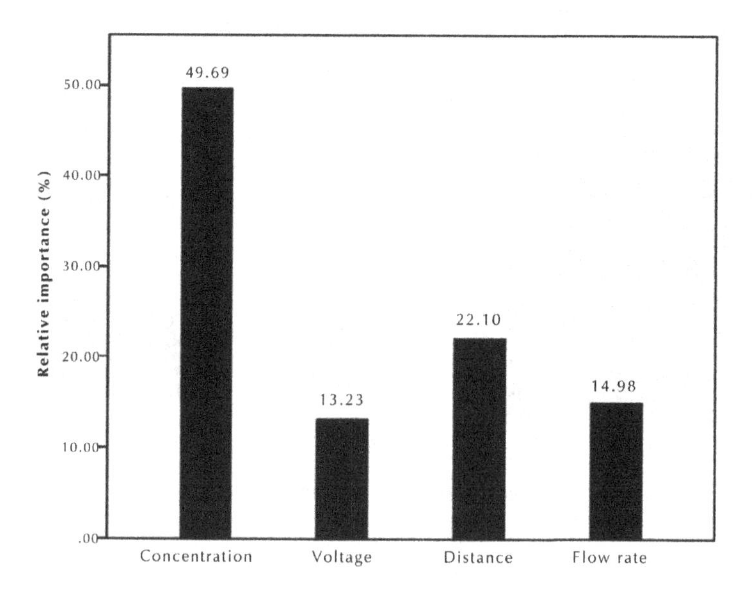

FIGURE 15.6 Relative importance of electrospinning parameters on the CA of electrospun fiber mat

15.4.4 OPTIMIZING THE CA OF ELECTROSPUN FIBER MAT

The optimal values of the electrospinning parameters were established from the quadratic form of the RSM. Independent variables (solution concentration, applied voltage, spinning distance, and volume flow rate) were set in range and dependent variable (CA) was fixed at minimum. The optimal conditions in the tested range for minimum CA of electrospun fiber mat are shown in Table 4.5. This optimum condition was a predicted value, thus to confirm the predictive ability of the RSM model for response, a further electrospinning and CA measurement was carried out according to the optimized conditions and the agreement between predicted and measured responses was verified. Figure 4.7 shows the SEM, average fiber diameter distribution and corresponding CA image of electrospun fiber mat prepared at optimized conditions.

TABLE 15.5 Optimum values of the process parameters for minimum CA of electrospun fiber mat

Solution concentration (wt.%)	Applied voltage (kV)	Spinning distance (cm)	Volume flow rate (ml/h)	Predicted CA (°)	Observed CA (°)
13.2	16.5	10.6	2.5	20	21

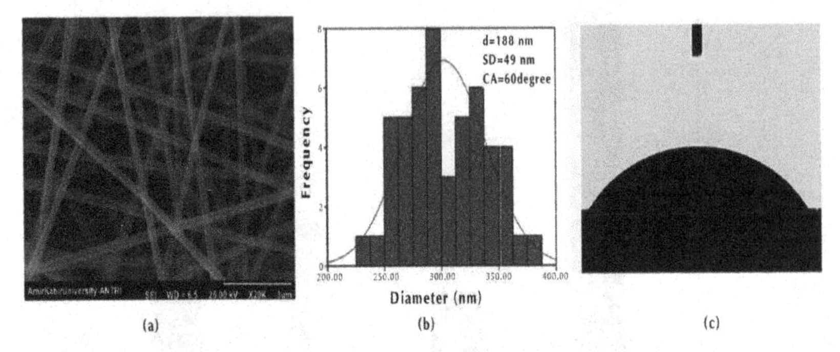

FIGURE 15.7 (a) SEM image, (b) fiber diameter distribution, and (c) CA of electrospun fiber mat prepared at optimized conditions

15.4.5 COMPARISON BETWEEN RSM AND ANN MODEL

Table 4.6 gives the experimental and predicted values for the CA of electrospun fiber mat obtained from RSM as well as ANN model. It is demonstrated that both models performed well and a good determination coefficient was obtained for both RSM and ANN. However, the ANN model shows higher determination coefficient ($R^2 = 0.965$) than the RSM model ($R^2 = 0.958$). Moreover, the absolute percentage error in the ANN prediction of CA was found to be around 5.94%, while for the RSM model, it was around 7.83%. Therefore, it can be suggested that the ANN model shows more accurately result than the RSM model. The plot of actual and predicted CA of electrospun fiber mat for RSM and ANN is shown in Figure 4.8.

TABLE 15.6 Experimental and predicted values by RSM and ANN models

No.	Experimental	Predicted		Absolute error (%)	
		RSM	ANN	RSM	ANN
1	44	47	48	6.41	9.97
2	54	54	54	0.78	0.46
3	61	59	61	3.70	0.42
4	65	66	61	2.06	6.06
5	38	39	38	2.37	0.54
6	49	47	49	5.10	0.68
7	51	51	51	0.35	0.45
8	56	58	56	4.32	0.17
9	48	45	60	6.17	24.37
10	30	31	27	4.93	9.35
11	35	36	31	2.34	11.15
12	22	22	21	1.18	4.15
13	30	30	32	1.33	6.04
14	33	28	33	13.94	0.60
15	25	24	25	5.04	0.87
16	26	26	26	0.27	1.33
17	29	26	26	10.10	9.16
18	28	26	26	6.89	5.91
19	25	26	26	4.28	5.38
20	24	26	26	8.63	9.77
21	21	26	26	24.14	25.45
22	31	30	31	2.26	0.57
23	35	31	35	10.34	0.66
24	33	36	32	8.18	2.18
25	37	37	37	0.59	0.34
26	19	29	21	52.11	10.23
27	28	30	30	7.07	8.20
28	39	34	31	12.05	21.30
29	36	35	36	1.72	0.04
30	20	25	20	26.30	2.27
R^2		0.958	0.965		
	Mean absolute error (%)			7.83	5.94

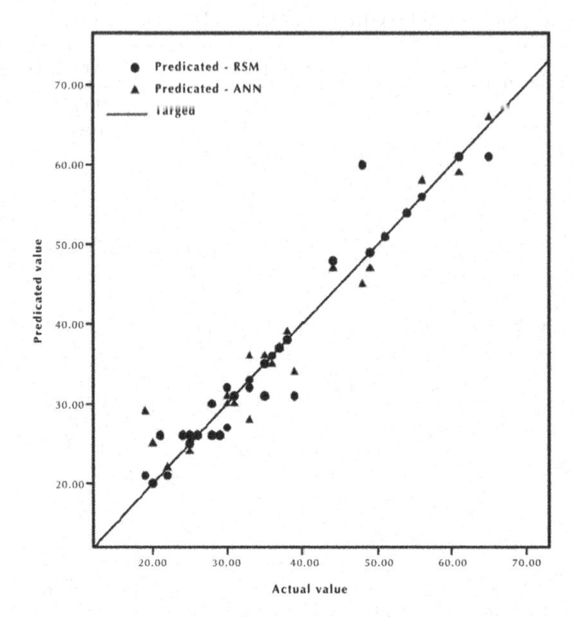

FIGURE 15.8 Comparison between the actual and predicted contact angle of electrospun nanofiber for RSM and ANN model

15.5 CONCLUSION

The morphology and properties of electrospun nanofibers depends on many processing parameters. In this work, the effects of four electrospinning parameters namely; solution concentration (wt.%), applied voltage (kV), tip to collector distance (cm), and volume flow rate (ml/hr) on CA of PAN nanofiber mat were investigated using two different quantitative models, comprising RSM and ANN. The RSM model confirmed that solution concentration was the most significant parameter in the CA of electrospun fiber mat. Comparison of predicted CA using RSM and ANN were also studied. The obtained results indicated that both RSM and ANN model shows a very good relationship between the experimental and predicted CA values. The ANN model shows higher determination coefficient ($R^2 = 0.965$) than the RSM model. Moreover, the absolute percentage error of prediction for the ANN model was much lower than that for RSM model, indicating that ANN model had higher modeling performance than RSM model. The minimum CA of electrospun fiber mat was estimated by RSM equation obtained at conditions of 13.2 wt.% solution concentration, 16.5kV of the applied voltage, 10.6 cm of tip to collector distance, and 2.5 ml/hr of volume flow rate.

15.6 APPENDIX

In 1966, Taylor discovered that finite conductivity enables electrical charge to accumulate at the drop interface, permitting a tangential electric stress to be generated. The tangential electric stress drags fluid into motion, and thereby generates hydrodynamic stress at the drop interface. The complex interaction between the electric and hydrodynamic stresses causes either oblate or prolate drop deformation, and in some special cases keeps the drop from deforming.

Feng has used a general treatment of Taylor–Melcher for stable part of electrospinning jets by one dimensional equations for mass, charge, and momentum. In this model a cylindrical fluid element is used to show electrospinning jet kinematic measurements.

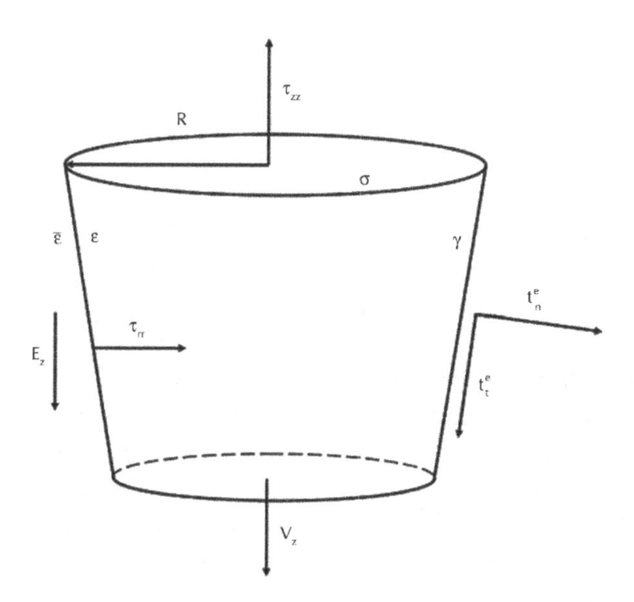

FIGURE 15.9 Scheme of the cylindrical fluid element used in electrohydrodynamic modeling

In (Figure 4.9) the essential parameters are: radius, R; velocity, v_z; electric field, E_z; total path length, L; interfacial tension, γ; interfacial charge, σ; tensile stress, τ; volumetric flow rate, Q; conductivity, K; density, ρ; dielectric constant, ε; and zero-shear rate viscosity, η_0. The most important equation that Feng used is

$$\tilde{R}^2\tilde{v}_z = 1$$

$$\tilde{R}^2\tilde{E}_z + Pe_e\tilde{R}\tilde{v}_z\tilde{\sigma} = 1$$

$$\tilde{\upsilon}_z\tilde{\upsilon}'_z = \frac{1}{Fr} + \frac{\tilde{T}'}{\mathrm{Re}_j\tilde{R}^2} + \frac{1}{We}\frac{\tilde{R}'}{\tilde{R}^2} + \varepsilon(\tilde{\sigma}\tilde{\sigma}' + \beta\tilde{E}_z\tilde{E}'_z + \frac{2\tilde{\sigma}\tilde{E}_z}{\tilde{R}})$$

$$\tilde{E}_z = \tilde{E}_0 - \ln\chi\left[(\tilde{\sigma}\tilde{R})' - \frac{\beta}{2}(\tilde{E}\tilde{R}^2)''\right]$$

$$E_0 = {}^{I_0 V_0}\!/_{R_0} \qquad \beta = ({}^{\varepsilon}\!/_{\bar{\varepsilon}}) - 1 \qquad \tau = R^2(\tilde{\tau}_{zz} - \tilde{\tau}_{rr})$$

Feng solved above equation under different fluid properties, particularly for non Newtonian fluids with extensional thinning, thickening, and strain hardening, but Helgeson et al. developed a simplified understanding of electrospinning jets based on the evolution of the tensile stress due to elongation.

The initialization of instability on the surfaces of liquids should be caused using the application of the external electric field that induces electric forces on surfaces of liquids. A localized approximation was developed to calculate the bending electric force acting on an electrified polymer jet, which is an important element of the electrospinning process for manufacturing of nanofibers. Using this force, a far reaching analogy between the electrically driven bending instability and the aerodynamically driven instability was established. The description of the wave's instabilities is expressed by equations called dispersion laws. The dependence of wavelength on the surface tension γ is almost linear and the wavelengths between jets are a little bit smaller for lower depths. Dependency of wavelength on electric field strength is exponential. The dispersion law is identified for four groups of dielectrics liquids with using of Clausius–Mossotti and Onsager's relation (nonpolar liquids with finite and infinite depth and weakly polar liquids with finite and infinite depth). According to these relations relative permittivity is a function of parameters, such as temperature, square of angular frequency, wave length number, and reflective index.

The best way for analyzing fluid mechanics problems is converting parameters to dimensionless form. By using this method the numbers of governing parameters for given geometry and initial condition reduce. The non dimensionalization of a fluid mechanics problem generally starts with the selection of a characteristic velocity then because the flow of non Newtonian fluids the stress depends non linearly on the flow kinematics, the deformation rate is a main quantity in the analysis of these flows. Next step after determining different parameters is evaluate Characteristic values of the parameters. Then the non dimensionalization procedure is to employ the selected characteristic quantities to obtain a dimensionless version of the conservation equations and get to certain numbers like Reynolds number and the Galilei number. The excessive number of governing dimensionless groups poses certain difficulties in the fluid mechanics analysis. Finally, by using these results in equation and applying boundary conditions it can be achieved to study different properties.

Electrospinning jets producing nanofibers from a polymer solution by electrical forces are fine cylindrical electrodes that create extremely high electric-field strength in their vicinity at atmospheric conditions. However, this quality of electrospinning is only scarcely investigated, and the interactions of the electric fields generated by them with ambient gases are nearly unknown. Pokorny et al. reported on the discovery that electrospinning jets generate X-ray beams up to energies of 20 keV at atmospheric conditions. Scientists investigated on the discovery that electrically charged polymeric jets and highly charged gold-coated nanofibrous layers in contact with ambient atmospheric air generate X-ray beams up to energies of hard X-rays.

The first detection of X-ray produced by nanofiber deposition was observed using radiographic films. The main goal of using these films was to understand Taylor cone creation. The X-ray generation is probably dependent on diameters of the nanofibers that are affected by the flow rate and viscosity. So, it is important to find the ideal concentration (viscosity) of polymeric solutions and flow rate to spin nanofibers as thin as possible. The X-ray radiation can produce black traces on the radiographic film. These black traces had been made by outer radiation generated by nanofibers and the radiation has to be in the X-ray region of electromagnetic spectra, because the radiation of lower energy is absorbed by the shield film. Radiographic method of X-ray detection is efficient and sensitive. It is obvious that this method did not tell us anything about its spectrum, but it can clearly show its space distribution. The humidity, temperature, and rheological parameters of polymer can affect on the X-ray intensity generated by nanofibers. The necessity of modeling in electrospinning process and a quick outlook of some important models will be discussed as follows.

Using theoretical prediction of the parameter effects on jet radius and morphology can significantly reduce experimental time by identifying the most likely values that will yield specific qualities prior to production. All models start with some assumptions and have short comings that need to be addressed. The basic principles for dealing with electrified fluids that Taylor discovered are impossible to account for most electrical phenomena involving moving fluids under the seemingly reasonable assumptions that the fluid is either a perfect dielectric or a perfect conductor. The reason is that any perfect dielectric still contains a non zero free charge density. The presence of both an axisymmetric instability and an oscillatory "whipping" instability of the centerline of the jet; however, the quantitative characteristics of these instabilities disagree strongly with experiments. During steady jetting, a straight part of the jet occurs next to the Taylor cone, where only axisymmetric motion of the jet is observed. This region of the jet remains stable in time. However, further along its path the jet can be unstable by non axisymmetric instabilities, such as bending and branching, where lateral motion of the jet is observed in the part near the collector.

Branching as the instability of the electrospinning jet can happen quite regularly along the jet if the electrospinning conditions are selected appropriately. Branching is a direct consequence of the presence of surface charges on the jet surface, as well as of the externally applied electric field. The bending instability leads to drastic stretching and thinning of polymer jets toward nano-scale in cross-section. Electrospun jets also caused to shape perturbations similar to undulations, which can be the source of various secondary instabilities leading to nonlinear morphologies developing on the jets. The bending instabilities that occur during electrospinning have been studied and mathematically modeled by Reneker et al by viscoelastic dumbbells connected together. Both electrostatic and fluid dynamic instabilities can contribute to the basic operation of the process.

Different stages of electrospun jets have been investigated by different mathematical models during last decade by one or three dimensional techniques.

Physical models which study the jet profile, the stability of the jet, and the cone-like surface of the jet have been develop due to significant effects of jet shape on fiber qualities. Droplet deformation, jet initiation and, in particular, the bending instability which control to a major extent fiber sizes and properties are controlled apparently predominantly by charges located within the flight jet. An accurate, predictive tool using a verifiable model that accounts for multiple factors would provide a means to run many different scenarios quickly without the cost and time of experimental trial and error.

The models typically treat the jet mechanics using the localized-induction approximation by analogy to aerodynamically driven jets. They include the viscoelasticity of the spinning fluid and have also been augmented to account for solvent evaporation in the jet. These models can describe the bending instability and fiber morphology. Because of difficulty in measure model variables they cannot accurately design and control the electrospinning process.

The principles for dealing with electrified fluids were summarized by many researchers. Their research showed that it is impossible to explain the most of the electrical phenomena involving moving fluids given the hypothesis that the fluid is either a perfect dielectric or a perfect conductor, since both the permittivity and the conductivity affect the flow. An electrified liquid always includes free charge. Although the charge density may be small enough to ignore bulk conduction effects, the charge will accumulate at the interfaces between fluids. The presence of this interfacial charge will result in an additional interfacial stress, especially a tangential stress, which in turn will modify the fluid dynamics.

The electro-hydrodynamic theory proposed by Taylor as the leaky dielectric model is capable of predicting the drop deformation in qualitative agreement with the experimental observations.

Although, Taylor's leaky dielectric theory provides a good qualitative description for the deformation of a Newtonian drop in an electric field, the validity of its analytical results is strictly limited to the drop experiencing small deformation in an infinitely extended domain. Extensive experiments showed a serious difference in this theoretical prediction.

Some investigations have been done to solve this defect. For example, to examine electrokinetic effects, the leaky dielectric model was modified by consideration the charge transport. When the conductivity is finite, the leaky dielectric model can be used. By use of this mean, the solution is weakly conductive so the jet carries electric charges only on its surface.

Comprehension of the drops behavior in an electric field is playing a critical role in practical applications. The electric field-driven flow is of practical importance in the processes in which improvement of the rate of mass or heat transfer between the drops and their surrounding fluid. Numerically investigations about the shape evolution of small droplets attached to a conducting surface depended on strong electric fields (weak, strong and super-electrical) have done and indicated that three different scenarios of droplet shape evolution are distinguished, based on numerical solution of the Stokes equations for perfectly conducting droplets by investigation of Maxwell stresses and surface tension. The advantages of this model are that the non-Newtonian effect on the drop dynamics is successfully identified on the basis of electro-hydrostatics at least qualitatively. In addition, the model showed that the deformation and breakup modes of the non-Newtonian drops are distinctively different from the Newtonian cases.

A simple two-dimensional model can be used for description of formation of barb electrospun polymer nano-wires with a perturbed swollen cross-section and the electric charges "frozen" into the jet surface. This model was integrated numerically using the Kutta–Merson method with the adoptable time step. The result of this modeling is explained theoretically as a result of relatively slow charge relaxation compared to the development of the secondary electrically driven instabilities which deform jet surface locally. When the disparity of the slow charge relaxation compared with the rate of growth of the secondary electrically driven instabilities becomes even more pronounced, the barbs transform in full scale long branches. The competition between charge relaxation and rate of growth of capillary and electrically driven secondary localized perturbations of the jet surface is affected not only by the electric conductivity of polymer solutions but also by their viscoelasticity. Moreover, a non linear theoretical model was able to resemble the main morphological trends recorded in the experiments.

There is not a theoretical model which can describe the electrospinning process under the multi-field forces so a simple model might be very useful to indicate the contributing factors. Modeling this process can be done in two ways: a) the deterministic approach which uses classical mechanics like Euler approach and Lagrange approach. b) The probabilistic approach uses E-infinite theory and quantum like properties. Many basic properties are harmonious by adjusting electrospinning parameters, such as voltage, flow rate, and others, and it can offer in-depth inside into physical understanding of many complex phenomena which cannot be fully explain.

KEYWORDS

- **Artificial neural network**
- **Chemical vapor deposition**
- **Contact angle**
- **Electrospun fiber matFirst, a central composite design**
- **Response surface methodology**

REFERENCES

1. Miwa, M., Nakajima, A., Fujishima, A., Hashimoto, K., & Watanabe, T. (2000). *Langmuir, 16,* 5754.

2. Öner, D. & McCarthy, T. J. (2000). *Langmuir, 16,* 7777.
3. Abdelsalam, M. E., Bartlett, P. N., Kelf, T., & Baumberg, J. (2005). *Langmuir, 21,* 1753.
4. Nakajima, A., Hashimoto, K., Watanabe, T., Takai, K., Yamauchi, G., & Fujishima, A. (2000). *Langmuir, 16,* 7044.
5. Zhong, W., Liu, S., Chen, X., Wang, Y., & Yang, W. (2006). *Macromolecules, 39,* 3224.
6. Shams Nateri, A. & Hasanzadeh, M. (2009). *Journal of Computational and Theoretical Nanoscience, 6,* 1542.
7. Kilic, A., Oruc, F., & Demir, A. (2008). *Textile Research Journal, 78,* 532.
8. Reneker, D. H., & Chun, I. (1996). *Nanotechnology, 7,* 216.
9. Shin, Y. M., Hohman, M. M., Brenner, M. P., & Rutledge, G. C. (2001). *Polymer, 42,* 9955.
10. Reneker, D. H., Yarin, A. L., Fong, H., & Koombhongse, S. (2000). *Journal of Applied Physics, 87,* 4531.
11. Zhang, S., Shim, W. S., & Kim, J. (2009). *Materials and Design, 30,* 3659.
12. Yördem, O. S., Papila, M., & Menceloğlu, Y. Z. (2008). *Materials and Design, 29,* 34.
13. Chronakis, I. S. (2005). *Journal of Materials Processing Technology, 167,* 283.
14. Dotti, F., Varesano, A., Montarsolo, A., Aluigi, A., Tonin, C., & Mazzuchetti, G. (2007). *Journal of Industrial Textiles, 37,* 151.
15. Lu, Y., Jiang, H., Tu, K., & Wang, L. (2009). *Acta Biomaterialia, 5,* 1562.
16. Lu, H., Chen, W., Xing, Y., Ying, D., & Jiang, B. (2009). *Journal of Bioactive and Compatible Polymers, 24,* 158.
17. Nisbet, D. R., Forsythe, J. S., Shen, W., Finkelstein, D. I., & Horne, M. K. (2009). *Journal of Biomaterials Applications, 24,* 7.
18. Ma, Z., Kotaki, M., Inai, R., & Ramakrishna, S. (2005). *Tissue Engineering, 11,* 101.
19. Hong, K. H. (2007). *Polymer Engineering and Science, 47,* 43.
20. Zhang, W., & Pintauro, P. N. (2011). *ChemSusChem, 4,* 1753.
21. Lee, S., & Obendorf, S. K. (2007). *Textile Research Journal, 77,* 696.
22. Myers, R. H., Montgomery, D. C., & Anderson-cook, C. M. (2009). *Response surface methodology: Process and product optimization using designed experiments,* (3rd edn.). USA: Wiley & Sons.
23. Gu, S. Y., Ren, J., & Vancso, G. J. (2005). *European Polymer Journal, 41,* 2559.
24. Dev, V. R. G., Venugopal, J. R., Senthilkumar, M., Gupta, D., & Ramakrishna, S. (2009). *Journal of Applied Polymer Science, 113,* 3397.
25. Galushkin, A. L. (2007). *Neural networks theory* (Moscow Institute of Physics & Technology). Moscow: Springer.
26. Ma, M., Mao, Y., Gupta, M., Gleason, K. K., & Rutledge, G. C. (2005). *Macromolecules, 38,* 9742
27. Haghi, A. K. & Akbari, M. (2007). *Physica Status Solidi (A), 204,* 1830.
28. Ziabari, M., Mottaghitalab, V., & Haghi, A. K. (2009). In W. N. Chang (Ed.), *Nanofibers: Fabrication, Performance, and Applications*. USA: Nova Science Publishers.
29. Ramakrishna, S., Fujihara, K., Teo, W. E., Lim, T. C., & Ma, Z. (2005). *An Introduction to Electrospinning and Nanofibers*. Singapore: World Scientific Publishing.
30. Zhang, S., Shim, W. S., & Kim, J. (2009). *Materials and Design Journal, 30,* 3659.
31. Kasiri, M. B., Aleboyeh, H., & Aleboyeh, A. (2008). *Environmental Science and Technology, 42,* 7970.

CHAPTER 16

NEW DEVELOPMENTS IN THE OPTIMIZATION OF ELECTROSPINNING PROCESS

M. HASANZADEH, B. HADAVI MOGHADAM,
M. H. MOGHADAM ABATARI, and A. K. HAGHI

CONTENTS

16.1 INTRODUCTION

Recently, it was demonstrated that electrospinning can produce superfine fiber ranging from micrometer to nanometer using an electric field force. In the electrospinning process, a strong electric field is applied between polymer solution contained in a syringe with a capillary tip and grounded collector. When the electric field overcomes the surface tension force, the charged polymer solution forms a liquid jet and travels towards collection plate. As the jet travels through the air, the solvent evaporates and dry fibers deposits on the surface of a collector [1-4].

The electrospun nanofibers have high specific surface area, high porosity, and small pore size. Therefore, they have been suggested as excellent candidate for many applications including filtration, multifunctional membranes, tissue engineering, protective clothing, reinforced composites, and hydrogen storage [5, 6].Studies have shown that the morphology and the properties of the electrospun nanofibers depend on many parameters including polymer solution properties (the concentration, liquid viscosity, surface tension, and dielectric properties of the polymer solution), processing parameters (applied voltage, volume flow rate, tip to collector distance, and the strength of the applied electric field), and ambient conditions (temperature, atmospheric pressure and humidity) [5-8].

Response surface methodology (RSM) is a combination of mathematical and statistical techniques used to evaluate the relationship between a set of controllable experimental factors and observed results. This optimization process is used in situations where several input variables influence some output variables of the system. The main goal of RSM is to optimize the response, which is influenced by several independent variables, with minimum number of experiments [9, 10]. Therefore, the application of RSM in electrospinning process will be helpful in effort to find and optimize the electrospun nanofibers properties.

In this chapter, a study has been conducted to investigate the relationship between four electrospinning parameters (solution concentration, applied voltage, tip to collector distance, and volume flow rate) and electrospun PAN nanofiber mat properties such as average fiber diameter (AFD) and contact angle (CA).

MAIN OBJECTIVES OF THIS CHAPTER:

Response surface methodology (RSM) based on central composite design (CCD) was employed to model and optimize the electrospinning parameters such as solution concentration (wt.%), applied voltage (kV), tip to collector distance (cm), and volume flow rate (ml/hr), that have important effects on average fiber diameter (AFD) and contact angle (CA) of nanofiber mat. It is observed that polymer solution played an important role to the AFD and CA of nanofibers. Analysis of variance (ANOVA) showed a high determination coefficient (R^2) value of 0.9640 and 0.9683 for AFD and CA respectively, which indicated that the both models have a good agreement with experimental data. According to model optimization of the process, the minimum CA of electrospun fiber mat is given by following conditions: 13.2wt.% solution concentration, 16.5kV of the applied voltage, 10.6cm of tip to collector distance, and 2.5ml/hr of volume flow rate.

16.2 MATERIALS AND METHODS

16.2.1 MATERIALS

Polyacrylonirile (PAN, M_w=100,000) was purchased from Polyacryle Co. (Iran) and *N-N*, dimethylformamide (DMF) was obtained from Merck Co. (Germany).

The polymer solutions with different concentration ranged from 10 to 14wt.% were prepared by dissolving PAN powder in DMF and were stirred for 24hrs at 50°C. These polymer solutions were used for electrospinning.

16.2.2 ELECTROSPINNING

A schematic of the electrospinning apparatus is shown in Figure 5.1. A polymer solution was loaded in a 5mL syringe connected to a syringe pump. The tip of the syringe was connected to a high voltage power supply (capable to produce 0–40kV). Under high voltage, a fluid jet was ejected from the tip of the needle and accelerated toward the grounded collector (aluminum foil). All electrospinnings were carried out at room temperature.

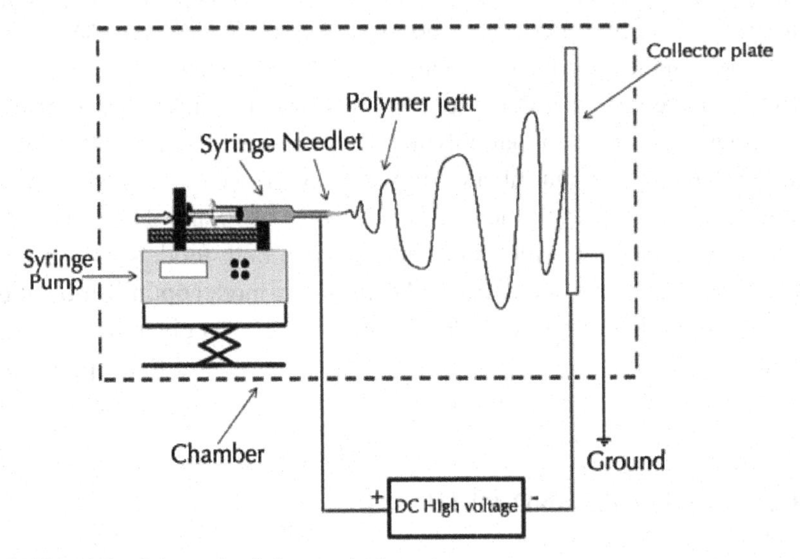

FIGURE 16.1 Schematic of electrospinning set up

16.2.3 MEASUREMENT AND CHARACTERIZATION

The electrospun nanofibers were sputter-coated with gold and their morphology was examined with a scanning electron microscope (SEM, Philips XL-30). Average diameter of electrospun nanofibers was determined from selected SEM image by measuring at least 50 random fibers using Image J software.

The wettability of electrospun fiber mat was determined by water contact angle measurement. Contact angles were measured by specially arranged microscope equipped with camera and PCTV vision software as shown in Figure 5.2. The volume of the distilled water for each measurement was kept at 1 μl.

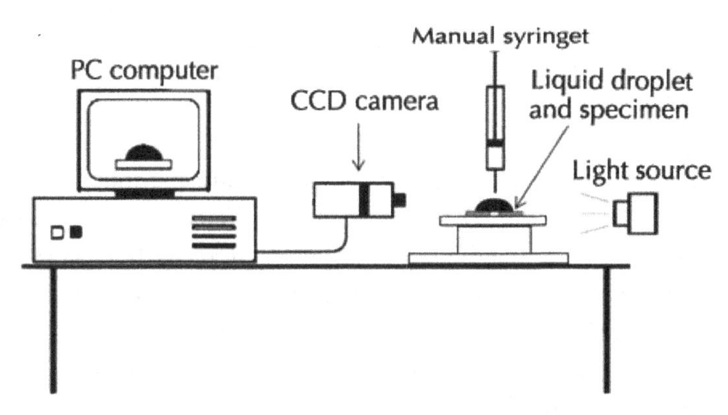

FIGURE 5.2 Schematic of contact angle measurement set up

16.2.4 EXPERIMENTAL DESIGN BY RSM

In this chapter, the effect of four electrospinning parameters on two responses, comprising the AFD and the CA of electrospun fiber mat, was evaluated using central composite design (CCD). The experiment was performed for at least three levels of each factor to fit a quadratic model. Polymer solution concentration (X_1), applied voltage (X_2), tip to collector distance (X_3), and volume flow rate (X_4) were chosen as independent variables and the AFD and the CA of electrospun fiber mat as dependent variables (responses). The experimental parameters and their levels are given in Table 5.1.

TABLE 16.1 Design of experiment (factors and levels)

Factor	Variable	Unit	Factor level		
			-1	0	1
X_1	Solution concentration	(wt.%)	10	12	14
X_2	Applied voltage	(kV)	14	18	22
X_3	Tip to collector distance	(cm)	10	15	20
X_4	Volume flow rate	(ml/hr)	2	2.5	3

A quadratic model, which also includes the linear model, is given below:

$$Y = \beta_0 + \sum_{i=1}^{4} \beta_i x_i + \sum_{i=1}^{4} \beta_{ii} x_i^2 + \sum_{i=1}^{3} \sum_{j=2}^{4} \beta_{ij} x_i x_j \tag{1}$$

where, Y is the predicted response, x_i and x_j are the independent variables, β_0 is a constant, β_i is the linear coefficients, β_{ii} is the squared coefficients and β_{ij} is the second-order interaction coefficients [9, 10].

The statistical analysis of experimental data was performed using Design-Expert software (Version 8.0.3, Stat-Ease, Minneapolis, MN, 2010) including analysis of variance (ANOVA). A design of 30 experiments for independent variables and responses for AFD and CA are listed in Table 5.2.

TABLE 16.2 The actual design of experiments and responses for AFD and CA

No.	Electrospinning Parameters				Responses	
	X^1 Concentration	X^2 Voltage	X^3 Distance	X^4 Flow rate	AFD (nm)	CA (°)
1	10	14	10	2	206±33	44±6
2	10	22	10	2	187±50	54±7
3	10	14	20	2	162±25	61±6
4	10	22	20	2	164±51	65±4
5	10	14	10	3	225±41	38±5
6	10	22	10	3	196±53	49±4
7	10	14	20	3	181±43	51±5
8	10	22	20	3	170±50	56±5
9	10	18	15	2.5	188±49	48±3
10	12	14	15	2.5	210±31	30±3
11	12	22	15	2.5	184±47	35±5
12	12	18	10	2.5	214±38	22±3
13	12	18	20	2.5	205±31	30±4
14	12	18	15	2	195±47	33±4
15	12	18	15	3	221±23	25±3

TABLE 16.2 *(Continued)*

No.	Electrospinning Parameters				Responses	
	X^1 Concentration	X^2 Voltage	X^3 Distance	X^4 Flow rate	AFD (nm)	CA (°)
16	12	18	15	2.5	199±50	26±4
17	12	18	15	2.5	205±31	29±3
18	12	18	15	2.5	225±38	28±5
19	12	18	15	2.5	221±23	25±4
20	12	18	15	2.5	215±35	24±3
21	12	18	15	2.5	218±30	21±3
22	14	14	10	2	255±38	31±4
23	14	22	10	2	213±37	35±5
24	14	14	20	2	240±33	33±6
25	14	22	20	2	200±30	37±4
26	14	14	10	3	303±36	19±3
27	14	22	10	3	256±40	28±3
28	14	14	20	3	283±48	39±5
29	14	22	20	3	220±41	36±4
30	14	18	15	2.5	270±43	20±3

16.3 RESULTS AND DISCUSSION

16.3.1 MORPHOLOGICAL ANALYSIS OF NANOFIBERS

The PAN solution in DMF were electrospun under different conditions, including various PAN solution concentrations, applied voltages, volume flow rates and tip to collector distances, to study the effect of electrospinning parameters on the morphology and properties of electrospun nanofibers.

Figure 5.3 shows the SEM images and fiber diameter distributions of electrospun fibers in different solution concentration as one of the most effective parameters to control the fiber morphology. As observed in Figure 5.3, the AFD increased with increasing solution concentration. It was suggested that the higher solution concentration would have more polymer chain entanglements and less chain mobility. This causes the hard jet extension and disruption during electrospinning process and producing thicker fibers.

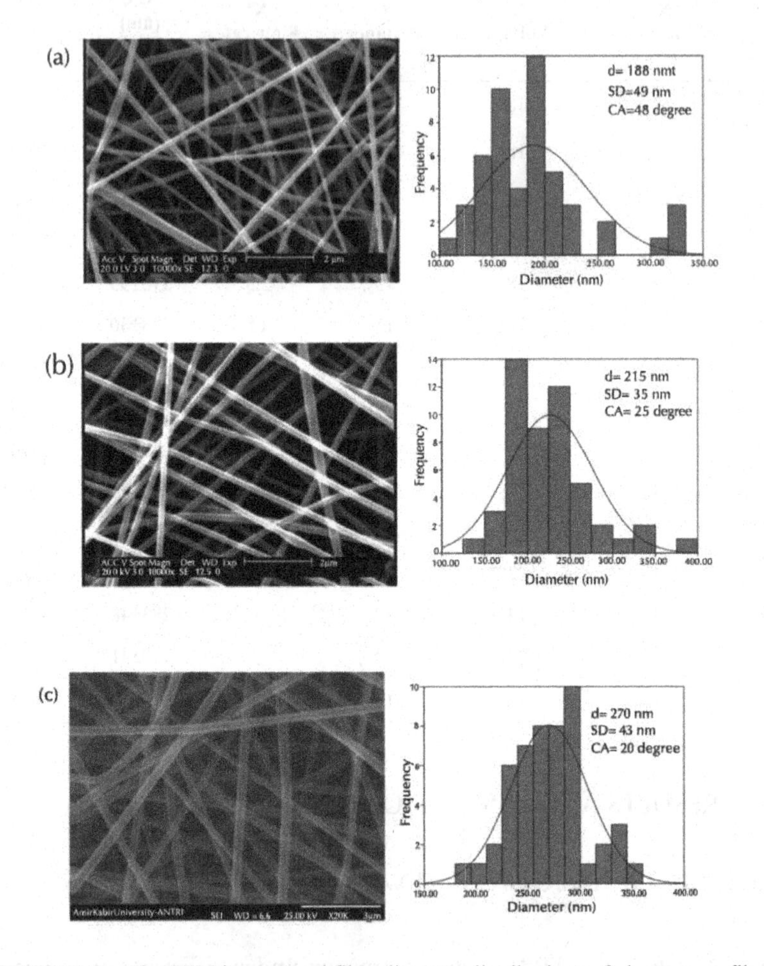

FIGURE 16.3 The SEM images and fiber diameter distributions of electrospun fibers in solution concentration of (a) 10wt.%, (b) 12wt.% and (c) 14wt.%

The SEM image and corresponding fiber diameter distribution of electrospun nanofiber in different applied voltage are shown in Figure 5.4. It is obvious that increasing the applied voltage cause an increase followed by a decrease in electrospun fiber diameter. As demonstrated by previous researchers [7, 8], increasing the applied voltage may decrease, increase or may not change the fiber diameter. In one hand, increasing the applied voltage will increase the electric field strength and higher electrostatic repulsive force on the jet, favoring the thinner fiber formation. On the other hand, more surface charge will introduce on the jet and the solution will be removed more quickly from the tip of needle. As a result, the AFD will be increased [8, 11].

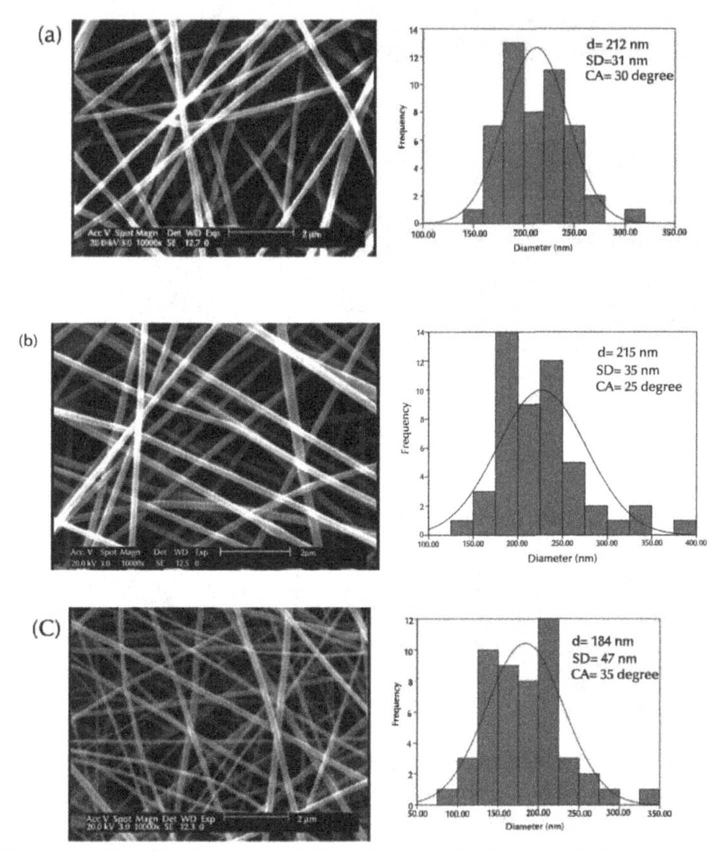

FIGURE 16.4 The SEM images and fiber diameter distributions of electrospun fibers in applied voltage of (a) 14kV, (b) 18kV and (c) 22kV

Figure 5.5 represents the SEM image and fiber diameter distribution of electrospun nanofiber in different spinning distance. It can be seen that the AFD decreased with increasing tip to collector distance. Because of the longer spinning distance could give more time for the solvent to evaporate, increasing the spinning distance will decrease fiber diameter [3, 8].

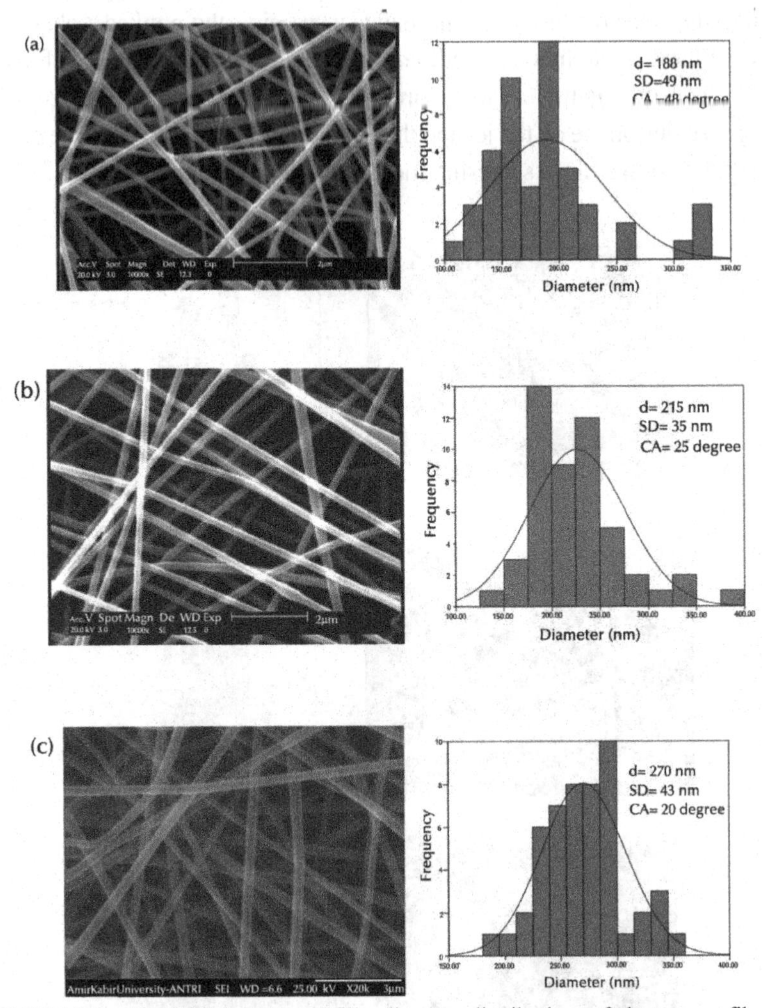

FIGURE 16.5 The SEM images and fiber diameter distributions of electrospun fibers in tip to collector distance of (a) 10cm, (b) 15cm and (c) 20cm.

The SEM image and fiber diameter distribution of electrospun nanofiber in different volume flow rate are illustrated in Figure 5.6. It is clear that increasing the volume flow rate cause an increase in average fiber diameter. Ideally, the volume flow rate must be compatible with the amount of solution removed from the tip of the needle. At low volume flow rates, solvent would have sufficient time to evaporate and thinner fibers were produced, but at high volume flow rate, excess amount of solution fed to the tip of needle and thicker fibers result [3, 12].

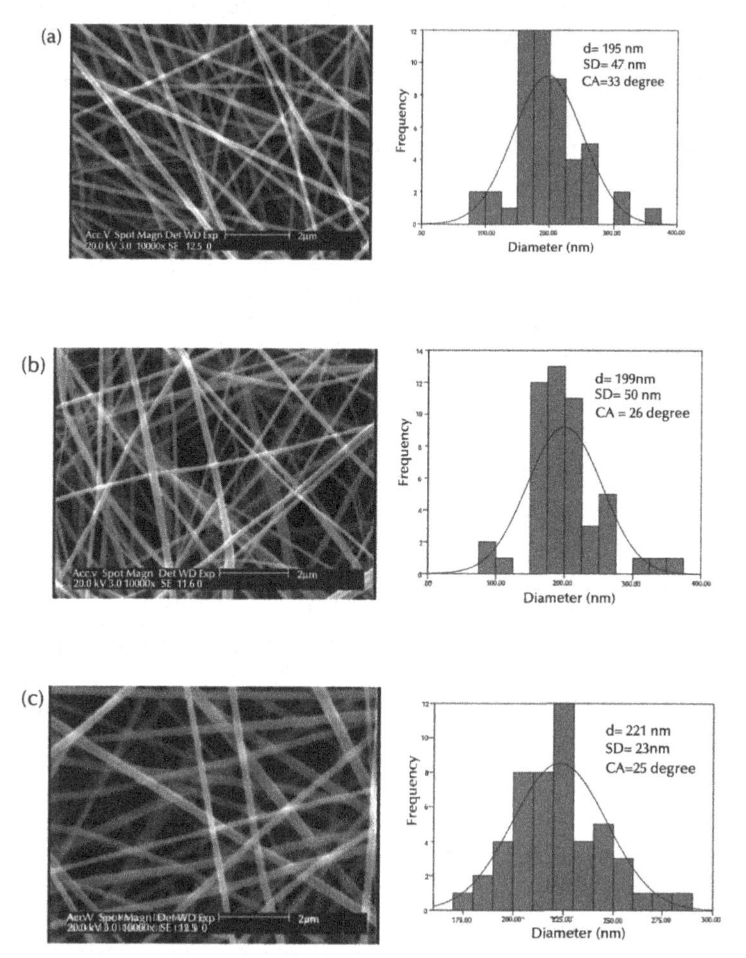

FIGURE 16.6 The SEM images and fiber diameter distributions of electrospun fibers in volume flow rate of (a) 2ml/hr, (b) 2.5ml/hr and (c) 3ml/hr

16.3.2 THE ANALYSIS OF VARIANCE (ANOVA)

The analysis of variance for AFD and CA of electrospun fibers has been summarized in Table 5.3 and Table 5.4 respectively, which indicated that the predictability of the models is at 95% confidence interval. Using 5% significance level, the factor is considered significant if the p-value is less than 0.05.

From the p-values presented in Table 5.3 and Table 5.4, it is obvious that p-values of terms X_3^2, $X_4^2, X_2X_3, X_1X_3, X_2X_4$ and X_3X_4 in the model of AFD and X_3^2, X_4^2, X_2X_3, X_2X_4 and X_3X_4 in the model of CA were not significant ($p > 0.05$).

The approximating function for AFD and CA of electrospun fiber obtained from Equation 2 and 3 respectively.

$$\begin{aligned} AFD = 211.89 + 31.17X_1 - 15.28X_2 - 12.78X_3 + 12.94X_4 \\ -8.44X_1X_2 + 6.31X_1X_4 \\ +18.15X_1^2 - 13.85X_2^2 \end{aligned} \quad (2)$$

$$\begin{aligned} CA = 26.07 - 9.89X_1 + 2.17X_2 + 4.33X_3 - 2.33X_4 \\ -1.63X_1X_2 - 1.63X_1X_3 + 1.63X_1X_4 \\ +9.08X_1^2 + 7.58X_2^2 \end{aligned} \quad (3)$$

TABLE 16.3 Analysis of variance for average fiber diameter (AFD)

Source	SS	DF	MS	F-value	Probe > F	Remarks
Model	31004.72	14	2214.62	28.67	<0.0001	Significant
X_1-Concentration	17484.50	1	17484.50	226.34	<0.0001	Significant
X_2-Voltage	4201.39	1	4201.39	54.39	<0.0001	Significant
X_3-Distance	2938.89	1	2938.89	38.04	<0.0001	Significant
X_4-Flow rate	3016.06	1	3016.06	39.04	<0.0001	Significant
Source	**SS**	**DF**	**MS**	**F-value**	**Probe > F**	**Remarks**
X_1X_2	1139.06	1	1139.06	14.75	0.0016	Significant
X_1X_3	175.56	1	175.56	2.27	0.1524	
X_1X_4	637.56	1	637.56	8.25	0.0116	Significant
X_2X_3	39.06	1	39.06	0.51	0.4879	
X_2X_4	162.56	1	162.56	2.10	0.1675	

TABLE 16.3 *(Continued)*

X_3X_4	60.06	1	60.06	0.78	0.3918	
X_1^2	945.71	1	945.71	12.24	0.0032	Significant
X_2^2	430.80	1	430.80	5.58	0.0322	Significant
X_3^2	0.40	1	0.40	0.0052	0.9433	
X_4^2	9.30	1	9.30	0.12	0.7334	
Residual	1158.75	15	77.25			
Lack of Fit	711.41	10	71.14	0.80	0.6468	

TABLE 16.4 Analysis of variance for contact angle (CA) of electrospun fiber mat

Source	SS	DF	MS	F-value	Probe > F	Remarks
Model	4175.07	14	298.22	32.70	<0.0001	Significant
X_1-Concentration	1760.22	1	1760.22	193.01	<0.0001	Significant
X_2-Voltage	84.50	1	84.50	9.27	0.0082	Significant
X_3-Distance	338.00	1	338.00	37.06	<0.0001	Significant
X_4-Flow rate	98.00	1	98.00	10.75	0.0051	Significant
X_1X_2	42.25	1	42.25	4.63	0.0481	Significant
X_1X_3	42.25	1	42.25	4.63	0.0481	Significant
X_1X_4	42.25	1	42.25	4.63	0.0481	Significant
X_2X_3	12.25	1	12.25	1.34	0.2646	
X_2X_4	6.25	1	6.25	0.69	0.4207	
X_3X_4	2.25	1	2.25	0.25	0.6266	
Source	SS	DF	MS	F-value	Probe > F	Remarks
X_1^2	161.84	1	161.84	17.75	0.0008	Significant
X_2^2	106.24	1	106.24	11.65	0.0039	Significant

TABLE 16.4 *(Continued)*

X_3^2	0.024	1	0.024	0.0026	0.9597
X_4^2	21.84	1	21.84	2.40	0.1426
Residual	136.80	15	9.12		
Lack of Fit	95.30	10	9.53	1.15	0.4668

Analysis of variance for AFD and CA showed that the models were significant ($p < 0.0001$), which indicated that the both models have a good agreement with experimental data. The value of determination coefficient (R^2) for AFD and CA was evaluated as 0.9640 and 0.9683 respectively.

The predicted versus actual response plots of AFD and CA are shown in Figure 5.7 and Figure 5.8 respectively. It can be observed that experimental values are in good agreement with the predicted values.

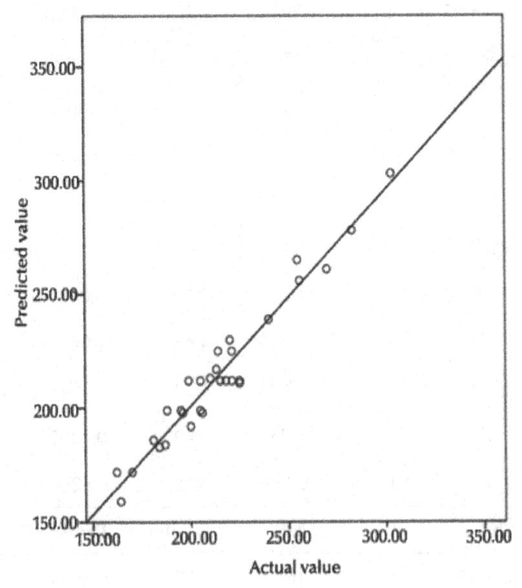

FIGURE 16.7 The predicted versus actual plot for AFD of electrospun fiber mat

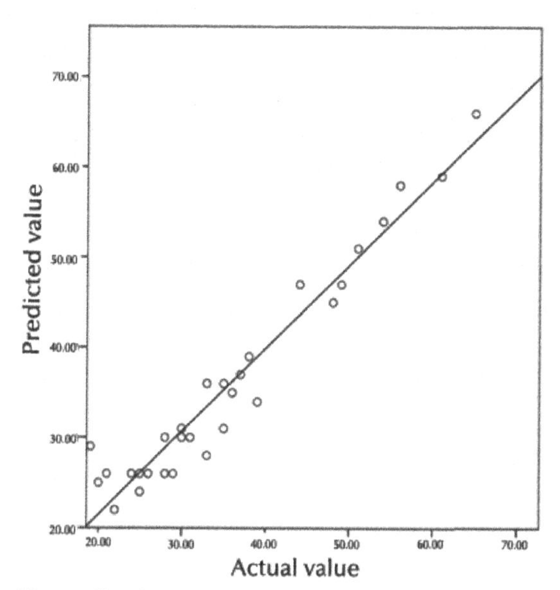

FIGURE 16.8 The predicted versus actual plot for CA of electrospun fiber mat

16.3.3 EFFECTS OF SIGNIFICANT PARAMETERS ON AFD

The response surface and contour plots in Figure 5.9a indicated that there was a considerable interaction between solution concentration and applied voltage at middle level of spinning distance (15cm) and flow rate (2.5ml/hr). It can be seen an increase in AFD with increase in solution concentration at any given voltage that is in agreement with previous observations [11, 12]. Generally, a minimum solution concentration is required to obtain uniform fibers from electrospinning. Below this concentration, polymer chain entanglements are insufficient and a mixture of beads and fibers is obtained. On the other hand, the higher solution concentration would have more polymer chain entanglements and less chain mobility. This causes the hard jet extension and disruption during electrospinning process and producing thicker fibers [7].

Figure 5.9b shows the response surface and contour plots of interaction between solution concentration and flow rate at fixed voltage (18kV) and spinning distance (15cm). It can be seen that at fixed applied voltage and spinning distance, an increase in solution concentration and volume flow rate results in fiber with higher diameter. As mentioned in the literature, the volume flow rate must be compatible with the amount of solution removed from the tip of the needle. At low volume flow rates, solvent would have sufficient time to evaporate and thinner fibers were produced, but at high volume flow rate, excess amount of solution fed to the tip of needle and thicker fibers were resulted [3, 8].

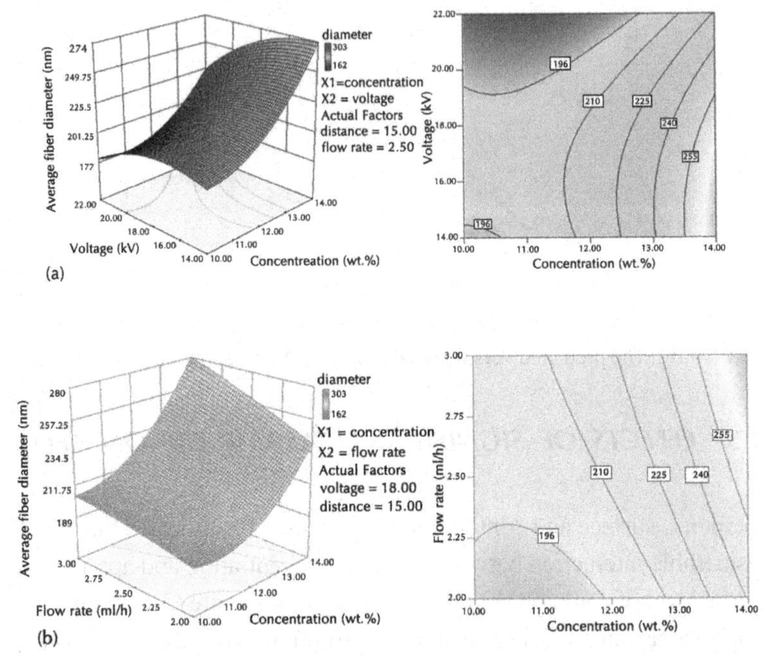

FIGURE 16.9 Response surface and contour plots of AFD showing the effect of: **a** solution concentration and applied voltage, **b** solution concentration and volume flow rate

16.3.4 EFFECTS OF SIGNIFICANT PARAMETERS ON CA

The response surface and contour plots in Figure 5.10a represented the CA of electrospun fiber mat at different solution concentration and applied voltage. It is obvious that at fixed spinning distance and volume flow rate, an increase in applied voltage and decrease in solution concentration result the higher CA. The tip to collector distance was found to be another important processing parameter as it influences the solvent evaporating rate and deposition time as well as electrostatic field strength. The impact of spinning distance on CA of electrospun fiber mat is illustrated in Figure 5.10b. Increasing the spinning distance causes the CA of electrospun fiber mat to increase. As demonstrated in Figure 5.10b, low solution concentration cause an increase in CA of electrospun fiber mat at large spinning distance. The response plots in Figure 5.10c shows the interaction between solution concentration and volume flow rate at fixed applied voltage and spinning distance. It is obvious that at any given flow rate, CA of electrospun fiber mat will increase as solution concentration decreases.

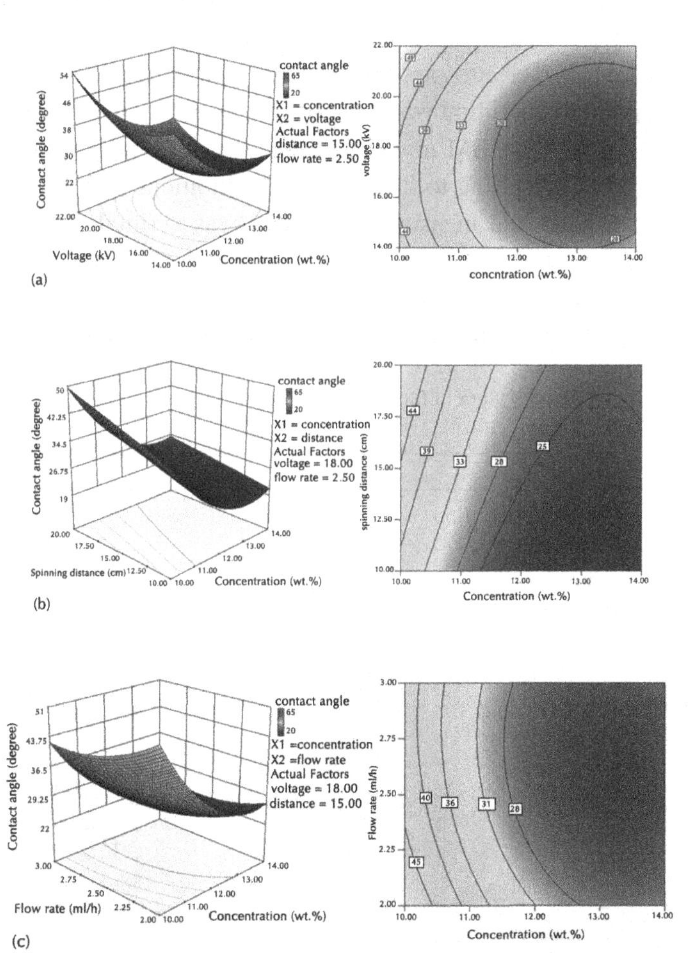

FIGURE 16.10 Response surface and contour plots of CA showing the effect of: **a** solution concentration and applied voltage, **b** solution concentration and spinning distance, **c** solution concentration and volume flow rate

16.3.5 DETERMINATION OF OPTIMAL CONDITIONS

It is well known that the value of CA for hydrophilic surfaces is less than 90°. Fabrication of these surfaces has attracted considerable interest for both fundamental research and practical studies. So, the goal of the present chapter is to minimize the CA of electrospun nanofibers. The optimal conditions of the electrospinning parameters were established from the quadratic form of the RSM. Independent variables namely, solution concentration, applied voltage, spinning distance, and volume flow rate were set in range and dependent variable (CA) was fixed at minimum. The optimal conditions in the tested range for minimum CA of electrospun fiber mat are shown in Table 5.5.

This optimum condition was a predicted value, thus to confirm the predictive ability of the RSM model for response, a further electrospinning was carried out according to the optimized conditions and the agreement between predicted and measured responses was verified. The measured CA of electrospun nanofiber mat (21°) was very close to the predicted value estimated to 20°. Figure 5.11 shows the SEM image and AFD distribution of electrospun fiber mat prepared at optimized conditions.

TABLE 16.5 Optimum values of the process parameters for minimum CA of electrospun fiber mat

Parameter	Optimum value
Solution concentration (wt.%)	13.2
Applied voltage (kV)	16.5
Spinning distance (cm)	10.6
Volume flow rate (ml/hr)	2.5

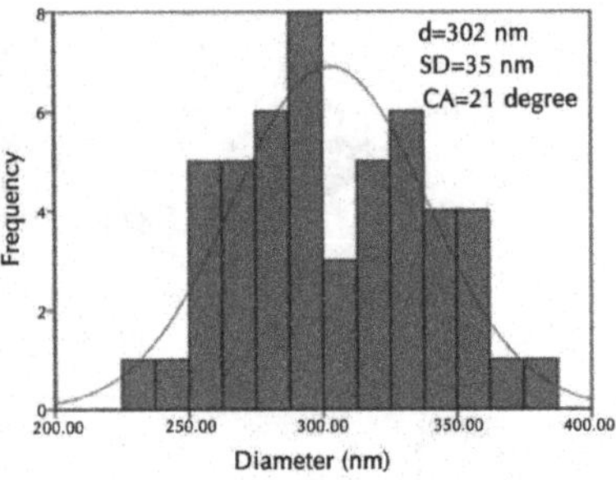

FIGURE 16.11 The SEM image and fiber diameter distribution of electrospun fiber mat prepared at optimized conditions

16.4 CONCLUSION

In this chapter, the effects of electrospinning parameters, comprising solution concentration (wt.%), applied voltage (kV), tip to collector distance (cm), and volume flow rate (ml/hr) on average diameter and CA of electrospun PAN nanofibers were investigated by statistical approach. Response surface methodology (RSM) was successfully employed to model and optimize the electrospun nanofibers diameter and CA. The response surface and contour plots of the predicted AFD and CA indicated that the nanofiber diameter and its CA are very sensitive to solution concentration changes. It was concluded that the polymer solution concentration was the most significant factor impacting the AFD and CA of electrospun fiber mat. The R^2 value was 0.9640 and 0.9683 for AFD and CA respectively, which indicates a good fit of the models with experimental data. The optimum value of the solution concentration, applied voltage, spinning distance, and flow rate were found to be 13.2wt.%, 16.5kV, 10.6cm and 2.5ml/hr, respectively, for minimum CA of electrospun fiber mat.

16.5 APPENDIX

16.5.1 THE BASICS OF ELECTROSPINNING MODELING

Modeling of the electrospinning process will be useful for the factors perception that cannot be measured experimentally. Although electrospinning gives continuous fibers, mass production and the ability to control nanofibers properties are not obtained yet. In electrospinning the nanofibers for a random state on the collector plate while in many applications of these fibers such as tissue engineering well-oriented nanofibers are needed. Modeling and simulations will give a better understanding of electrospinning jet mechanics. The development of a relatively simple model of the process has been prevented by the lack of systematic, fully characterized experimental observations suitable to lead and test the theoretical development. The governing parameters on electrospining process which are investigated by modeling are solution volumetric flow rate, polymer weight concentration, molecular weight, the applied voltage and the nozzle to ground distance. The macroscopic nanofiber properties can be determined by multiscale modeling approach. For this purpose, at first the effective properties determined by using modified shear lag model then by using of volumetric homogenization approach, the macro scale properties concluded.

Till date two important modeling zones have been introduced. These zones are: a) The zone close to the capillary (jet initiation zone) outlet of the jet and b) The whipping instability zone where the jet spirals and accelerates towards the collector plate.

The parameters influence the nature and diameter of the final fiber so obtaining the ability to control them is a major challenge. For selected applications it is desirable to control not only the fiber diameter, but also the internal morphology. An ideal operation would be: the nanofibers diameter to be controllable, the surface of the fibers to be intact and a single fiber would be collectable. The control of the fiber diameter can be affected by the solution concentration, the electric field strength, the feeding rate at the needle tip and the gap between the needle and the collecting screen (Figure 5.12).

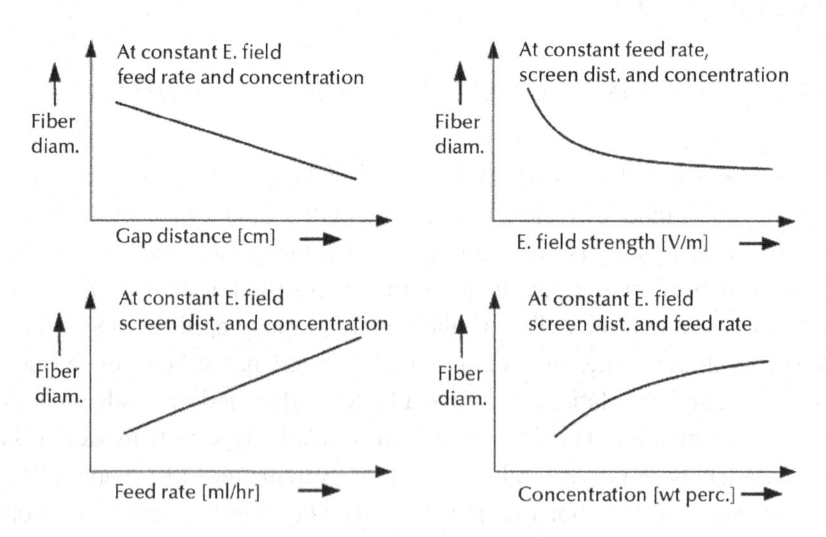

FIGURE 16.12 Effect of process parameters on fiber diameter

Control over the fiber diameter remains a technological bottleneck. However A cubic model for mean fiber diameter was developed for samples scientists. A suitable theoretical model of the electrospinning process is one that can show a strong-moderate-minor rating effects of these parameters on the fiber diameter. Some disadvantages of this method are low production rate, non oriented nanofiber production, difficulty in diameter prediction and controlling nanofiber morphology, absence of enough information on rheological behavior of polymer solution and difficulty in precise process control that emphasis necessity of modeling.

Recently, significant progress has been made in the development of numerical algorithms for the stable and accurate solution of viscoelastic flow problems, which exits in processes like electrospinning process. A limitation is made to mixed finite element methods to solve viscoelastic flows using constitutive equations of the differential type.

The analysis of viscoelastic flows includes the solution of a coupled set of partial differential equations: The equations depicting the conservation of mass, momentum and energy, and constitutive equations for a number of physical quantities present in the conservation equations such as density, internal energy, heat flux, stress, and so on depend on process.

There are fundamentally different approaches like: A mixed formulation may be adopted different parameters like velocity, pressure, and so on including the constitutive equation, is multiplied independently with a weighting function and transformed in a weighted residual form. The constitutive equation may be transformed into an ordinary differential equation (ODE). For transient problems this can, for instance, be achieved in a natural manner by adopting a Lagrangian formulation.

By introducing a selective implicit/explicit treatment of various parts of the equations, a certain separating at each time step of the set of equations may be obtained to improve computational efficiency. This suggests the possibility to apply devoted solvers to sub-problems of each fractional time step.

Due to the reason that nanofibers are made of polymeric solutions forming jets, it is necessary to have a basic knowledge of hydrodynamics. According to the effort of finding a fundamental description of fluid dynamics, the theory of continuity was implemented. The theory describes fluids as small elementary volumes which still are consisted of many elementary particles.

The equation of continuity:

$$\frac{\partial \rho_m}{\partial t} + div(\rho_m v) = 0 \text{ (For incompressible fluids } div(v) = 0 \text{)}$$

The Euler's equation simplified for electrospinning:

$$\frac{\partial v}{\partial t} + \frac{1}{\rho_m} \nabla p = 0$$

The equation of capillary pressure:

$$P_c = \frac{\gamma \partial^2 \zeta}{\partial x^2}$$

The equation of surface tension:

$$\Delta P = \gamma \left(\frac{1}{R_X} + \frac{1}{R_Y} \right)$$ R_x and R_y are radii of curvatures

The equation of viscosity:

$$\tau_{ij} = \eta \left(\frac{\partial v_i}{\partial x_j} + \frac{\partial v_j}{\partial x_i} \right)$$ (For incompressible fluids, $\tau_{i,j}$ = Stress tensor)

$$V = \frac{\eta}{\rho_m} \text{ (Kinematic viscosity)}$$

Symbols	Definition
ρ_m	Mass density
P	Pressure
t	Time
γ	Surface tension
η	Viscosity
τ	Stress tensor
ζ	Surface deflection

Symbols	Definition
Q	constant volume flow rate
v	Jet velocity
I	The current carried by the jet
K	Electrical conductivity
E	Electric field
σ	Surface charge density
ε	The dielectric constants of the jet
$\tilde{\varepsilon}$	The dielectric constants of the ambient air
r_i	The position of bead i
F_C	The net coulomb force
F_E	The electric field force
F_{ve}	The viscoelastic force
F_B	The surface tension force
F_q	The Lorenz force

KEYWORDS

- **Electrospinning process**
- **Response surface methodology**
- **Electrospun nanofibers**
- **Central composite design**
- **Contact angle**

REFERENCES

1. Nateri, A. S., & Hasanzadeh, M. (2009). *Journal of Computational and Theoretical Nanoscience,* 6(1542).
2. Kilic, A., Oruc, F., & Demir, A. (2008). *Textile Research Journal, 78*(532).

3. Ramakrishna, S., Fujihara, K., Teo, W. E., Lim, T. C., & Ma, Z. (Eds.) (2005). *An Introduction to Electrospinning and Nanofibers*. Singapore: World Scientific Publishing.
4. Brown. P. J., & Stevens. K. (Eds.) (2007). *Nanofibers and Nanotechnology in Textiles*. Cambridge, UK: Woodhead Publishing
5. Shin, Y. M., Hohman, M. M., Brenner, M. P., & Rutledge, G. C. (2001). *Polymer, 42*(9955).
6. Yördem, O. S., Papila, M., & Menceloğlu, Y. Z. (2008). *Materials & Design, 29*(34).
7. Haghi, A. K., & Akbari, M. (2007). *Physica Status Solidi A, 204* (1830).
8. Ziabari, M., Mottaghitalab, V., & Haghi, A. K. (2009). In Chang, W. N. (Ed.) *Nanofibers: Fabrication, Performance, and Applications* (pp. 153–182). USA: Nova Science Publishers.
9. Myers, R. H., Montgomery, D. C., & Anderson-cook, C. M. (2009). *Response surface methodology: process and product optimization using designed experiments* (3rd Edn.). USA: John Wiley and Sons.
10. Gu, S. Y., Ren, J., & Vancso, G. J. (2005). *European Polymer Journal, 41*(2559).
11. Zhang, C., Yuan, X., Wu, L., Han, Y., & Sheng, J. (2005). *European Polymer Journal, 41*(423).
12. Zhang, S., Shim, W. S., & Kim, J. (2009). *Material Design, 30*(3659).

CHAPTER 17

A STUDY ON APPLICATION OF BIOACTIVE SUBSTANCES: PART I

I. S. BELOSTOTSKAJA, E. B. BURLAKOVA, V. M. MISIN,
G. A. NIKIFOROV, N. G. KHRAPOVA, and V. N. SHTOL'KO

CONTENTS

17.1 EFFECT OF POSITION AND KIND OF SUBSTITUTES ON PHENOL TOXICITY

Notwithstanding general point of view, peroxide oxidation of lipids is a physiological process of normal metabolizing tissues and follows the same laws as lipids oxidation does *in vitro* in liquid phase. Fine regulation of lipid oxidation in cell takes place both by chemical ways and with the help of enzymes. The rate of lipid oxidation is supported by a few systems, each of which is working within strict limits on relating stages of the process, having its own limiting reactions [1].

The lipid oxidation rate depends mainly on concentration changes of the substances, breaking the oxidation chains. Only natural antioxidants (AO) (basically phenols) have the ability to react with peroxide radicals. There is no interaction with peroxide radicals of lipids in any other system of defense. None of the enzymes with antioxidant action break oxidation chain. Only natural AO are inhibitors of radical processes. They can destroy redundant peroxide radicals of lipids. Lipid oxidation rate is significantly affected by AO more than by other systems. It defines the unique role of natural AO in regulation the intensity of lipid oxidation [2].

The low level quantity of lipids peroxide in normal tissues is explained by well-balanced processes of peroxide formation and expenses. Breaking this balance results in changing the antioxidant status of the organism and may be the cause of some pathology [2, 3]. In some studies it was established that AO therapy was necessary for treating many diseases, and free radical processes do play a certain role. For inhibition reactions of chain oxidation *in vivo* as substances, phenol derivatives are widely used .They possess physiological activity in mammals. There appeared biologically active supplements and medicines of the antioxidant effect, which are of great help to treat many diseases, for example: dibunoli, mexidol, emoxipin, trimexidini, probucoli [4-10]. In Russia the substances group of the antioxidant effect was chosen from biologically active supplements [11].

For many substances toxicity was carefully studied; clinical, preclinical tests and experiments were carried out. However, in literature there is a little information on toxicity of numerous synthesized phenols. They are joined into different groups and have various phenols and having substituents in position 2, 4, 6 to the group OH. For example, in [12], pulmonary toxicity of 24 phenols was studied on mice, which have various lipophil alkyl substituents in positions 2, 4, 6, and injected I.P in mammals.

In this study systematic competitive investigations were carried out for the estimation toxicity of 27 phenol compounds, having in p- position substituents with different heteroatoms. They have the same formula:

R^2, R^4, R^6 – various substituents; R^3=R^5=H

17.1.1 EXPERIMENTAL PART

The synthesis of studied phenols was described in [4, 13-14]. Toxicity of all the compounds was estimated by a value LD_{50}, measured in mg/kg weight of animal. For some compounds, values as much as possible transferable dose (MTD) and LD_{100} were additionally determined. Toxicity was determined on mice (males) of the line *Bulb* (mass18-22 gr) with a single I.P. These mammals were housed in standard conditions of vivarium. Water-soluble preparations were injected as solutions of distilled water; liposoluble - 10% in solution Twin-80, used as solvent. Each doze off this preparation was tested on not less than 4 mammals. For this calculation, Beren's method was used (the method "frequency accumulation"), because it was simple and rather reliable [15].

17.1.2 DISCUSSION AND RESULT

All investigated substances are resulted in Table 17.1. For better comparison of the obtained results on toxicity, investigated compounds were grouped separately in Table 17.2–17.9. It was necessary to estimate toxicity of the substituted phenols depending on:

TABLE 17.1 Toxicity of the phenols containing various substituents. R3 = R5 = H

№ patt.	substituents			MTD, mg/kg	LD50, mg/kg	LD100, mg/kg
	R2	R4	R6			
1	$C(CH3)_3$	CH_3	$C(CH_3)3$	250	375	650
2	$C(CH3)_3$	NH_2	$C(CH_3)3$	40	80	100
3	$C(CH3)_3$	CH_2NH_2	$C(CH_3)3$	30	70	130
4	$C(CH3)_3$	$(CH_2)2NH_2$	$C(CH_3)3$	60	75	80
5	$C(CH3)_3$	$(CH_2)3NH_2$	$C(CH_3)3$	50	60	-
6	$C(CH3)_3$	$(CH_2)4NH_2$	$C(CH_3)3$	-	50	-
7	$C(CH3)_3$	$CH(CH_3)NH_2$	$C(CH_3)3$	80	105	130
8	$CH(CH_3)_2$	$CH(CH_3)NH_2$	$C(CH_3)3$	30	80	100
9	$CH(CH_3)_2$	$CH(CH_3)NH_2$	$CH(CH_3)2$	27	35	45
10	CH_3	$CH(CH_3)NH_2$	$C(CH_3)3$	-	50	100
11	$C(CH_3)_3$	$CH(C_2H_5)NH_2$	$C(CH_3)3$	40	70	110
12	H	$(CH_2)3NH_2$	$C(CH_3)3$	100	165	-

TABLE 17.1 *(Continued)*

13	$C(CH_3)_3$	$CH_2NHCOCH_3$	C(CH3)3	350	425	-
14	$C(CH_3)_3$	(CH_2)2NHCOCH_3	$C(CH_3)_3$	-	175	-
15	$C(CH_3)_3$	(CH_2)3NHCOCH_3	$C(CH_3)_3$	75	125	-
16	$C(CH_3)_3$	CN	$C(CH_3)_3$	300	450	525
17	$C(CH_3)_3$	CH_2CN	$C(CH_3)_3$	50	95	180
18	$C(CH_3)_3$	(CH_2)2CN	$C(CH_3)_3$	200	360	475
19	H	(CH_2)2CN	$C(CH_3)_3$	250	285	350
20	H	(CH_2)2CN	H	200	302	400
21	CH_3	CH_2CN	$C(CH_3)_3$	50	152	250
22	CH_3	CN	$C(CH_3)_3$	50	150	-
23	H	CN	$C(CH_3)_3$	250	185	-
24	H	CN	H	200	300	-
25	$C(CH_3)_3$	OH	$C(CH_3)_3$	-	390	-
26	$C(CH_3)_3$	(CH_2)2OH	$C(CH_3)_3$	-	300	-
27	$C(CH_3)_3$	(CH_2)3OH	$C(CH_3)_3$	-	225	-

- The type of p-substituents under identical 2,6-di-tert-butyl substituents;
- Spacer length (CH2)n between phenol ring and p-substituents under identical 2,6-di-tert-butyl substituents;
- The type of both o-substituents under identical p-substituents.

First, the phenol toxicity was compared for various p-substituents and identical 2, 6-di-tert-butyl substituents, possessing maximum screening effect (table 17.2). Toxicity estimated by the value LD_{50} was found to decrease in a row (in brackets values LD_{50} were given) of 2, 6-di-tert-butylphenols having next n- substituents.

$$-NH_2\ (80) \approx -CH_2NH_2\ (70) > -CH_3\ (375) > -OH\ (390) \approx -C\ (CH_3)_3\ (400) > -CN\ (450)$$

TABLE 17.2 Dependence of toxicity of the phenols having identical assistants $R^2=R^6=C(CH_3)_3$, from type of p- substituent's; $R^3=R^5=H$

Substituents			MTD, mg/kg	LD_{50}, mg/kg	LD_{100}, mg/kg
R^2	R^4	R^6			
$C(CH_3)_3$	NH_2	$C(CH_3)_3$	40	80	100
$C(CH_3)_3$	CH_2NH_2	$C(CH_3)_3$	30	70	130
$C(CH_3)_3$	CH_3	$C(CH_3)_3$	250	375	650
$CH(CH_3)_2$	OH	$CH(CH_3)_2$	-	390	-
$C(CH_3)_3$	$C(CH_3)_3$	$C(CH_3)_3$	250	400	-
$C(CH_3)_3$	CN	$C(CH_3)_3$	300	450	525

Toxicity decreasing of some phenol compounds from this row was approved with increasing the value of another parameter toxicity-MTD (in brackets the value MTD is given).

$$-NH_2\ (40) \approx -CH_2NH_2\ (30) >> -CH_3\ (250) = -C\ (CH_3)_3\ (250) > -CN\ (300\text{-}350)$$

The same value decrease in toxicity was being observed for parameter LD_{100} in a row.

$$-NH_2\ (110) > -CH_2NH_2\ (130) >> -CN\ (525) > -CH_3\ (650)$$

TABLE 17.3 Dependence of toxicity of phenols on remoteness of p-substituents on a benzene ring. R4 = (CH2)n X ; R2=R6= C(CH3)3 ; R3=R5=H

n	X=CN			X=NH2			X=NHCOCH3			X=OH
	MTD, mg/kg	LD50, mg/kg	LD100, mg/kg	MTD, mg/kg	LD50, mg/kg	LD100, mg/kg	MTD, mg/kg	LD50, mg/kg	LD100, mg/kg	LD100, mg/kg
0	300	450	525	40	80	100	-	-	-	390
1	50	95	180	30	70	130	350	425	500	-
2	200	360	475	60	75	80	-	175	-	300
3	-	-	-	50	60	-	75	125	200	225
4	-	-	-	-	50	-	-	-	-	-

For phenols, having identical 2, 6-di-tert-butyl substituent's, dependence of their toxicity on the distance of the functional p-substituents from benzene ring was investigated (table 17.3). The tendency was found to increase phenol toxicity connected with the spacer length -$(CH_2)_n$- for all investigated types of substituents (–NH_2 , –$NHCOCH_3$, –OH , –CN) both electron donors and electron acceptors (table 17.3).

However, for 2, 6-di-tert-butylphenol with p-substituent - CN, separated only by one group -CH_2- (n=1) from benzene ring, a sharp toxicity increase, as compared with phenol toxicity, having n=0, was being observed. Further, for substances with the increasing bridges - $(CH)_n$- toxicity sharply diminishes. But it remained greater than the phenol toxicity did, having p-substituent just near benzoyl ring (n=0). It is interesting that the observed extreme dependence was repeating for all values, characterizing the toxicity of these compounds (MTD, LD_{50}, LD_{100}).

The observed occurrence of extreme phenol toxicity, having methylene (or methyl) group in p-position, could possibly be explained by greater ability of these phenols in comparison with other phenols. Thus there can be reactions of appearing of phenol radicals with the following reaction of demerization or reaction of disproportionate with formation of cyclohexadienone-2, 5 [4, 16-17].

For 2, 6-di-*tert*-butylphenol with p-substituent -NH_2 the same tendency of toxicity increase, according to increase of spacer length -$(CH_2)_n$- has been observed (table 17.3). Toxicity increase were observed for this phenol (n=1) with the help of values MTD and LD_{50}. It was interesting that sharper toxicity increase has been observed for similar phenol, but with CN substituent.

Identical dependence of sufficient toxicity increase with the spacer length - $(CH_2)_n$- was observed for 2,6-di-*tert*-butylphenol having p-substituent's -$(CH_2)_n NHCOCH_3$ (table 17.3).

Symmetric 2, 6-di-*tert*-butylphenol (as an example), containing group -CH_2NH_2 in p-position, toxicity dependence on the type of α-substituents in p-methylene group was investigated (table 17.4). It was found phenol toxicity to decrease after introduction of CH_3 group in α-position of the substitute CH_2NH_2. The introduction of C_2H_5 group returns the substituted phenol toxicity to its initial meaning.

TABLE 17.4 Toxicity of the phenols containing various R^4 substituents. $R^3 = R^5 = H$

No patt.	substituents R^2	R^4	R^6	MTD, mg/kg	LD_{50}, mg/kg	LD_{100}, mg/kg
1	$C(CH_3)_3$	CH_2NH_2	$C(CH_3)_3$	30	70	130
2	$C(CH_3)_3$	$CH(CH_3)NH_2$	$C(CH_3)_3$	80	105	130
3	$C(CH_3)_3$	$CH(C_2H_5)NH_2$	$C(CH_3)_3$	40	70	110

It was noted that the introduction of acetyl group to N-position of symmetrical 4-aminomethyl-2, 6-di-tert-butylphenol has decreased the substance toxicity according to the value LD_{50} in 6 times (table 17.5).

TABLE 17.5 Toxicity of the phenols containing various R^4 substituent's. $R^3 = R^5 = H$

No patt.	Substituents R^2	R^4	R^6	MTD, mg/kg	LD_{50}, mg/kg	LD_{100}, mg/kg
1	$C(CH_3)_3$	CH_2NH_2	$C(CH_3)_3$	30	70	130
2	$C(CH_3)_3$	$CH_2NHCOCH_3$	$C(CH_3)_3$	350	425	500

It was interesting to study the dependence of phenol toxicity on the type of substituents in o-positions.

For phenols (table 17.6), having identical p-substituent $-CH(CH_3)NH_2$, it was investigated phenol toxicity influenced by the substituents of the types R_2 and R_6 (tert-butyl, iso-propyl, methyl). Over all the values, describing toxicity (MTD, LD_{50}, LD_{100}), phenol toxicity increases with the quantity decrease of tert-butyl groups, which are in both o-positions of these phenol in a row

tert-butyl + tert-butyl < tert-butyl + iso-propyl < tert-butyl + methyl < iso-propyl + iso-propyl

TABLE 17.6 Toxicity of the phenols containing various R^2, R^6 substituent's. $R^3 = R^5 = H$

No patt.	Substituents			MTD, mg/kg	LD_{50}, mg/kg	LD_{100}, mg/kg
	R^2	R^4	R^6			
1	$C(CH_3)_3$	$CH(CH_3)NH_2$	$C(CH_3)_3$	80	105	130
2	$CH(CH_3)_2$	$CH(CH_3)NH_2$	$C(CH_3)_3$	30	80	100
3	CH_3	$CH(CH_3)NH_2$	$C(CH_3)_3$	-	50	100
4	$CH(CH_3)_2$	$CH(CH_3)NH_2$	$CH(CH_3)_2$	25	35	45

Such toxicity increase can be explained both by decrease of steric hindrance, created to electron donor group -OH by the volume tert-butyl substituents, and by hyper conjugation effect of o-substituents in OH-group [18].

For n-CN of the substituted phenols (table 17.7) after decreasing the number of tert-butyl substituents in o-position it was found the sharp toxicity increase with the following decrease of it. Nevertheless the toxicity of CN-substituted 2, 6-di-*tert*-butylphenol is less than toxicity of others CN-substituted phenol derivatives.

TABLE 17.7 Toxicity of the phenols containing various R^2, R^6 substituent's. $R^3 = R^5 = H$

No patt.	substituents			MTD, mg/kg	LD_{50}, mg/kg	LD_{100}, mg/kg
	R^2	R^4	R^6			
1	$C(CH_3)_3$	CN	$C(CH_3)_3$	300	450	525
2	CH_3	CN	$C(CH_3)_3$	50	150	-
3	H	CN	$C(CH_3)_3$	250	185	-
	H	CN	H	200	300	-

The analogous dependence has been observed under investigation according to values MTD, LD_{50} and LD_{100} of three phenols, having group-CH_2CH_2CN in p-position. They differ in quantity of tert-butyl substituents in o-position (table 17.8). Toxicity increasing as found when the quantity of tert-butyl substituents decrease. However, extreme dependence of mono-tert-butyl-substituted phenol toxicity has been observed.

TABLE 17.8 Toxicity of the phenols containing various R^2, R^6 substituent's. $R^3 = R^5$ =H

No patt.	substituents			MTD, mg/kg	LD50, mg/kg	LD100, mg/kg
	R2	R4	R6			
1	$C(CH_3)_3$	$(CH_2)_2CN$	$C(CH_3)_3$	200	360	475
2	H	$(CH_2)_2CN$	$C(CH_3)_3$	250	285	350
3	H	$(CH_2)_2CN$	H	200	300	400

So, the least toxicity of sterically hindered 2, 6-di-*tert*-butilphenols has been observed for a few phenol groups with different p-substitutes .It can be explained by their worst bioaccessibility [4].

However, only for 2, 6-di-*tert*-butyl-4-cyanophenol and 2-*tert*-butyl-6-methyl-4-cyanophenol (table 17.9) reverse dependence was being observed, the change of one tert-butyl group to methyl group resulted in essential toxicity decrease. This was proved by the growth of both measured values LD_{50} and LD_{100}.

TABLE 17.9 Toxicity of the phenols containing various R^2, R^6 substituent's. $R^3 = R^5$ =H

No patt.	substituents			MTD, mg/kg	LD_{50}, mg/kg	LD_{100}, mg/kg
	R^2	R^4	R^6			
1	$C(CH_3)_3$	CH_2CN	$C(CH_3)_3$	50	95	180
2	CH_3	CH_2CN	$C(CH_3)_3$	50	150	250

The observed property dependence on the structure for various o-substituted phenols with polar substitutes in p-position does not correlate with the same dependence observed for those with nonpolar alkyl substitutes [12]. Perhaps after introducing the substituents of various types to phenols, the competition of a few tendencies is being observed, each of them influencing on selected toxicity of the investigated substances [19].

First, the change of lipophility takes place, which results in the change of phenol transport through lipid membrane layers to the following receptors. For example, in [20] the influence of AO on the activity of lipodependent proteinkinase was noted not only through lipid membrane changes, but through direct interaction with enzyme. As the substituted phenols are accumulated in lipids, so lipophility dropping of substituted phenol should promote it to supply those fields of lipid membranes which are enriched with oxidized lipids and the products of their metabolism [19].

Second, the structure change results in the reaction ability change in substituted phenol groups, responsible for metabolizing with these phenols. OH - groups, particularly, are responsible for conjugation reactions glucuronic acid and sulfates [21, 22]. In turn heteroatom groups in p-position take place participate in the processes on the first stage of metabolism with the following stage of metabolism by conjugation with possible metabolites – 2,5-cyclohexadienones formation [4, 12].

Third, the structure change of substituted phenols results in constant of ionization change of OH-group. It is sufficiently influences upon selective toxicity, connected with substrate ionization [19].

So, the above stated factors make difficult to find the true phenol toxicity dependence on structure. Besides it is difficult to compare these results with those obtained for phenols with lipophilic alkyl substituents [12]. Indeed, from these results in [12], it follows that 2, 6-di-*tert*-butyl-4-methylphenol is one of the strongest toxicants. Similar principal result difference may be explained that strong phenol toxicity has been studied on mice in this study, and in [12] specific mice organs (lungs) have been investigated there; its affects on lungs.

For full explanation obtained, it is desirable to investigate specific affect of these phenols on various targets in some organs with help of pharmacokinetics. Nevertheless, the obtained results in this study may be of help in pharmacological investigations and planning studies connected with synthesis of new, nontoxical, biologically active phenol compounds.

17.1.3 CONCLUSION

- Phenols having two sterically hindered tert-butyl substitutes in o-position have the least toxicity.

- Toxicity of 2, 6-di-*tert*-butylphenol derivatives increases with removing the functional p-substituents of benzene ring.
- The declinations from the law of nature of toxicity changes were observed for p-substituted phenols, in which functional group is selected from benzene ring only by one methylene group.
- The obtained results may be of help in pharmacological investigations and planning studies connected with synthesis of new, nontoxical, biologically active phenol compounds.

KEYWORDS

- **Analogous Dependence**
- **Antioxidants**
- **Lipid Oxidation Rate**
- **Phenol Toxicity**

REFERENCES

1. Burlakova, E. B. & Khrapova, N. G. (1985) Peroxide oxidation of lipids of membranes and natural antioxidants. *Russian Chemical Reviews.*, 54(9), 1540–1558.
2. Khrapova, N. G. (2005). Peroxide oxidation of lipids of Biological membranes and Food Additives, *In Chemical and Biological Kinetics. New Horizons. V. 2. Biological Kinetics*, E. B. Burlakova (Ed.), Khimiya, Moscow, pp. 46-60.
3. Burlakova, E. B., Alesenko, A. V., Molochkina, E. M. et. al. (1975). *Bioantioxidant in Radiation Sickness and Malignant Disease*, E. B. Burlakova (Ed.), Nauka, Moscow.
4. Ershov, V. V., Nikiforov, G. A., & Volodkin, A. A. (1972). *Sterically hindred phenols*, Khimiya, Moscow,.
5. Zarudij, F. S., Gilmutdinov, G. Z., Zarudy, R. F. et.al.. (2001). 2,6-di-tert-butyl-4-methyl phenol (Dibunoli, Ionoli, Tonafol) is a classical antioxidant. *Khimiko-Farmacevtichesky Zh.*, 35(3), 42-48.
6. Zorkina, A. V., Kostin, J. V., Inchina, V. I. et.al.. (1998) Antioxidative both hypolipidemical properties of mexidol and emoxipin at long immobili stress. *Khimiko-Farmacevtichesky Zh.*, 32(5), 3-6.
7. Kotljarov, A. A., Smirnov, L. D., Smirnova, L. E. et. al. (2002) Research of joint application mexidol with antiarrhytmic preparations. *Experimental and clinical pharmacology: Two-month scientific-theoretical magazine.* 65(5), 31-34.

8. Zenkov, N. K., Kandalintseva, N. V., Lankin, V. Z. et. al. (2003). *Phenolic bioantioxidants*, SO RAMN, Novosibirsk,.
9. Burlakova, E. B. (2005). Bioantioxidants: yesterday, today, tomorrow. *In Chemical and Biological Kinetics. New Horizons. V. 2. Biological Kinetics.* V. 2. E. B. Burlakova (Ed.), Khimiya, Moscow, pp. 10-45.
10. Kravchuk, E. A., Keselyova, T. N., Ostrovsky, M. A. et. al. (2008). Influence an antioxidant preparation trimeksidin on a current of an early stage experimental uveitis. *Refraction surgery and ophthalmology*, 8(1), 36-41.
11. The Russian federal register of biologically active additives to food, 2th edition, processed and added. (2001).
12. Mizutani, T., Ishida, I., Yamamoto, K., & Tajima, K. (1982). Pulmonary Toxicity of butylated hydroxytoluene and related alkylphenols: structural requirements for toxic potency in mice. *Toxicology and Applied Pharmacology*, 62, 273-281.
13. Ershov, V. V. & Belostotskaja, I. S. (1965). Di-tert-butylspirocyclodienones. *Izv.Akad. Nauk SSSR, Ser. Khim.*, (7), 1301-1303.
14. Belostotskaja, I. S., Volodkin, A. A., Ostapets-Sveshnikova, G. D., & Ershov, V. V. (1966). Synthesis α-replaced hydroxyphenylacetic acids and phenylethylamines of some the sterically hindred phenols. *Izv.Akad. Nauk SSSR, Ser. Khim.* (10), 1833-1835.
15. Belen'ky, M. L. (1959). Elements of a quantitative estimation of pharmacological effect, Acad. (Ed.), *Sci. of Latv.*, SSR,
16. Roginsky, V. A. (1988). *Phenolic antioxidants: Reactionary ability and efficiency*, Nauka, Moscow.
17. Takahashi, O. & Hiraga, K. (1997). 2,6-Di-tert-butyl-4-methylene-2,5-cyclohexadienone: a hepatic metabolite of butylated hydroxytoluene in rats. *Fd. Cosmet. Toxicol*, 17, 451-454.
18. Temnikova, T. I. (1968). Course of theoretical bases of organic chemistry, Khimiya, Leningrad.
19. Albert, A. (1985). Selective Toxicity, 7th edition. Chapman and Hall, London.
20. Hohlov, A. P. (1988). Mechanisms of regulation of activity of membrane receptors synthetic antioxidants from a class sterically hindred phenols. *The bulletin of experimental biology and medicine*. (10), 440-444.
21. Kabiev, O. K. & Balmuhanov, S. B. (1975). Natural phenols – a perspective class of antineoplastic and radio protective connections, *Meditsina*, Moscow..
22. Vergejchik, T. H. (2009). *Toxicological chemistry*, MEDpress-inform, Moscow.

CHAPTER 18

A STUDY ON APPLICATION OF BIOACTIVE SUBSTANCE: PART II

F. F. NIYAZY, N. V. KUVARDIN, and E. A. FATIANOVA

CONTENTS

18.1 CHANGE OF SOME PHYSICO -CHEMICAL PROPERTIES OF ASCORBIC ACID AND PARACETAMOL HIGH-DILUTED SOLUTIONS AT THEIR JOINT PRESENCE

During last decades there is a tendency of the growing interest to the study of high-diluted solutions of bioactive substances. Besides, concentration ranges under study are related to the category of super-small or, in other words, 'illusory' concentrations. Such solutions, unlike more saturated ones, but with pre-therapeutic content of active substance, may possess high biological activity.

Use of bio-objects to reveal super-small doses effect (SSD) in the substances is the most exact method today allowing not only to define the effect existence and to find out how it shows itself, but to determine concentration ranges of its action. However, the use of this method will entail great difficulties. In this connection it is necessary to search for other methods, including physico-chemical ones, allowing to define presence of SSD effect in the compounds only at the stage of preliminary tests. Study of physico-chemical bases of SSD effect display is one of the most interesting questions in the given sphere of research and attracts attention of many scientists [1-4].

Antineoplastic and antitumorous agents, radioprotectors, neutropic preparations, neupeptides, hormones, adaptogenes, immunomodulators, antioxidants (AO), detoxicants, stimulants and inhibitors of plants growth and so on are included into the group of bioactive substances possessing SSD effect. Study of high-diluted solutions of bioactive compounds was carried out on one-component solution that is ones containing only one solute. But now, mainly multicomponent medical preparations, possessing several therapeutic actions, are used in medicine. So, preparations of analgesic-antipyretic action are possibly used at sharp respiratory illnesses, accompanied by muscular pain and rise of temperature. It is possible that effects of multicomponent medical preparation in super-small concentrations and its separate components will differ.

We studied some physico-chemical properties of high-diluted aqueous solutions of paracetamol and ascorbic acid with the purpose of finding out peculiarities of their change at solutions dilution and also of definition of possible concentration ranges of SSD effect action. Paracetamol and ascorbic acid are the components of combined analgesic-antipyretic and anti-inflammatory preparations [5]. Ascorbic acid is used as fortifying and stimulating remedy for immune system. Paracetamol (acetominophene) has anaesthetic and febrifugal effect.

18.1.1 EXPERIMENTAL

We prepared one-component solutions of ascorbic acid and paracetamol, two-component solutions of paracetamol with ascorbic acid (relation of dissolute compounds in solutions is 1:1), in the following concentrations of dissolute substances (mole/l): 10^{-1}, 10^{-3}, 10^{-5}, 10^{-7},10^{-9}, 10^{-11}, 10^{-13}, 10^{-15}, 10^{-17}, 10^{-19}, 10^{-21}, 10^{-23}. Water cleansed by reverse osmosis was used as solvent. Solutions were prepared by successive dilution by 100 times using classical methods. Initial solution was 0,1M one. Before choosing of solution portion for the following dilution the sample was subjected to taking antilog.

Prepared solutions were studied by cathetometric method of substances screening, the ones acting in super-small concentrations, and also by method of electronic spestroscopy.

Cathetometric method of substances screening is based on the study of the change of solution meniscus height in capillary [6]. Results of measuring meniscus height of ascorbic acid, paracetamol, paracetamol with ascorbic acid solutions are given in Figures 18.1, 18.2, and 18.3, accordingly.

Meniscus height of dilution with ascorbic acid concentration of 10^{-3} mole/l was 0.8 mm (Figure 18.1). During further dilution value of meniscus height in the capillary reduces, but changes are not uniquely defined. The most lowering of meniscus height is observed in samples in which content of ascorbic acid is 10^{-9}, 10^{-13}, 10^{-15}, and 10^{-17} mole/l.

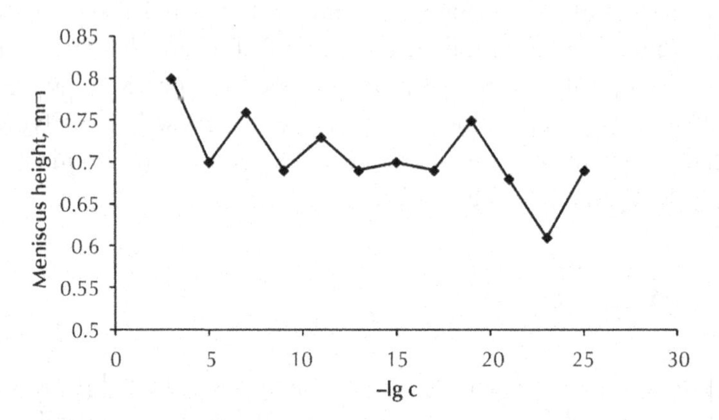

FIGURE 18.1 Values of meniscus height in the capillary of ascorbic acid solutions (concentration, mole/l)

Meniscus height reduces on an average by 13.75%. Lowering of meniscus height by 23.7%, in comparison with more concentrated solution, is also observed for the sample with ascorbic acid content of 10^{-23} mole/l.

Equivalent lowering of meniscus in the capillary has been stated for solutions with ascorbic acid concentration of 10^{-9} mole/l and 10^{-13}–10^{-17} mole/l. In between these concentration ranges there is concentration range in which there are no essential changes. Being based on literature data and also on cathetometric method for screening of substances activity in super-small doses, we can assume that dilutions of ascorbic acid in concentrations of 10^{-9} mole/l and 10^{-13}–10^{-17} mole/l show biological activity regarding bio-objects.

While studying solutions of paracetamol by cathetometric method it can be observed that in first dilutions by 100 times (paracetamol concentrations being 10^{-3}–10^{-7} mole/l) height of meniscus reduces slightly, maximum 5.7%, regarding meniscus height of the first dilution. This slight lowering of meniscus height in the capillary is caused by rather large dose of active substance in these dilutions. However, sudden lowering of meniscus height in the capillary up to 0.71mm, which is by 19.3% lower than meniscus height of the first dilution, is observed at diluting paracetamol solution up to the concentration of 10^{-9} mole/l (Figure 18.2).

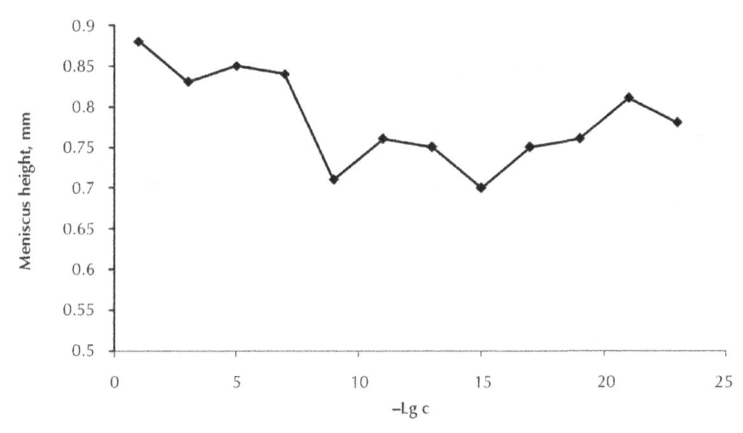

FIGURE 18.2 Values of meniscus height in the capillary of paracetamol solution (concentration, mole/l)

The same dependence is observed for dilution of paracetamol solution with concentration of 10^{-15} mole/l. So, at this concentration meniscus height is 0.7mm.

Growth of meniscus height in the capillary is observed further for solutions with the following dilution. This process is motivated by very high dilution, which is the solution, according to its composition and properties, tries to attain the state of pure solvent.

These changes are polymodal dependence effect concentration, described for different substances and different properties in domestic and world scientific literature. From the Figure and its description it is clearly seen that there are concentrations of paracetamol solution of 10^{-9} mole/l and 10^{-15} mole/l, for which there has been stated essential lowering of meniscus in the capillary regarding other concentrations. Between these concentration ranges there is a concentration range in which there are no essential changes, this range being the so-called "dead zone".

We have studied solutions of ascorbic acid and paracetamol mixture. It is necessary to note that not uniquely defined change of meniscus height is observed in solutions containing simultaneously two active substances. Meniscus of one-component solutions is narrower than that of water, but in two-component solutions both reduction and increase of meniscus height values are possible in comparison with water (Figure 18.3).

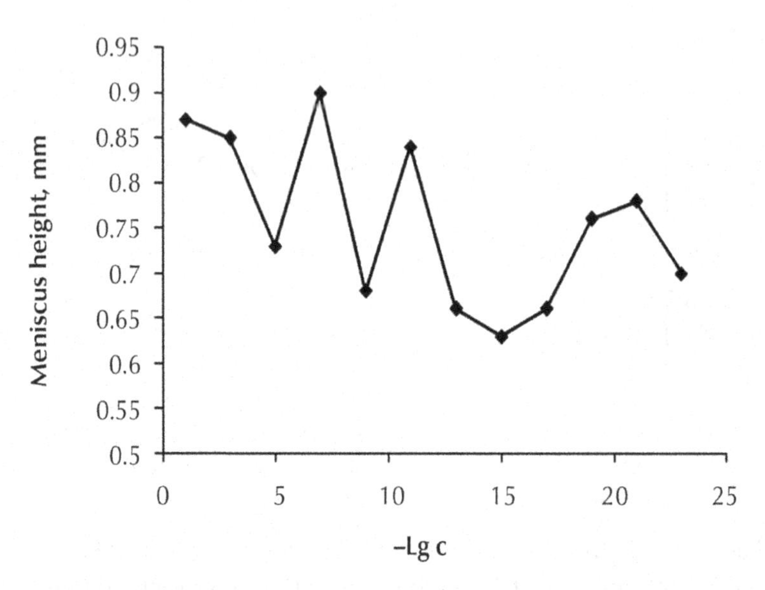

FIGURE 18.3 Values of meniscus height in the capillary of paracetamol solutions with ascorbic acid

Lowering of meniscus height is observed in solutions with concentration of paracetamol with ascorbic acid 10^{-5}, 10^{-9}, 10^{-13}, 10^{-15}, 10^{-17} mole/l on 11.5 %, 17.5%, 20%, 23.6%, and 20% correspondingly. It is possible to distinguish two concentration ranges in the field of super-small concentrations where reduction of meniscus height in the capillary takes place: This is – 10^{-9} mole/l and 10^{-13} – 10^{-17} mole/l. Received concentration ranges coincide with data of cathetometric studies of ascorbic acid and paracetamol solutions (Table 18.1).

Increase of meniscus height in comparison with the solvent is observed in solutions with concentrations of diluted compounds of 10^{-1}, 10^{-3}, 10^{-7}, 10^{-11} mole/l. Growth of meniscus height above such value of the solvent can be explained by the change in SSD effect display.

We have analyzed ultra-violet spectra of solutions of ascorbic acid, paracetamol, paracetamol with ascorbic acid. All the spectra were read from spectrophotometer Cary 100, UV-Visible Spectrophotometer in the range of 200–350 nm.

The most absorption in all solutions takes place in short wave's part of the effective wave band. More concentrated solutions (10^{-1}, 10^{-3} mole/l) of ascorbic acid have maximum absorption in the length interval of 220–280 nm, of paracetamol 220–310 nm, paracetamol with ascorbic acid 200–320 nm. Tendency to narrowing of absorption field, reduction of optical density value and quantity of tops up to one or two is registered in all solutions, under study, as content of diluted substance reduces.

TABLE 18.1 Results of study of ascorbic acid, paracetamol, paracetamol with ascorbic acid solutions of wide concentration range by method of electronic spectroscopy and cathetometric method.

Method of study	Concentration ranges, mole/l		
	Ascorbic acid solutions	**Paracetamol solutions**	**Solutions of paracetamol with ascorbic acid**
Cathetometric method of screening	10^{-9}, 10^{-13} – 10^{-17}, 10^{-23}	10^{-9}, 10^{-15}	10^{-9}, 10^{-13} – 10^{-17}
Electronic spectroscopy	10^{-9}, 10^{-15}, 10^{-19}	10^{-9}, 10^{-13}, 10^{-15}, 10^{-21}	10^{-9}, 10^{-13}, 10^{-15}, 10^{-21}

Solutions spectra with ascorbic acid concentration of 10^{-15} and 10^{-19} mole/l are equal in form and differ from spectra of other dilutions. Lowering of ascorbic acid concentration is not accompanied by uninterrupted reduction of optical density value. Irregular growth of absorption in comparison with more concentrated solutions is observed in solutions with ascorbic acid concentrations of 10^{-9}, 10^{-15}, 10^{-19} mole/l (Table 18.2).

TABLE 18.2 Maximum values of absorption of ascorbic acid solutions in ultra-violet field

Concentration, mole/l	Absorption maximum	
	Wave length, nm	**A**
10^{-3}	230-270	0,79
10^{-5}	265	0,04
10^{-7}	220-230	0,012
10^{-9}	225	0,03
10^{-11}	235	0,01

TABLE 18.2 *(Continued)*

10^{-13}	220	0,015
10^{-15}	220	0,04
10^{-17}	220-240	0,005
10^{-19}	220	0,03
10^{-21}	220	0,005
10^{-23}	220	0

Community in given structures may be assumed taking into account coincidence of waves length and values of optical density for solutions with concentrations 10^{-9}, 10^{-15}, 10^{-19} mole/l.

Gradual reduction of paracetamol concentration in solutions is not accompanied by the same reduction of optical density value. Growth of absorption in comparison with more concentrated solutions is observed for solutions with concentrations 10^{-9}, 10^{-13}, 10^{-15}, and 10^{-21} mole/l (Table 18.3). All this allows assuming the rise of structural changes in these solutions.

TABLE 18.3 Maximum values of absorption of paracetamol solutions in ultra-violet field.

Concentration, mole/l	Absorption maximum	
	Wave length, nm	A
10^{-1}	230	0,58
10^{-3}	270	0,53
10^{-5}	245	0,15
10^{-7}	240	0,02
10^{-9}	235	0,015
10^{-11}	240	0,007
10^{-13}	245	0,007
10^{-15}	220-245	0,01
10^{-17}	235	0,003
10^{-19}	290	0,001
10^{-21}	225	0,02
10^{-23}	230	0,005

Spectrum of ascorbic acid and paracetamol solution in concentrations of 10^{-1} mole/l is characterized by wide absorption band in the field of 200–320nm. Dilution of 0.1M solution by 100 times is accompanied by reduction of absorption field width up to 200–280nm without changing of spectrum form and intensity of absorption on peaks.

Changing of spectrum form accompanied by essential reduction of absorption from 4 to 0.095 with the maximum on the length 243nm takes place while diluting solution of ascorbic acid with paracetamol up to the concentration 10^{-5} mole/l (Table 18.4).

TABLE 18.4 Maximum values of absorption of ascorbic acid with paracetamol solutions (1:1) in ultra-violet field.

Concentration, mole/l	Absorption maximum	
	Wave length, nm	A
10^{-1}	237	4,651
10^{-3}	235	4,208
10^{-5}	243	0,095
10^{-7}	205	0,242
10^{-9}	205	0,293
10^{-11}	207	0,084
10^{-13}	206	0,367
10^{-15}	205	0,186
10^{-17}	207	0,086
10^{-19}	207	0,089
10^{-21}	206	0,261
10^{-23}	207	0,105

Two peaks on lengths 205–206 nm and 270–273 nm are shown on spectra of solutions with paracetamol and ascorbic acid concentrations 10^{-7}, 10^{-9}, 10^{-13}, 10^{-15}, and 10^{-21} mole/l. One peak on the length 207 nm is shown on spectra of solutions with paracetamol and ascorbic acid concentrations 10^{-11}, 10^{-17}, 10^{-19}, and 10^{-23} mole/l. Display of maximum on the length 270 nm occurred under conditions that absorption on maximum 205 nm was not less than 0.1.

Increase of optical density is observed in spectra of solutions with paracetamol and ascorbic acid concentrations 10^{-7}, 10^{-9}, 10^{-13}, 10^{-19}, and 10^{-21} mole/l.

18.1.2 CONCLUSION

As a result of this work we, by cathetometric method, have defined concentration ranges of possible display of medical compounds biological activity that forms 10^{-9} mole/l and 10^{-13}–10^{-17} mole/l for ascorbic acid and 10^{-9}, and 10^{-15} of paracetamole, and also 10^{-9}, and 10^{-13}–10^{-17} for the mixture of ascorbic acid paracetamol. This allows assuming compatibility of components data in given concentration ranges.

While studying solutions in super-small doses of ascorbic acid, paracetamol and also at their joint presence by method of electronic spectroscopy it has been found out that irregular growth of absorption is observed for solutions with ascorbic acid concentrations 10^{-9}, 10^{-15}, and 10^{-19} mole/l in comparison with more concentrated solutions and for paracetamol solutions with concentrations 10^{-9}, 10^{-13}, 10^{-15}, and 10^{-21} mole/l. But for the mixture ascorbic acid paracetamol such interval is 10^{-9}, 10^{-13}, 10^{-15}, and 10^{-21} mole/l.

This fact allows speaking about appearance of medical compounds structures with water in these solutions which differ by relatively high absorption.

Information received by method of ultra-violet spectroscopy, agrees with data of cathetometric screening both for solutions of ascorbic acid and of paracetamol.

KEYWORDS

- **Cathetometric Method**
- **Dead Zone**
- **Paracetamol**
- **Super-small Doses Effect**
- **Ultra-violet Spectroscopy**

REFERENCES

1. Konovalov, A. I. (2009). Physico-chemical mystery of super-small doses. *Chemistry and life.*, 2, 5-9.
2. Kuznetsov, P. E., Zlobin, V. A., & Nazarov, G. V. On the question about physical nature of superlow concentrations action. Heads of reports at III International Symposium "Mechanisms of super-small doses action", Moscow, December, 3–6, 2002, p.229
3. Chernikov, F. R. (1999). Method for evaluation of homoeopathic preparations and its physico-chemical foundations. *Matherials of Congress of homoeopathists of Russia.* Novosibirsk, p.73.
4. Pal'mina, N. P. (2009). Mechanisms of super-small doses action. *Chemistry and life.*, N2, p.10.
5. Maslikovsky, M. D. (1995). *Combined analgesic-febrifugal and anti-inflammatory preparations.* M., p. 208
6. Niyazi, F. F. & Kuvardin, N. V. *Method for determining abilities of bioactive substances to display "super-small doses" effect.* Patent No. 2346260 of 10.02.2009

CHAPTER 19

PRODUCTION OF ELECTROSPUN NANOFIBERS: AN INVESTIGATION ON GOVERNING PARAMETERS

A. K. HAGHI and GENNADY E. ZAIKOV

CONTENTS

19.1 INTRODUCTION

There is an enormous requirement for cleaner air around the world which has activated immense interests in the development of high efficiency filters or face masks. Nonwoven nanofibrous media have low basis weight, high permeability, and small pore size that make them appropriate for a wide range of filtration applications. In addition, nanofibers web offers unique properties like high specific surface area (ranging from 1 to 35 m^2/g depending on the diameter of fibers), good interconnectivity of pores and the potential to incorporate active chemistry or functionality on nanoscale. To date, the most successful method of producing nanofibers is through the process of electrospinning. The electrospinning process uses high voltage to create an electric field between a droplet of polymer solution at the tip of a needle and a collector plate [1-3]. When the electrostatic force overcomes the surface tension of the drop, a charged, continuous jet of polymer solution is ejected. As the solution moves away from the needle and toward the collector, the solvent evaporates and jet rapidly thins and dries. On the surface of the collector, a nonwoven web of randomly oriented solid nanofibers is deposited. It has been found that the morphology of the web, such as fiber diameter and its uniformity of the electrospun nanofibers, are dependent on many processing parameters. These parameters can be divided into three main groups:

- Solution properties,
- Processing conditions,
- Ambient conditions. Each of the parameters has been found to affect the morphology of the electrospun fibers [4].

19.1.1 SOLUTION PROPERTIES

Parameters such as viscosity of solution, solution concentration, molecular weight (MW) of polymer, electrical conductivity, elasticity, and surface tension have an important effect on the morphology of nanofibers.

VISCOSITY

The viscosity range of a different nanofiber solution which is spinnable is different. One of the most significant parameters influencing the fiber diameter is the solution viscosity. A higher viscosity results in a large fiber diameter. Beads and beaded fibers are less likely to be formed for the more viscous solutions. The bead diameter becomes bigger and the average distance between beads on the fibers become longer as the viscosity increases.

SOLUTION CONCENTRATION

In the electrospinning process, for fiber formation to occur a minimum solution concentration is required. As the solution concentration increases, a mixture of beads and fibers is obtained. The shape of the beads changes from spherical to spindle-like when the solution concentration varies from low to high levels. The fiber diameter increases with increasing solution concentration because of the higher viscosity resistance. Nevertheless, at higher concentration the viscoelastic force which usually resists rapid changes in fiber shape may result in uniform fiber formation. However, it is impossible to electrospinning if the solution concentration or the corresponding viscosity become too high due to the difficulty in liquid jet formation.

MOLECULAR WEIGHT

The MW also has a significant effect on the rheological and electrical properties such as viscosity, surface tension, conductivity and dielectric strength. It has been observed that too low MW solution tends to form beads rather than fibers and high molecular weight nanofiber solution gives fibers with larger average diameter.

SURFACE TENSION

The surface tension of a liquid is often defined as the force acting at right angles to any line of unit length on the liquid surface. By reducing surface tension of a nanofiber solution, fibers could be obtained without beads. The surface tension seems more likely to be a function of solvent compositions, but is negligibly dependent on the solution concentration. Different solvents may contribute different surface tensions. However, a lower surface tension of a solvent will not necessarily always be more suitable for electrospinning. Generally, surface tension determines the upper and lower boundaries of electrospinning window if all other variables are held constant. The formation of droplets, bead and fibers can be driven by the surface tension of solution, and lower surface tension of the spinning solution helps electrospinning to occur at lower electric field.

SOLUTION CONDUCTIVITY

There is a significant drop in the diameter of the electrospun nanofibers when the electrical conductivity of the solution increases. Beads may also be observed due to low conductivity of the solution, which results in insufficient elongation of a jet by electrical force to produce uniform fiber. In general, electrospun nanofibers with the smallest fiber diameter can be obtained with the highest electrical conductivity. This indicates that the drop in the size of the fibers is due to the increased electrical conductivity.

19.1.2 PROCESSING CONDITIONS

Another important parameter that affects the electrospinning process is the various external factors exerting on the electrospinning jet. This includes the Applied voltage, the feed rate, diameter of nuzzle/needle and distance between the needle and collector.

APPLIED VOLTAGE

In the case of electrospinning, the electric current due to the ionic conduction of charge in the nanofiber solution is usually assumed small enough to be negligible. The only mechanism of charge transport is the flow of solution from the tip to the target. Thus, an increase in the electrospinning current generally reflects an increase in the mass flow rate from the capillary tip to the grounded target when all other variables (conductivity, dielectric constant, and flow rate of solution to the capillary tip) are held constant. Increasing the applied voltage (*i.e.*, increasing the electric field strength) will increase the electrostatic repulsive force on the fluid jet, which favors the thinner fiber formation. On the other hand, the solution will be removed from the capillary tip more quickly as jet is ejected from Taylor cone. This results in the increase of the fiber diameter [2].

FEED RATE

The morphological structure can be slightly changed by changing the solution flow rate. When the flow rate exceeded a critical value, the delivery rate of the solution jet to the capillary tip exceeded the rate at which the solution was removed from the tip by the electric forces. This shift in the mass-balance resulted in sustained but unstable jet and fibers with big beads formation. In the first part of this study, the production of electrospun nanofibers was investigated [5]. In another part, a different case study was presented to show how nanofibers can be laminated for application in filter media.

DISTANCE BETWEEN THE NEEDLE AND COLLECTOR

When the distance between the tip of needle and the collector is reduced, the jet will have a shorter distance to travel before it reaches the collector plate. Moreover, the electric field strength will also increase at the same time and this will increase the acceleration of the jet to the collector. As a result, there may not have enough time for the solvents to evaporate when it hits the collector [6].

DIAMETER OF NEEDLE

The internal diameter of needle has a certain effect on the electrospinning process. A smaller internal diameter was found to reduce the amount of beads on the electrospun web and was also found to cause a reduction in the diameter of fibers.

19.1.3 AMBIENT CONDITIONS

Since electrospinning is influenced by external electric field, any changes in the electrospinning environment will also affect the electrospinning process. Any interaction between the surrounding and the polymer solution may have an effect on the electrospun fiber morphology. For example, high humidity was found to cause the formation of pores on the surface of the fibers.

CHITOSAN

Over the recent years, interest in the application of naturally occurring polymers such as polysaccharides and proteins, owing to their abundance in the environment, has grown considerably. Chitin, the second most abundant polysaccharide found on earth next to cellulose, is a major component of the shells of crustaceans such as crab, shrimp, and crawfish. The structural characteristics of chitin are similar to those of glycosaminoglycans. When chitin is deacetylated over about 60% it becomes soluble in dilute acidic solutions and is referred to chitosan or poly(*N*-acetyl-D-glucosamine). Chitosan and its derivatives have attracted much research because of their unique biological properties such as antibacterial activity, low toxicity, and biodegradability.

Depending on the chitin source and the methods of hydrolysis, chitosan varies greatly in its MW and degree of deacetylation (DDA). The

MW of chitosan can vary from 30 kDa to well above 1000 kDa and its typical DDA is over 70%, making it soluble in acidic aqueous solutions. At a pH of about 6–7, the biopolymer is a polycations and at a pH of 4.5 and below, it is completely protonated. The fraction of repeat units which are positively charged is a function of the DDA and solution pH. A higher DDA would lead to a larger number of positively charged groups on the chitosan backbone.

As mentioned, chitosan has several unique properties such as the ability to chelate ions from solution and to inhibit the growth of a wide variety of fungi, yeasts and bacteria. Although the exact mechanism witch chitosan exerts these properties is currently unknown, it has been suggested that the polycationic nature of this piopolymer that forms from acidic solutions below pH 6.5 is a crucial factor. Thus, it has been proposed that the positively charged amino groups of the glucosamine units interact with negatively charged components in microbial cell membranes altering their barrier properties, thereby preventing the entry of nutrients or causing the leakage of intracellular contents.

ELECTROSPINNING OF CHITOSAN

Chitosan is insoluble in water, alkali, and most mineral acidic systems. However, though its solubility in inorganic acids is quite limited, chitosan is in fact soluble in organic acids, such as dilute aqueous acetic, formic, and lactic acids. Chitosan also has free amino groups, which makes it a positively charged polyelectrolyte. This property makes chitosan solutions highly viscous and complicates its electrospinning. Furthermore, the formation of strong hydrogen bonds in a 3-D network prevents the movement of polymeric chains exposed to the electrical field. Different strategies were used for bringing chitosan in nanofiber form. The three top most abundant techniques include blending of favorite polymers for electrospinning process with chitosan matrix, alkali treatment of chitosan backbone to improve electro spinnability through reducing viscosity and employment of concentrated organic acid solution to produce nanofibers by decreasing of surface tension. Electrospinning of polyethylene oxide(PEO)/chitosan and polyvinyl alcohol(PVA)/ chitosan blended nanofiber are two recent studies based on first strategy. In the second protocol, the MW of chitosan decreases through alkali treatment. Solutions of the treated chitosan in aqueous 70–90% acetic acid produce nanofibers with appropriate quality and processing stability. Using concentrated organic acids such as acetic acid and triflouroacetic acid (TFA) with and without dichloromethane (DCM) has been reported exclusively for producing neat chitosan nanofibers. They similarly reported the decreasing of surface tension and at the same time enhancement of charge density of chitosan solution without significant effect on viscosity. This new method suggests significant influence of the concentrated acid solution on the reducing of the applied field required for electrospinning.

ELECTROSPUN NANO-WEB LAMINATION

While electrospun webs suggest exciting characteristics, it has been reported that they have limited mechanical properties [3-6]. To compensate this drawback in order to use of them in filtration applications, electrospun nanofibers web could be laminated *via* an adhesive into a multilayer system. The adhesives are as solvent/water-based adhesive or as hot-melt adhesive. At the first group, the adhesives are as solution in solvent or water, and solidify by evaporating of the carrying liquid. Solvent-based adhesives could 'wet' the surfaces to be joined better than water-based adhesives, and also could solidify faster. But unfortunately, they are environmentally unfriendly, usually flammable and more expensive than those. Of course it does not mean that the water-based adhesives are always preferred for laminating, since in practice, drying off water in terms of energy and time is expensive too. Besides, water-based adhesives are not resisting to water or moisture because of their hydrophilic nature. At the second group, hot-melt adhesives are environmentally friendly, inexpensive, require less heat and energy, and so are now more preferred. Generally there are two procedures to melt these adhesives; static hot-melt laminating that accomplish by flat iron or Hoffman press and continuous hot-melt laminating that uses the hot calendars. In addition, these adhesives are available in several forms; as a web, as a continuous film, or in powder form. The adhesives in film or web form are more expensive than the corresponding adhesive powders. The web form are discontinuous and produce laminates which are flexible, porous and breathable, whereas, Continuous film adhesives cause stiffening and produce laminates which are not porous and permeable to both air and water vapour. This behaviour attributed to impervious nature of adhesive film and its shrinkage under the action of heat. Thus, the knowledge of laminating skills and adhesive types is very essential to producing appropriate multilayer structures [7-12]. Specifically, this subject becomes more highlight as we will laminate the ultrathin nanofibers web, because the laminating process may be adversely influenced on the nanofibers web properties.

In this chapter, new multilayer structures were made by laminating of nanofibers web into cotton fabric *via* hot-melt method at different temperatures. Effects of laminating temperature on the nanofibers web morphology, air transport properties, and the adhesive force were discussed.

19.2 EXPERIMENTAL

19.2 1 PREPARATION OF CHITOSAN-MWNT SOLUTION

Chitosan with DDA of 85% and MW of 5×10^5 was supplied by Sigma-Aldrich. The MWNTs, supplied by Nutrino, have an average diameter of 4 nm and purity of about 98%. All of the other solvents and chemicals were commercially available and used as received without further purification. A Branson Sonifier 250 operated at 30W used to prepare the MWNT dispersion in chitosan/organic acid solution (70/30 TFA/DCM). First, 3mg of as received MWNTs was dispersed into deionized water or DCM using solution sonicating for 10 min. Next, chitosan was added to MWNTs dispersion to preparing a 10 wt% solution after sonicating for 5 min. Finally, in order to obtain a chitosan/MWNT solution, organic acid solution was added and the dispersion was stirred for 10 hrs.

19.2.2 ELECTROSPINNING

After the preparation of spinning solution, it was inserted into a syringe with a stainless steel nozzle and then the syringe was placed in a metering pump from WORLD PRECISION INSTRUMENTS (Florida, USA). Next, this set installed on a homemade plate which it could traverse to left-right direction along drum collector (Figure 19.1). The electrospinning conditions and layers properties for laminating are summarized in Table 1 (refer to part 1 of this article). The electrospinning process was carried out for 8h and the electrospun fibres were collected on an aluminium-covered rotating drum which was previously covered with a Poly-Propylene Spunbond Nonwoven (PPSN) substrate. After removing of PPSN covered with electrospun fibres from drum and attaching another layer of PPSN on it, this set was incorporated between two cotton weft-warp fabrics as a structure of fabric-PPSN-nanofibers web-PPSN-fabric (Figure 19.2). Finally, hot-melt laminating performed using a simple flat iron for 1min, under a pressure of 9gf/cm^2 and at temperatures 85,110,120,140, and 150°C (above softening point of PPSN) to form multilayer structures.

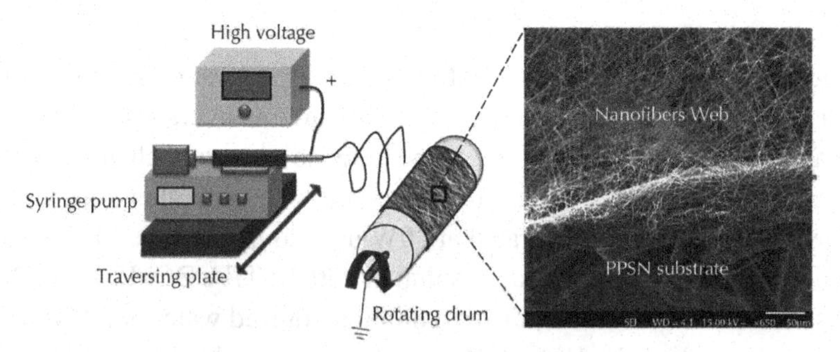

FIGURE 19.1 Electrospinning setup and an enlarged image of Nanofibers Web on PPSN.

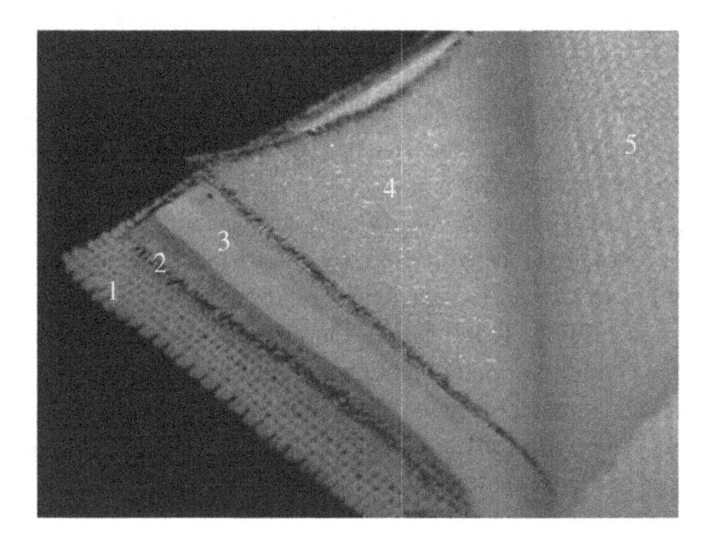

FIGURE 19.2 Multilayer components: (1) Fabric, (2) PPSN, (3) Nanofibers Web, (4) PPSN, (5) Fabric.

19.2.3 CHARACTERIZATIONS

THE MORPHOLOGY OF ELECTROSPUN FIBERS

The electrospun fibers were characterized using scanning electron microscope (SEM, Seron Technology, AIS-2100, Korea) to study the fiber morphology. The sample was sputter coated with Au/Pd to prevent charging during SEM imaging. Image processing software (MICROSTRUCTURE MEASUREMENT) was used to measure the fiber diameter from the SEM micrograph. Fiber diameter was measured at 50 different points for determining the average and distribution of diameter. The Mass per unit area (g/m^2 or gsm) of the electrospun web was measured by dividing the mass of the web by its area.

THE MORPHOLOGY OF MULTILAYER STRUCTURES

A piece of each multilayer was freeze fractured in liquid nitrogen and after sputter-coating with Au/Pd, a cross-section image of them captured using a scanning electron microscope.

Also, to consider the nanofiber web surface after hot-melt laminating, other laminations were prepared by a non-stick sheet made of Teflon (0.25 mm thickness) as a replacement for one of the fabrics (fabric/PPSN/nanofibers web/PPSN/Teflon sheet). Laminating process was carried out at the same conditions which mentioned to produce primary laminations. Finally, after removing of Teflon sheet, the nanofibers layer side was observed under an optical microscope (MICROPHOT-FXA, Nikon, Japan) connected to a digital camera.

MEASUREMENT OF AIR PERMEABILITY

The air permeability of multilayer structures, which is a measure of the structural porosity, was measured by air permeability tester (TEXTEST FX3300, Zürich, Switzerland). It was tested five pieces of each sample under air pressure of 125 Pa, at ambient condition (16°C, 70%RH), and then the average air permeability was calculated.

19.3 DISCUSSION AND RESULTS

The goal of this research is to develop chitosan based multilayer structures in order to use of them in filtration applications. In electrospinning phase, PPSN was chosen as a substrate to provide strength to the nanofiber web and to prevent of its destruction in removing from the collector. In Figure 19.1, an ultrathin layer of nanofibers web on PPSN layer is illustrated, which conveniently shows the relative fiber size of nanofiber (326 ± 68 nm) web compared to PPSN fibers. Also, this Figure shows that the macropores of PPSN substrate is covered with numerous electrospun nanofibers, which will create innumerable microscopic pores in this system.

But in laminating phase, this substrate acts as an adhesive and causes to bond the nanofiber web to the fabric. In general, it is relatively simple to create a strong bond between these layers, which guarantees no delamination or failure in multilayer structures; the challenge is to preserve the original properties of the nanofiber web and fabrics. In other words, the application of adhesive should have minimum affect on the fabric or nanofiber web structure. In order to achieve to this aim, it is necessary that: a) the least amount of a highly effective adhesive applied, b) the adhesive correctly cover the widest possible surface area of layers for better linkage between them and c) the adhesive penetrate to a certain extent of the nanofiber web/fabric. Therefore, we selected PPSN, which is a hot-melt adhesive in web form. As mentioned chapter, the perfect use of web form adhesive can be lead to produce multilayer fabrics which are porous, flexible, and permeable to both air and water vapour. On the other hand, since the melting point of PPSN is low, hot-melt laminating can perform at lower temperatures. Hence, the probability of shrinkage that may happen on layers in effect of heat becomes smaller; Of course, the thermal degradation of chitosan nanofiber web begins above 250°C and the cotton fabrics are intrinsically resistant to heat too. By this description, laminating process performed at five different temperatures to consider the effect of laminating temperature on the nanofiber web/multilayer properties.

The SEM images of multilayer fabric cross-section after laminating at different temperatures shown in Figure 19.3. It is obvious that these images cannot deliver useful information about nanofibers web morphology in multilayer structure, so it becomes impossible to consider the effect of laminating temperature on nanofibers web. Hence, in a novel way, we decided to prepare a secondary multilayer by substitution of one of the fabrics with Teflon sheet. By this replacement, the surface of nanofiber web will become accessible after laminating; because Teflon is a non-stick material and easily separates from the adhesive.

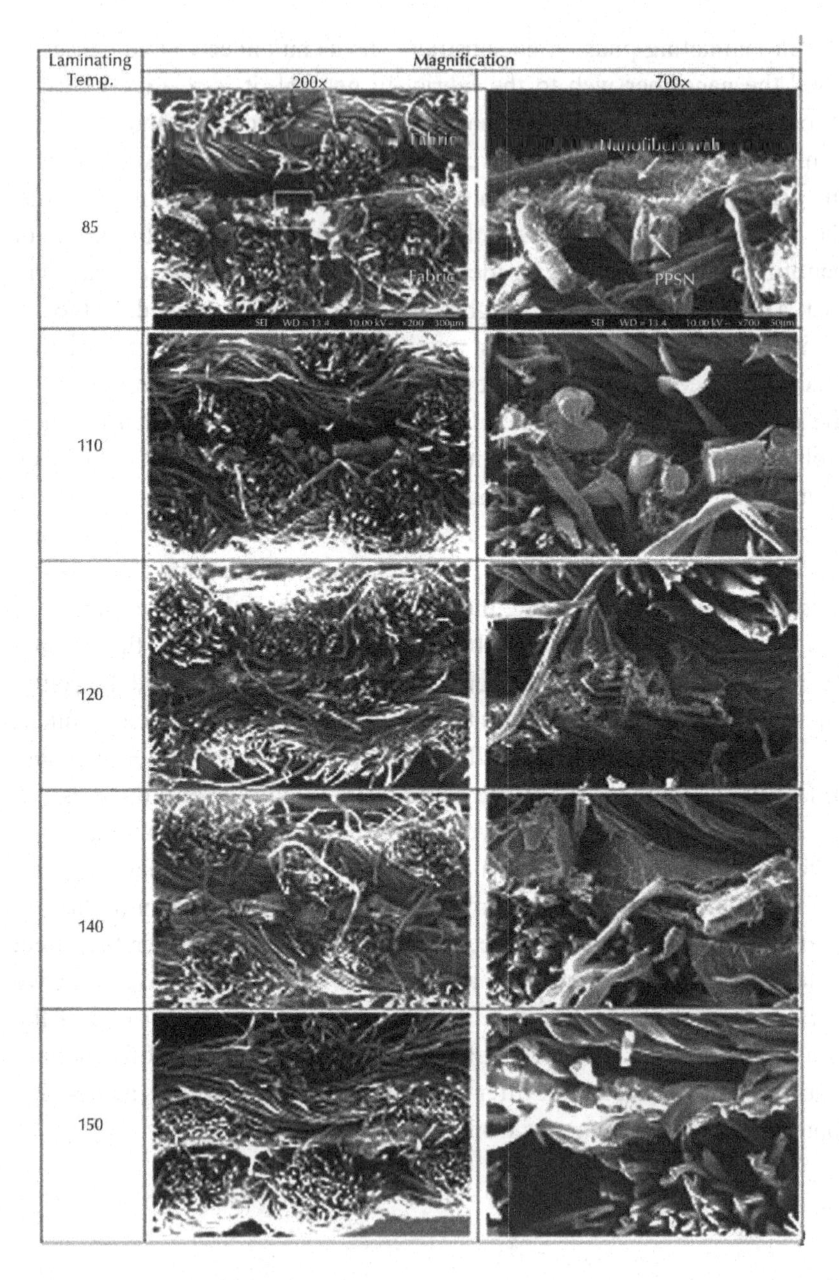

FIGURE 19.3 The SEM images of multilayer fabric cross-section at different laminating temperatures.

Figure 19.4 presents optical microscope images of nanofiber web and adhesive after laminating at different temperatures. It is apparent that the adhesive gradually flattened on nanofiber web (Figure 19.4(a-c)) when laminating temperature increased to melting point of adhesive (140°C). This behavior is attributed to increment in plasticity of adhesive because of temperature rise and the pressure applied from the iron weight. But, by selection of melting point as laminating temperature, the adhesive completely melted and began to penetrate into the nanofiber web structure instead of spread on it (Figure 19.4(d)). This penetration, in some regions, was continued to some extent that the adhesive was even passed across the web layer. The dark crisscross lines in Figure 19.4(d) obviously show wherever this excessive penetration is occurred. The adhesive penetration could intensify by increasing of laminating temperature above melting point; because the fluidity of melted adhesive increases by temperature rise. Figure 19.4(e) clearly shows the amount of adhesive diffusion in the web which was laminated at 150°C. At this case, the whole diffusion of adhesive lead to create a transparent film and to appear the fabric structure under optical microscope.

It is obvious in Figure 19.4(a-c) that the pore size of adhesive layer becomes smaller in effect of this transformation. Therefore, we can conclude that the adhesive layer, as a barrier, resists to convective air flow during experiment and finally reduces the air permeability of multilayer fabric according to the pore size decrease. But, this conclusion is unacceptable for the samples laminated at melting point (140°C); since the adhesive was missed self layer form in effect of penetration into the web/fabric structures (Figure 19.4d). At these samples, the adhesive penetration leads to block the pores of web/fabrics and to prevent of the air pass during experiment. It should be noted that the adhesive was penetrated into the web much more than the fabric because, PPSN structurally had more surface junction with the web (Figure 19.3). Therefore, at here, the nanofiber web absorbs the adhesive and forms an impervious barrier to air flow.

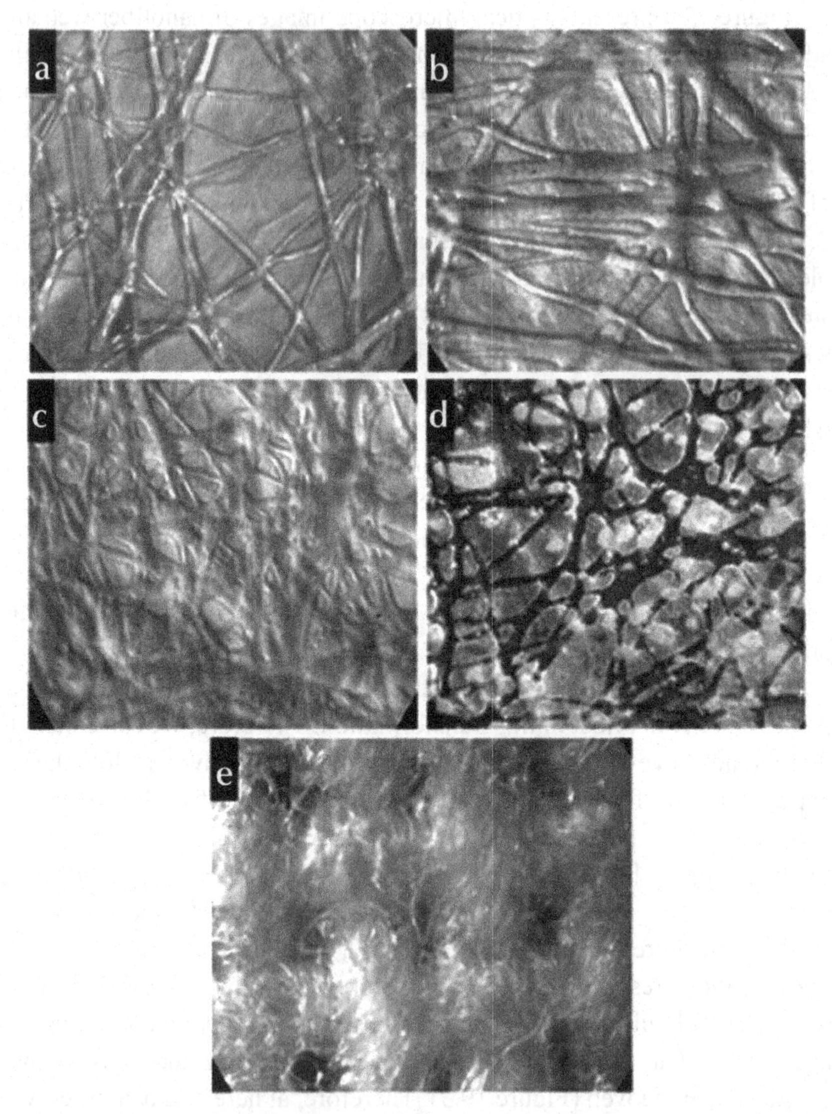

FIGURE 19.4 The optical microscope images of nanofiber web surface at different laminating temperatures (100x magnification), a) 85°C, b) 110°C, c) 120°C, d) 140°C, and e) 150°C

Furthermore, we only observed that the adhesive force between layers was improved according to temperature rise. For example, the samples laminated at 85°C were exhibited very poor adhesion between the nanofiber web and the fabrics as much as they could be delaminated by light abrasion of thumb, as illustrated at Figure 19.3. Generally, it is essential that no delamination occurs during use of this multilayer structure, because the nanofiber web might be destroyed due to abrasion of other layers. Before melting point, improving the adhesive force according to temperature rise is simultaneously attributed to the more penetration of adhesive into layers and the expansion of bonding area between them, as already discussed. Also at melting point, the deep penetration of adhesive into the web/fabric leads to increase in this force.

19.4 CONCLUSION

In the second parts of this study, production of a new laminated chitosan nanofiber with particular application in filtration is introduced. Meanwhile, effect of laminating temperature on the nanofibers web/multilayer structures properties investigated to make next generation of filter media. First, we demonstrate that it is impossible to consider the effect of laminating temperature on the nanofiber web morphology by a SEM image of multilayer cross-section. Thus, we prepared a surface image of nanofiber web after laminating at different temperature using an optical microscope. It was observed that nanofiber web was approximately unchanged when laminating temperature was below PPSN melting point. In addition, to compare air transport properties of multilayer fabrics, air permeability tests were performed. It was found that by increasing laminating temperature, air permeability was decreased. Furthermore, it only was observed that the adhesive force between layers in multilayer was increased with temperature rise. These results indicate that temperature is an effective parameter for laminating of nanofiber web to make a new type of antibacterial filter media.

KEYWORDS

- **Ambient Conditions**
- **Diameter of Needle**
- **Electrospinning Process**
- **Processing Conditions**
- **Solution Properties**
- **Surface Tension**

REFFERENCES

1. Tan, S. H. & Inai, R. (2005). Systematic parameters study for ultra-fine fibre fabrication *via* electrospinning Process. *Polymer*, 46, 6128.
2. itzel, De. J. M. (2001). Controlled deposition of electrospun Polyethylene Oxide fibres. *Polymer*, 42, 8163.
3. Zarkoob, S. (2004). Structure and morphology of electrospun silk nanofibres. *Polymer*, 45, 3973.
4. Atheron, S. (2004). Experimental investigation of the governing parameters in the electrospinning of polymer solutions. *polymer*, 45, 2017.
5. Ziabari, M., Mottaghitalab, V., McGove rn, S. T., & Haghi, A. K. (2008). Measuring Electrospun Nanofibre Diameter: A Novel Approach. *Chin. Phys. Lett.*, 25(8), 3071.
6. Ziabari, M., Mottaghitalab, V., McGove rn, S. T., & Haghi, A. K. (2007). A new image analysis based method for measuring electrospun nanofibre diameter. *Nanoscale Research Letter*, 2, 297.
7. Ziabari, M., Mottaghitalab, V., & Haghi, A. K. (2008). Simulated image of electrospun nonwoven web of PVA and corresponding nanofibre diameter distribution. *Korean J. of Chemical Engng.*, 25(4), 919.
8. Ziabari, M., Mottaghitalab, V., & Haghi, A. K. (2008). Evaluation of electrospun nanofibre pore structure parameters. *Korean J. of Chemical Engng.*, 25(4), 923.
9. Ziabari, M., Mottaghitalab, V., & Haghi, A. K. (2008). Distance transform algoritm for measuring nanofibre diameter. *Korean J. of Chemical Engng.*, 25(4), 905.
10. Haghi, A. K. & Ak bari, M. (2007). Trends in electrospinning of natural nanofibres. *Physica Status Solidi*, 204(6), 1830.
11. Kanafchian, M., Valizadeh, M., & Haghi, A. K. (2011) Prediction of nanofiber diameter for improvements in incorporation of multilayer electrospun nanofibers, *Korean J. Chem. Eng.*, 28(3), 751.
12. Kanafchian, M., Valizadeh, M., & A. K. Haghi. (2011). Fabrication of nanostructured and multicompartmental fabrics base on electrospun nanofibers, *Korean J. Chem. Eng.*, 28(3), 763.

CHAPTER 20

A NEW CLASS OF POLYMER NANOCOMPOSITES

A. A. OLKHOV, D. J. LIAW, G. V. FETISOV,
M. A. GOLDSCHTRAKH, N. N. KONONOV, A. A. KRUTIKOVA,
P. A. STOROZHENKO, G. E. ZAIKOV, and A. A. ISCHENKO

CONTENTS

20.1 INTRODUCTION

In recent years, considerable efforts have been devoted for search new functional nanocomposite materials with unique properties that are lacking in their traditional analogues. Control of these properties is an important fundamental problem. The use of nanocrystals as one of the elements of a polymer composite opens up new possibilities for targeted modification of its optical properties because of a strong dependence of the electronic structure of nanocrystals on their sizes and geometric shapes. An increase in the number of nanocrystals in the bulk of composites is expected to enhance long range correlation effects on their properties. Among the known nanocrystals, nanocrystalline silicon (nc-Si) attracts high attention due to its extraordinary optoelectronic properties and manifestation of quantum size effects. Therefore, it is widely used for designing new generation functional materials for nanoelectronics and information technologies. The use of nc-Si in polymer composites calls for a knowledge of the processes of its interaction with polymeric media. Solid nanoparticles can be combined into aggregates (clusters), and when the percolation threshold is achieved, a continuous cluster is formed.

An orderly arrangement of interacting nanocrystals in a long range potential minimum leads to formation of periodic structures. Because of the well-developed interface, an important role in such systems belongs to adsorption processes, which are determined by the structure of the nanocrystal surface. In a polymer medium, nanocrystals are surrounded by an adsorption layer consisting of polymer, which may change the electronic properties of the nanocrystals. The structure of the adsorption layer has an effect on the processes of self-organization of solid-phase particles, as well as on the size, shape, and optical properties of resulting aggregates. According to data obtained for metallic [1] and semiconducting [2] clusters, aggregation, and adsorption in three-phase systems with nanocrystals have an effect on the optical properties of the whole system. In this context, it is important to reveal the structural features of systems containing nanocrystals, characterizing aggregation, and adsorption processes in these systems, which will make it possible to establish a correlation between the structural and the optical properties of functional nanocomposite systems.

Silicon nanoclusters embedded in various transparent media are a new and interesting object for physicochemical investigation. For example, for particles smaller than 4 nm in size, the quantum size effects become significant. It makes possible to control the luminescence and absorption characteristics of materials based on such particles using of these effects [3, 4]. For nanoparticles about 10 nm in size or larger (containing ~10^4 Si atoms), the absorption characteristics in the UV and visible ranges are determined in many respects by properties like typical of massive crystalline or amorphous silicon samples. These characteristics depend on a number of factors: the presence of structural defects and impurities, the phase state, and so on. [5, 6]. For effective practical application and creation on a basis nc-Si the new polymeric materials possessing useful properties: sun-protection films [7], coverings [8] photoluminescent and electroluminescent composites [9, 10], stable to light dyes [11], embedding of these nanosized particles in polymeric matrixes becomes an important synthetic problem.

The method of manufacturing of silicon nanoparticles in the form of a powder by plasma chemical deposition, which was used in this study, makes possible to vary the chemical composition of their surface layers. As a result, another possibility of controlling their spectral characteristics arises, which is absent in conventional methods of manufacturing of nanocrystalline silicon in solid matrices (for example, in a-SiO_2) by implantation of charged silicon particles [5] or radio frequency deposition of silicon [2]. Polymer composites based on silicon nanopowder are a new object for comprehensive spectral investigation. At the same time, detailed spectral analysis has been performed for silicon nanopowder prepared by laser induced decomposition of gaseous SiH_4 (see, for example, [6, 12]). It is of interest to consider the possibility of designing new effective UV protectors based on polymer containing silicon nanoparticles [13]. An advantage of this nanocomposite in comparison with other known UV protectors is its environmental safety, that is, the ability to hinder the formation of biologically harmful compounds during UV-induced degradation of components of commercial materials. In addition, changing the size distribution of nanoparticles and their concentration in a polymer and correspondingly modifying the state of their surface, one can deliberately change the spectral characteristics of nanocomposite as a whole. In this case, it is necessary to minimize the transmission in the wavelength range below 400 nm (which determines the properties of UV-protectors [13]) by changing the characteristics of silicon powder.

High-strength polyethylene films containing 0.5–1.0 wt% of nanocrystalline silicon (nc-Si) were synthesized. Samples of nc-Si with an average core diameter of 7–10 nm were produced by plasmochemical method and by laser-induced decomposition of monosilane. Spectral studies revealed almost complete (up to ~95%) absorption of UV radiation in 200–400 nm spectral regions by 85 micron thick film if the nc-Si content approaches to 1.0 wt%. The density function of particle size in the starting powders and polymer films containing immobilized silicon nanocrystallites were obtained using the modeling a complete profile of X-ray diffraction patterns, assuming spherical grains and the lognormal distribution. The results of X-ray analysis shows that the crystallite size distribution function remains almost unchanged and the crystallinity of the original polymer increases to about 10% with the implantation of the initial nc-Si samples in the polymer matrix.

20.2 OBJECTS OF RESEARCH

In this chapter, the possibilities of using polymers containing silicon nanoparticles as effective UV protectors are considered. First, the structure of nc-Si obtained under different conditions and its aggregates, their adsorption and optical properties was studied in order to find the ways to control the UV spectral characteristics of multiphase polymer composites containing nanocrystalline silicon. Also, the purpose of this work was to investigate the effect of the concentration of silicon nanoparticles embedded in polymer matrix and the methods of preparation of these nanoparticles on the spectral characteristics of such nanocomposites. On the basis of the data obtained, recommendations for designing UV protectors based on these nanocomposites were formulated.

The nc-Si consists of core–shell nanoparticles in which the core is crystalline silicon coated with a shell formed in the course of passivation of nc-Si with oxygen and nitrogen. The nc-Si samples were synthesized by an original procedure in an argon plasma in a closed gas loop. To do this, we used a plasma vaporizer/condenser operating in a low-frequency arc discharge. A special consideration was given to the formation of a nano-crystalline core of specified size. The initial reagent was a silicon powder, which was fed into a reactor with a gas flow from a dosing pump. In the reactor, the powder vaporized at 7000–10000°C. At the outlet of the high-temperature plasma zone, the resulting gas-vapor mixture was sharply cooled by gas jets, which resulted in condensation of silicon vapor to form an aerosol. The synthesis of nc-Si in a low-frequency arc discharge was described in detail in [3].

The microstructure of nc-Si was studied by transmission electron microscopy (TEM) on a Philips NED microscope. X-ray powder diffraction analysis was carried out on a Shimadzu Lab XRD-6000 diffractometer. The degree of crystallinity of nc-Si was calculated from the integrated intensity of the most characteristic peak at $2\theta = 28^{\circ}$. Low-temperature adsorption isotherms at 77.3 K were measured with a Gravimat-4303 automated vacuum adsorption apparatus. The FTIR spectra were recorded on, in the region of 400–5000 cm^{-1} with resolution of about 1 cm^{-1}.

Three samples of nc-Si powders with specific surfaces of 55, 60, and 110 m^2/g were studied. The *D* values for these samples calculated by Eq. (2) are 1.71, 1.85, and 1.95, respectively; that is they are lower than the limiting values for rough objects. The corresponding *D* values calculated by Eq. (3) are 2.57, 2.62, and 2.65, respectively. Hence, the adsorption of nitrogen on nc-Si at 77.3 K is determined by capillary forces acting at the liquid-gas interface. Thus, in argon plasma with addition of oxygen or nitrogen, the ultra-disperse silicon particles are formed, which consist of a crystalline core coated with a silicon oxide or oxynitride shell. This shell prevents the degradation or uncontrollable transformation of the electronic properties of nc-Si upon its integration into polymer media. Solid structural elements (threads or nanowires) are structurally similar, which stimulates self-organization leading to fractal clusters. The surface fractal dimension of the clusters determined from the nitrogen adsorption isotherm at 77.3K is a structurally sensitive parameter, which characterizes both the structure of clusters and the morphology of particles and aggregates of nanocrystalline silicon.

As the origin materials for preparation film nanocomposites served polyethylene of low density (LDPE) marks 10803-020 and ultradisperse crystal silicon. Silicon powders have been received by a method plazmochemical recondensation of coarse-crystalline silicon in nanocrystalline powder. Synthesis nc-Si was carried out in argon plasma in the closed gas cycle in the plasma evaporator the condenser working in the arc low-frequency category. After particle synthesis nc-Si were exposed microcapsulating at which on their surfaces the protective cover from SiO_2, protecting a powder from atmospheric influence and doing it steady was created at storage. In the given work powders of silicon from two parties were used: nc-Si-36 with a specific surface of particles ~36 m^2/g and nc-Si-97 with a specific surface ~97 m^2/g.

Preliminary mixture of polyethylene with a powder nc-Si firms "Brabender" (Germany) carried out by means of closed hummer chambers at temperature 135 ± 5°C, within 10 min and speed of rotation of a rotor of 100 min^{-1}. Two compositions LDPE + nc-Si have been prepared: (1) composition PE + 0.5% nc-Si-97 on a basis nc-Si-97, containing 0.5 weights silicon %; (2) composition PE + 1% nc-Si-36 on a basis nc-Si-36, containing 1.0 weights silicon %.

Formation of films by thickness 85 ± 5 micron was spent on semi-industrial extrusion unit ARP-20-150 (Russia) for producing the sleeve film. The temperature was 120-190 °C on zones extruder and extrusion die. The speed of auger was 120 min^{-1}. The technological parameters of nanocomposites chooses, proceeding from conditions of thermostability and the characteristic viscosity recommended for processing polymer melting.

20.3 EXPERIMENTAL METHODS

The mechanical properties and an optical transparency of polymer films, their phase structure, and crystallinity, and also communication of mechanical and optical properties with a microstructure of polyethylene and granulometric structure of modifying powders nc-Si were observed.

Physicomechanical properties of films at a stretching (extrusion) measured in a direction by means of universal tensile machine EZ-40 (Germany) in accordance with Russian State Standard GOST-14236-71. Tests are spent on rectangular samples in width of 10 mm, and a working site of 50 mm. The speed of movement of a clip was 240 mm/min. The five parallel samples were tested.

Optical transparency of films was estimated on absorption spectra. Spectra of absorption of the obtained films were measured on spectrophotometer SF-104 (Russia) in a range of wavelengths 200–800 nm. Samples of films of polyethylene and composite films PE + 0.5% nc-Si-36 and PE + 1% nc-Si-36 in the size 3 × 3 cm were investigated. The special holder was used for maintenance uniform a film tension.

X-ray diffraction analysis by wide-angle scattering of monochromatic X-rays data was applied for research phase structure of materials, degree of crystallinity of a polymeric matrix, the size of single-crystal blocks in powders nc-Si and in a polymeric matrix, and also functions of density of distribution of the size crystalline particles in initial powders nc-Si.

X-ray diffraction measurements were observed on Guinier diffractometer: chamber G670 Huber [14] with bent Ge (111) monochromator of a primary beam which are cutting out line $K\alpha_1$ (length of wave λ = 1.5405981 Å) characteristic radiation of x-ray tube with the copper anode. The diffraction picture in a range of corners 2θ from 3° to 100° was registered by the plate with optical memory (IP-detector) of the camera bent on a circle. Measurements were spent on original powders nc-Si-36 and nc-Si-97, on the pure film LDPE further marked as PE, and on composite films PE + 0.5% nc-Si-97 and PE + 1.0% nc-Si-36. For elimination of tool distortions effect diffractogram standard SRM660a NIST from the crystal powder LaB_6 certificated for these purposes by Institute of standards of the USA was measured. Further it was used as diffractometer tool function.

Samples of initial powders nc-Si-36 and nc-Si-97 for X-ray diffraction measurements were prepared by drawing of a thin layer of a powder on a substrate from a special film in the thickness 6 microns (MYLAR, Chemplex Industries Inc., Cat. No: 250, Lot No: 011671). Film samples LDPE and its composites were established in the diffractometer holder without any substrate, but for minimization of structure effect two layers of a film focused by directions extrusion perpendicular to each other were used.

Phase analysis and granulometric analysis was spent by interpretation of the X-ray diffraction data. For these purposes the two different full-crest analysis methods [15, 16] were applied: (1) method of approximation of a profile diffractogram using analytical functions, polynoms and splines with diffractogram decomposition on making parts; (2) method of diffractogram modeling on the basis of physical principles of scattering of X-rays. The package of computer programs WinXPOW was applied to approximation and profile decomposition diffractogram ver. 2.02 (Stoe, Germany) [17], and diffractogram modeling at the analysis of distribution of particles in the sizes was spent by means of program PM2K (version 2009) [18].

20.4 DISCUSSION AND RESULT

The results of mechanical tests of the prepared materials are presented to Table 20.1 from which it is visible that additives of particles nc-Si have improved mechanical characteristics of polyethylene.

TABLE 20.1 Mechanical characteristics of nanocomposite films based of LDPE and nc-Si.

Sample	Tensile strength, kg/cm^2	Relative elongation-at-break, %
PE	100 ± 12	200–450
PE + 1% ncSi-36	122 ± 12	250–390
PE + 0.5% ncSi-97	118 12	380–500

The results presented in the table shows that the additives of powders of silicon raise the mechanical characteristics of films, and the effect of improvement of mechanical properties is more expressed in case of composite PE + 0.5% nc-Si-97 at which in comparison with pure polyethylene relative elongation-at-break has essentially grown.

Transmittance spectra of the investigated films are shown on Figure 20.1.

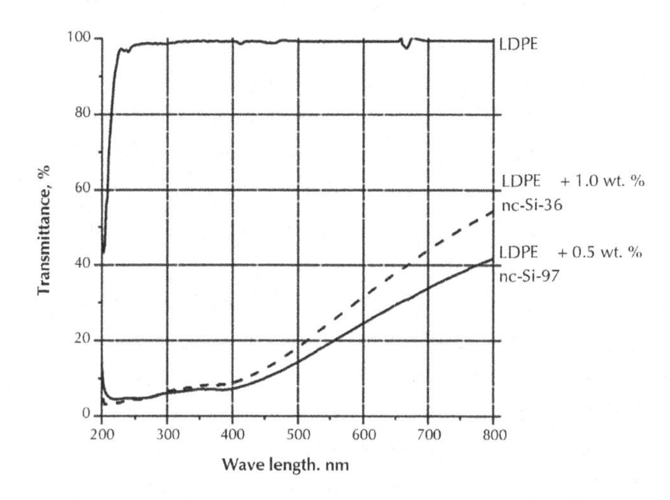

FIGURE 20.1 Transmittance spectra of the investigated films LDPE and nanocomposite films PE + 0.5% nc-Si-97 and PE + 1.0% nc-Si-36.

It is visible that the additives of powders nc-Si reduces the transparency of films in all investigated range of wavelengths, but especially strong decrease transmittance (almost in 20 times) is observed in a range of lengths of waves of 220–400 nm, that is. in UV areas.

The wide-angle scattering of X-rays data were used for the observing phase structure of materials and their component. Measured x-ray diffractograms of initial powders nc-Si-36 and nc-Si-97 on intensity and Bragg peaks position completely corresponded to a phase of pure crystal silicon (a cubic elementary cell of type of diamond – spatial group $Fd\overline{3}m$, cell parameter a_{Si} = 0.5435 nm).

For the present research granulometric structure of initial powders nc-Si is of interest. Density function of particle size in a powder was restored on X-ray diffractogram a powder by means of computer program PM2K [18] in which the method [19] modeling of a full profile diffractogram based on the theory of physical processes of diffraction of X-rays is realized. Modeling was spent in the assumption of the spherical form of crystalline particles and logarithmically normal distributions of their sizes. Deformation effects from flat and linear defects of a crystal lattice were considered. The received function of density of distribution of the size crystalline particles for initial powders nc-Si are represented graphically on Figure 20.2 in the signature to which statistical parameters of the found distributions are resulted. These distributions are characterized by such important parameters, as Mo(d) – position of maximum (a distribution mode); $<d>_V$ – average size of crystalline particles based on volume of the sample (the average arithmetic size) and Me(d) – the median of distribution defining the size d, specifying that particles with diameters less than this size make half of volume of a powder.

The results represented on Figure 20.2, show that initial powders nc-Si in the structure have particles with the sizes less than 10 nm which especially effectively absorb the UV radiation. The both powders modes of density function of particle size are very close, but median of density function of particle size of a powder nc-Si-36,it is essential more than at a powder nc-Si-97. It suggests that the number of crystalline particles with diameters is less 10 nanometers in unit of volume of a powder nc-Si-36 much less, than in unit of volume of a powder nc-Si-97. As a part of a powder nc-Si-36 a lot of particles with a diameter more than 100 nm and even there are particles more largely 300 nm whereas the sizes of particles in a powder nc-Si-97 do not exceed 150 nm and the basic part of crystalline particles has diameter less than 100 nm.

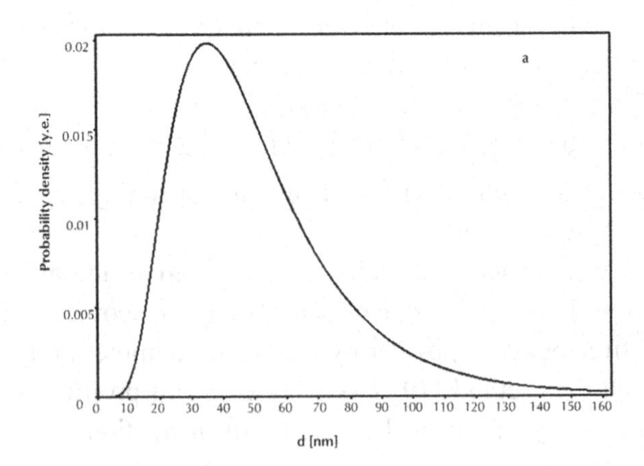

FIGURE 20.2 *(Continued)*

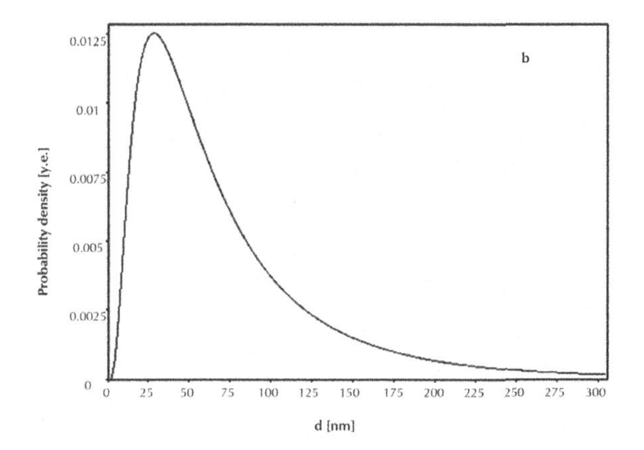

FIGURE 20.2 Density function of particle size in powders nc-Si, received from X-ray diffractogram by means of program PM2K.

(a) – nc-Si-97	Mo(d) = 35 nm	Me(d) = 45 nm	$<d>_V = 51$ nm;
(б) – nc-Si-36	Mo(d) = 30 nm	Me(d) = 54 nm	$<d>_V = 76$ nm.

The phase structure of the obtained film was estimated on wide-angle scattering diffractogram only qualitatively. Complexity of diffraction pictures of scattering and structure do not poses the quantitative phase analysis of polymeric films [20]. At the phase analysis of polymers often it is necessary to be content with the comparative qualitative analysis which allows watching evolution of structure depending on certain parameters of technology of production. Measured wide-angle X-rays scattering diffractograms of investigated films are shown on Figure 20.3. Diffractogramms have a typical form for polymers. As a rule, polymers are the two-phase systems consisting of an amorphous phase and areas with distant order, conditionally named crystals. Their diffractograms represent [20] superposition of intensity of scattering by the amorphous phase which looks like wide halo on the small-angle area (in this case in area 2θ between 10° and 30°), and intensity Bragg peaks scattering by a crystal phase.

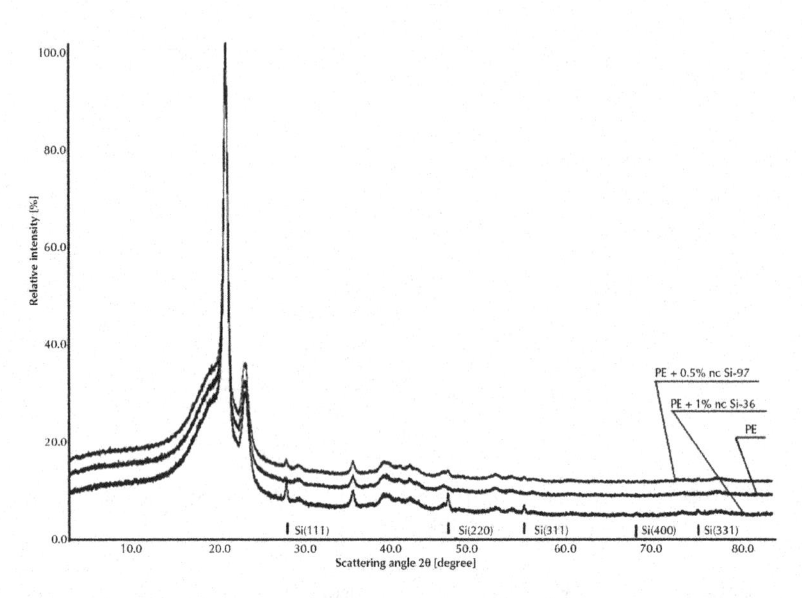

FIGURE 20.3 Diffractograms of the investigated composite films in comparison with diffractogram of pure polyethylene. Below vertical strokes specify reference positions of diffraction lines of silicon with their interference indexes (hkl).

Data on Figure 20.3 are presented in a scale of relative intensities (intensity of the highest peak is accepted equal 100%). For convenience of consideration curves are represented with displacement on an axis of ordinates. The scattering plots without displacement represents completely overlapping of diffractogram profiles of composite films with diffractogram of a pure LDPE film, except peaks of crystal silicon which were not present on PE diffractogram. It testifies that additives of powders nc-Si practically have not changed the crystal structure of polymer.

The peaks of crystal silicon are well distinguishable on diffractograms of films with silicon. Heights of the peaks of silicon with the same name (that is peaks with identical indexes) on diffractograms of the composite films PE + 0.5% nc-Si-97 and PE + 1.0% nc-Si-36 differ approximately twice that corresponds to a parity of mass concentration Si set at their manufacturing.

The degree of crystallinity of polymer films (a volume fraction of the crystal ordered areas in a material) was defined by diffractograms (Figure 20.3), for a series of samples only semi-quantitative (more/less). The essence of the method of crystallinity definition consists in analytical division of a diffractogram profile on the Bragg peaks from crystal areas and diffusion peak of an amorphous phase [20], as is shown in Figure 20 4.

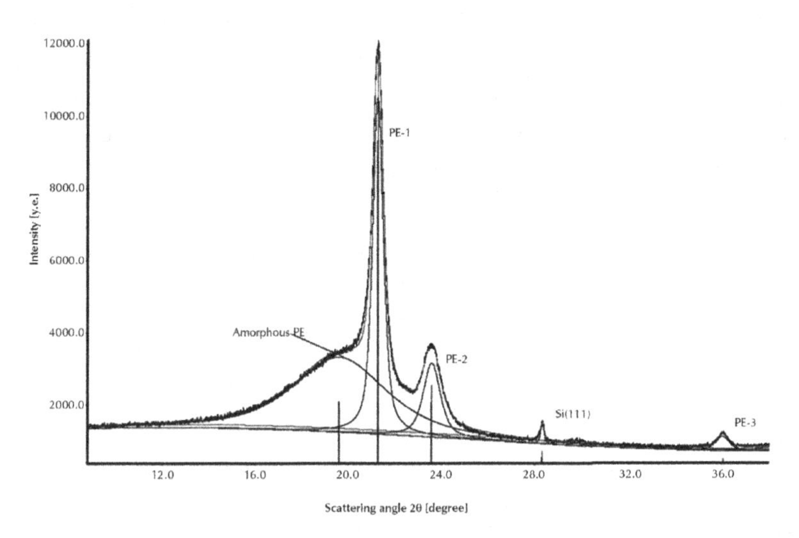

FIGURE 20.4 Diffractogram decomposition on separate peaks and a background by means of approximation of a full profile by analytical functions on an example of the data from sample PE + 1% nc-Si-36 (Figure 20 3). PE-n designate Bragg peaks of crystal polyethylene with serial numbers n from left to right. Si (111) – Bragg silicon peak nc-Si-36. Vertical strokes specify positions of maxima of peaks.

Peaks profiles of including peak of an amorphous phase, were approximated by function pseudo-Voigt, a background 4 order polynoms of Chebysheva. The nonlinear method of the least squares minimized a difference between intensity of points experimental and approximating curves. The width and height of approximating functions, positions of their maxima and the integrated areas, and also background parameters were thus specified. The relation of integrated intensity of a scattering profile by an amorphous phase to full the integrated intensity of scattering by all phases except for particles of crystal silicon gives a share of amorphy of the sample, and crystallinity degree turns out as a difference between unit and an amorphy fraction.

It was supposed that one technology of film obtaining allowed an identical structure. It proved to be true by coincidence relative intensities of all peaks on diffractograms Figure 20.3, and the sample consist only crystal and amorphous phases of the same chemical compound. Therefore, the received values of degree of crystallinity should reflect correctly a tendency of its change at modification polyethylene by powders nc-Si though because of a structure of films they can quantitatively differ considerably from the valid concentration of crystal areas in the given material. The found values of degree of crystallinity are represented in Table 20.2.

TABLE 20.2 Characteristics of the ordered (crystal) areas in polyethylene and its composites with nc-Si

PE			**PE + 1% nc-Si-36**			**PE + 0.5% nc-Si-97**		
Crystallinity	46%		47,5%			48%		
2θ [˚]	d [E]	ε	2θ [°]	d [E]	ε	2θ [°]	d [E]	ε
21.274	276	8.9	21.285	229	7.7	21.282	220	7.9
23.566	151	12.8	23.582	128	11.2	23.567	123	11.6
36.038	191	6.8	36.035	165	5.8	36.038	162	5.8
Average values	206	9.5×10^{3}		174	8.2×10^{-3}		168	8.4×10^{-3}

One more important characteristic of crystallinity of polymer is the size *d* of the ordered areas in it. For definition of the size of crystalline particles and their maximum deformation ε *in* X-ray diffraction analysis [21] Bragg peaks width on half of maximum intensity (Bragg lines half-width) is often used. In the given research the sizes of crystalline particles in a polyethylene matrix calculated on three well expressed diffractogram peaks Figure 20.3. The peaks of polyethylene located at corners 2θ approximately equal 21.28°, 23.57° and 36.03° (peaks PE-1, PE-2 and PE-3 on Figure 20.4) were used. The ordered areas size *d* and the maximum relative deformation ε of their lattice were calculated by the joint decision of the equations of Sherrera and Wilson [21] with use of half-width of the peaks defined as a result of approximation by analytical functions, and taking into account experimentally measured diffractometer tool function. The calculations were spent by means of program *WinX^{POW} size/strain*. Received d and ε, and also their average values for investigated films are presented in Table 20.2. The updated positions of maxima of diffraction peaks used at calculations are specified in the table.

The offered technology allowed the obtaining of films LDPE and composite films LDPE + 1% nc-Si-36 and LDPE + 0.5% nc-Si-97 an identical thickness (85 microns). Thus concentration of modifying additives nc-Si in composite films corresponded to the set structure that is confirmed by the X-ray phase analysis.

By direct measurements it is established that additives of powders nc-Si have reduced a polyethylene transparency in all investigated range of lengths of waves, but especially strong transmittance decrease (almost in 20 times) is observed in a range of lengths of waves of 220-400 nm, that is in UV areas. Especially strongly effect of suppression of UV radiation is expressed in LDPE film + 0.5% nc-Si-97 though concentration of an additive of silicon in this material is less. It is possible to explain this fact that according to experimentally received function of density of distribution of the size, the quantity of particles with the sizes is less 10 nm on volume/weight unit in a powder nc-Si-97 more than in a powder nc-Si-36.

Direct measurements define mechanical characteristics of the received films – durability at a stretching and relative lengthening at disrupture (Table 19.1). The received results shows that additives of powders of silicon raise durability of films approximately on 20% in comparison with pure polyethylene. Composite films in comparison with pure polyethylene also have higher lengthening at disrupture, especially this improvement is expressed in case of composite PE + 0.5% nc-Si-97. The observable improvement of mechanical properties correlates with degree of crystallinity of films and the average sizes of crystal blocks in them (Table 19.2). By results of the X-ray analysis the highest crystallinity at LDPE film + 0.5% nc-Si-97, and at it the smallest size the crystal ordered areas that should promote durability and plasticity increase.

20.5 ACKNOWLEDGMENT

This work is supported by grants RFBR № 10-02-92000 and RFBR № 11-02-00868 also by grants FCP "Scientific and scientific and pedagogical shots of innovative Russia", contract № 2353 from 17.11.09 and contract № 2352 from 13.11.09.

KEYWORDS

- **Nanocrystalline Silicon**
- **Polyethylene**
- **Polymer Nanocomposites**
- **Spectroscopy**
- **UV-Protective Film**
- **X-ray Diffraction Analysis**

REFERENCES

1. Karpov, S. V. & Slabko, V. V. (2003). *Optical and Photophysical Properties of Fractally Structured Metal Sols*, Sib. Otd. Ross. Akad. Nauk, Novosibirsk.
2. Varfolomeev, A. E., Volkov, A. V., Godovskii, D. Yu., *et al.*, (1995) *Pis'ma Zh. Eksp. Teor. Fiz.*, 62, 344.
3. Delerue, C., Allan, G., & Lannoo, M. (1999) *J. Lumin.*, 80, 65.
4. Soni, R. K., Fonseca, L. F., Resto, O., et al. (1999). *J. Lumin.*, 83–84, 187.
5. Altman, I. S., Lee, D., Chung, J. D. et al. (2001). *Phys. Rev. B: Condens. Matter Mater. Phys.* 63, 161402.
6. Knief, S. & Niessen, W. von. (1999). *Phys. Rev. B: Condens. Matter Mater. Phys.* 59, 12940.
7. Olkhov, A. A., Goldschtrakh, M. A., & Ischenko, A. A. (2009). RU Patent № 2009145013.
8. Bagratashvili, V. N., Tutorskii, I. A., Belogorokhov, A. I. et al. (2005). // Reports of Academy of Sciences. *Physical Chemistry*. 405, 360.
9. V. Kumar (Ed.), (2008). *Nanosilicon*. Elsevier Ltd. 13, 368.
10. *Nanostructured Materials. Processing, Properties, and Applications*. Carl C. Koch. (Ed.) William Andrew Publishing, New Yark. 752 (2009).
11. Ischenko, A. A., Dorofeev, S. G., Kononov, N. N. et al. (2009). RU Patent №2009146715.

12. Kuzmin, G. P., Karasev, M. E., Khokhlov, E. M. et al. (2000). *Laser Phys.* 10, 939.
13. Beckman, J. & Ischenko, A. A. (2003). RU Patent No. 2 227 015.
14. Stehl, K. (2000). The Huber G670 imaging-plate Guinier camera tested on beamline I711 at the MAX II synchrotron. *J. Appl. Cryst.,* 33, 394-396.
15. Fetisov, G. V. (2010) The X-ray phase analysis. Chapter 11, pp.153-184. Analytical chemistry and physical and chemical methods of the analysis. T. 2. A. A. Ischenko. M.: ITc Academy, p. 416.
16. Scardi, P. & Leoni, M. (2006). Line profile analysis: pattern modelling versus profile fitting. *J. Appl. Cryst.*, 39, 24-31.
17. WINX^{POW} Version 1.06. STOE & CIE GmbH Darmstadt/Germany.(1999).
18. Leoni, M., Confente, T., & Scardi, P. (2006). PM2K: a flexible program implementing Whole Powder Pattern Modelling., *Z. Kristallogr. Suppl.*, 23, 249-254.
19. Scardi, P. (2008). Recent advancements in whole powder pattern modeling. *Z. Kristallogr. Suppl.* 27, 101-111.
20. Strbeck, N. (2007). *X-ray scattering of soft matter.* Springer-Verlag Berlin Heidelberg.. 20, 238.
21. Iveronova, V. I. & Revkevich, U. P. (1978) The theory of scattering of X-rays. M.: MGU. p. 278

INDEX

C

F

G

H

I

K

L

M

N

O

P

V

W

X

Y

Z

Printed in the United States
by Baker & Taylor Publisher Services